# Development of the Vertebrate Retina

# PERSPECTIVES IN VISION RESEARCH

Series Editor: Colin Blakemore
University of Oxford
Oxford, England

---

**Development of the Vertebrate Retina**
Edited by Barbara L. Finlay and Dale R. Sengelaub

**Parallel Processing in the Visual System**
THE CLASSIFICATION OF RETINAL GANGLION CELLS AND
ITS IMPACT ON THE NEUROBIOLOGY OF VISION
Jonathan Stone

---

A Continuation Order Plan is available for this series. A continuation order will bring delivery of each new volume immediately upon publication. Volumes are billed only upon actual shipment. For further information please contact the publisher.

# Development of the Vertebrate Retina

Edited by
## Barbara L. Finlay
*Cornell University*
*Ithaca, New York*

### and

## Dale R. Sengelaub
*Indiana University*
*Bloomington, Indiana*

Plenum Press • New York and London

Library of Congress Cataloging in Publication Data

Development of the vertebrate retina / edited by Barbara L. Finlay and Dale R.
Sengelaub.
       p.      cm. —(Perspectives in vision research)
    Based on papers presented at a conference held at Cornell University, Ithaca, N.Y., in
June 1987.
    Includes bibliographies and index.
    ISBN-13: 978-1-4684-5594-6        e-ISBN-13: 978-1-4684-5592-2
    DOI: 10.1007/978-1-4684-5592-2

    1. Retina—Growth—Congresses. 2. Retina—Differentiation—Congresses. I. Finlay,
Barbara L. (Barbara La Verne), 1950-    . II. Sengelaub, Dale Robert, 1956-    . III.
Series.
    [DNLM: 1. Retina—growth & development—congresses. 2. Vertebrates—congresses.
WW 270 D489 1987]
QP479.D48   1989
596'.03'32—dc19
DNLM/DLC                                                                              88-38829
for Library of Congress                                                                   CIP

© 1989 Plenum Press, New York
Softcover reprint of the hardcover 1st edition   1989
A Division of Plenum Publishing Corporation
233 Spring Street, New York, N.Y. 10013

For William, Laura, and Caitlin

# Contributors

**Lyn D. Beazley** • Department of Psychology, University of Western Australia, Nedlands Western Australia 6009, Australia

**Lee-Ann Coleman** • Department of Psychology, University of Western Australia, Nedlands Western Australia 6009, Australia

**Ursula C. Drager** • Department of Neurobiology, Harvard Medical School, Boston, Massachusetts 02115

**Sarah A. Dunlop** • Department of Psychology, University of Western Australia, Nedlands Western Australia 6009, Australia

**Russell D. Fernald** • Institute of Neuroscience, University of Oregon, Eugene, Oregon 97403

**Barbara L. Finlay** • Department of Psychology, Cornell University, Ithaca, New York 14853

**Richard L. Fortney** • Department of Neurological Sciences, Rush Medical College, Chicago, Illinois 60612, and Department of Anatomy and Cell Biology, University of Illinois, College of Medicine, Chicago, Illinois 60680

**Alison M. Harman** • Department of Psychology, University of Western Australia, Nedlands Western Australia 6009, Australia

**Sally G. Hoskins** • Department of Biological Sciences, Columbia University, New York, New York 10027; *present address:* Department of Biology, City College of New York, New York, New York 10031

**Howard C. Howland** • Section of Neurobiology and Behavior, Cornell University, Ithaca, New York 14853

**Audie G. Leventhal** • Department of Anatomy, University of Utah, School of Medicine, Salt Lake City, Utah 84132

**V. Hugh Perry** • Department of Experimental Psychology, University of Oxford OX1 3UD, England

**Edward H. Polley** • Departments of Anatomy and Cell Biology, and Ophthalmology, University of Illinois, College of Medicine, Chicago, Illinois 60680

**Thomas A. Reh**  •  Department of Medical Physiology, University of Calgary, Calgary, Alberta, Canada T2N 4N1

**Frank Schaeffel**  •  Section of Neurobiology and Behavior, Cornell University, Ithaca, New York 14853

**Jeffrey D. Schall**  •  Department of Brain and Cognitive Science, Massachusetts Institute of Technology, Cambridge, Massachusetts 02139

**Jerry Silver**  •  Neuroscience Program, Department of Developmental Genetics, School of Medicine, Case Western Reserve University, Cleveland, Ohio 44106

**Alan D. Springer**  •  Department of Anatomy, New York Medical College, Valhalla, New York 10595

**Maree J. Webster**  •  Neuroscience Program, Department of Developmental Genetics, School of Medicine, Case Western Reserve University, Cleveland, Ohio 44106

**Kenneth C. Wikler**  •  Section of Neuroanatomy, Yale University, School of Medicine, New Haven, Connecticut 06510

**Roger P. Zimmerman**  •  Departments of Neurological Sciences and Physiology, Rush Medical College, Chicago, Illinois 60612, and Department of Anatomy and Cell Biology, University of Illinois, College of Medicine, Chicago, Illinois, 60680

# Preface

The vertebrate retina has a form that is closely and clearly linked to its function. Though its fundamental cellular architecture is conserved across vertebrates, the retinas of individual species show variations that are also of clear and direct functional utility. Its accessibility, readily identifiable neuronal types, and specialized neuronal connectivity and morphology have made it a model system for researchers interested in the general questions of the genetic, molecular, and developmental control of cell type and shape. Thus, the questions asked of the retina span virtually every domain of neuroscientific inquiry—molecular, genetic, developmental, behavioral, and evolutionary. Nowhere have the interactions of these levels of analysis been more apparent and borne more fruit than in the last several years of study of the development of the vertebrate retina.

Fields of investigation have a natural evolution, moving through periods of initial excitement, of framing of questions and controversy, to periods of synthesis and restatement of questions. The study of the development of the vertebrate retina appeared to us to have reached such a point of synthesis. Descriptive questions of how neurons are generated and deployed, and questions of mechanism about the factors that control the retinal neuron's type and distribution and the conformation of its processes have been posed, and in good part answered. Moreover, the integration of cellular accounts of development with genetic, molecular, and whole-eye and behavioral accounts has begun.

The task undertaken by the contributors to this volume was therefore to summarize the work of the past several years on the development of the cells and systems of the retina, and to pose unanswered questions or new directions for the field. We restricted the field of inquiry to forces acting on the generation and development of retinal neurons and their processes: the cells and systems of the neural retina. The new directions suggested by this work often are at a different level of analysis. New molecular tools combined with a satisfactory systems-level description of how the retina develops now allow functional questions to be asked at a genetic and molecular level. The number of species now studied gives an adequate comparative base to frame general questions about the mechanisms of evolution of the eye. Both approaches are represented by the contributors to this volume.

Assembly of this volume followed a conference, "Development of the

Vertebrate Retina," held at Cornell University, Ithaca, New York, in June of 1987. We are indebted to the Arts College of Cornell University, and in particular the Department of Psychology, which provided much of the funds to underwrite the conference. We acknowledge the intellectual contributions of conference attendees not represented in this volume, namely, Jonathan Stone, Leo Chalupa, Rob Williams, and Josh Wallman. We thank Lacey Burmingham for her excellent secretarial assistance throughout the organization of the conference and production of this volume.

Barbara L. Finlay
Dale R. Sengelaub

*Ithaca and Bloomington*

# Contents

## II. PHYLOGENETIC, EVOLUTIONARY, AND FUNCTIONAL ASPECTS OF RETINAL DEVELOPMENT

## 9. Development of Cell Density Gradients in the Retinal Ganglion Cell Layer of Amphibians and Marsupials: Two Solutions to One Problem

**Lyn D. Beazley, Sarah A. Dunlop, Alison M. Harman, and
Lee-Ann Coleman**

# Development of the Vertebrate Retina

# Cellular Aspects of Retinal Development

I

The organization of the mature vertebrate retina emerges from the development of individual cells and their processes. In this section, retinal development is addressed at the cellular level: the regulation and timing of the production of particular cell classes; the local and intrinsic factors that control dendritic and somal morphology; and the temporal and spatial factors controlling axonal outgrowth and projection specificity. While each paper emphasizes a particular cellular feature, together these papers demonstrate the multiple and interleaving factors that control each one of these features and the critical role of developmental timing in orchestrating them.

The problem of specification of retinal cell types and their proportions is addressed in the first three chapters. Polley, Zimmerman, and Fortney describe at both the light and electron microscope levels the sequential and overlapping patterns of neurogenesis in the cat retina that give rise to the cell classes and characteristic lamination of the adult retina. Fernald discusses the particular problem of rod neurogenesis, using a teleost fish whose protracted retinal development allows an especially clear resolution of the mode of rod generation from a separate germinal zone. New information on the regulation of generation and differentiation of retinal cell classes by molecular cues existing in the local microenvironment of the developing retina is reviewed by Reh.

The next three chapters discuss the factors controlling axon outgrowth, from initial events forming the optic nerve to the control of projection laterality and topographic map formation. The multiple uses of the information supplied by the timing of developmental events are particularly evident in this group of chapters. Webster, Drager, and Silver describe the initial development of the optic stalk in normal animals and hypopigmented mutants, and they argue that retinal pigmentation is a time-dependent cue important both in the establishment of lateralized projections in the optic stalk and in the establishment of normal cell density gradients. Springer reviews how spatiotemporal gradients in retinogenesis contribute to the organization of the optic nerve in various species, and he considers how this information might be employed to contribute to the solution of retinotopic mapping problems as axons contact their central targets. Hoskins examines the interaction of cell lineage and time in the particular case of the production of ipsilaterally projecting ganglion cells in frogs on metamorphosis, in a general review of the factors that determine projection laterality.

The final two chapters in this section review factors intrinsic and extrinsic to retinal cells which determine the nature of the retinal mosaic. Perry describes the regulation of the number and specialized distribution of retinal cells by dendritic competition for retinal afferents. Leventhal and Schall further consider the interaction of particular ganglion cell classes with their local environment which produces mature dendritic and somal array, with attention to the special problem of the primate fovea.

# Neurogenesis and Maturation of Cell Morphology in the Development of the Mammalian Retina

1

Edward H. Polley, Roger P. Zimmerman, and Richard L. Fortney

## 1.1. INTRODUCTION

The cellular organization and morphology of the adult mammalian retina results from a complex series of interacting developmental events. These events may occur sequentially or simultaneously, alone or in combination. Each individual event of development may have pervasive influences that modify or direct morphogenesis of the retina as a whole, or it may have local effects on birth and differentiation of individual populations of neural and glial cells. The time period during which individual modulatory phenomena may be effective can vary with the stage of development. In this review we limit our discussion to the production of neuroblasts, the maturation of their cellular morphology, and the contributions of their processes to the structure and the circuitry of the adult mammalian retina. The role of cell death in the development of the retina is considered elsewhere in this volume (Chapters 7, 9, and 10).

**Edward H. Polley** • Departments of Anatomy and Cell Biology, and Ophthalmology, University of Illinois, College of Medicine, Chicago, Illinois 60680. **Roger P. Zimmerman** • Departments of Neurological Sciences and Physiology, Rush Medical College, Chicago, Illinois 60612, and Department of Anatomy and Cell Biology, University of Illinois, College of Medicine, Chicago, Illinois 60680. **Richard L. Fortney** • Department of Neurological Sciences, Rush Medical College, Chicago, Illinois 60612, and Department of Anatomy and Cell Biology, University of Illinois, College of Medicine, Chicago, Illinois 60680.

There is a large body of classical neuroanatomical literature that describes the processes of neurogenesis and morphological differentiation in the retina (Ramon y Cajal, 1893, 1929; Polyak, 1941; Morest, 1970; Sidman, 1961; Hinds and Hinds, 1974, 1978, 1983). Most of these early descriptive studies of retinal development were based on rodent animal models, although similar information exists for the primate retina (Polyak, 1941; Barber, 1955; Duke-Elder and Cook, 1963; Mann, 1969). More recent work has built upon this classical foundation and addresses specific genetic, inductive, or trophic mechanisms that might direct the design and construction of the mature retina (Sorge *et al.*, 1984; Barnstable *et al.*, 1985; Adler, 1986; Constantine-Paton *et al.*, 1986; Turner and Cepko, 1987; Wetts and Fraser, 1988). These studies, employing techniques of immunocytochemistry and molecular biology, have also primarily concentrated on rodent models.

We have investigated the sequence of neurogenesis and the maturation of cellular morphology in retinal development of the cat and ferret using the techniques of autoradiography and scanning electron microscopy (Polley *et al.*, 1982, 1985, 1986; Zimmerman *et al.*, 1985, 1987, 1988). Tritiated thymidine provided to ventricular cells at the time of DNA replication prior to the final mitotic division permitted us to label selectively the daughter cells. Autoradiography then allowed us to identify the tritium-labeled cells and describe their morphology and their distribution in the maturing and adult retina (Hickey *et al.*, 1983; Walsh and Polley, 1985; Polley *et al.*, 1986).

Scanning electron microscopy has enabled us to examine and describe the changing cellular architecture of the retina at various stages of development. The high resolution and depth of field afforded by this technique revealed the changing cellular morphology of newly generated cells, as well as the appearance and elaboration of their axonal and dendritic processes (see Hilfer, 1983, and Borwein, 1985).

We have studied early retinal growth and development using these research tools in the context of reexamining and extending some classical descriptions of retinal development. For the purposes of this discussion, we concentrate on two episodes in neurogenesis: an early period of ganglion cell production, E (embryonic day) 19–E30, and a later period of neurogenesis in the first postnatal week. We describe the time of birth and distribution of the labeled cells that make up the cellular laminae of the adult retina, as well as the development of the plexiform and optic nerve fiber layers whose appearance and presence reflect the elaboration of the dendritic and axonal processes of the retinal neurons. We focus on neurogenesis and differentiation of specific classes of neurons, the migration of committed cells away from the ventricular zone, and the dilution in cellular density of the pre- and postmitotic cells formed earlier. Finally, we describe a model of the adult retina in which the emerging cytoarchitecture reflects the temporal sequence of cell birth, migration, and insertion of newly generated cells into the developing retina. Possible mechanisms that may act during retinal development to produce the characteristic laminar organization, mosaic cellular patterns, and patterns of cellular distribution are also discussed.

## 1.2.  BACKGROUND

Studies of mammalian retinal neurogenesis and development have shown that while the overall process of development is relatively similar in the various species, the time course of the developmental events leading to maturity varies widely. The time course for the developmental events of retinal morphogenesis and neurogenesis in any species is determined by the length of the period of gestation. Studies of retinal development in rodents (mouse, rat, hamster) show an apparent compression of developmental events relative to larger animals (cat, ferret, monkey) (see Chapter 10). Except for the duration of gestation, the timing of major events of retinal development are closely similar in the cat and ferret.

Despite the temporal compression of the neurogenetic sequences in smaller animals, the changing morphology of the developing eye is remarkably similar for most vertebrates. The compression of events in smaller animals tends to obscure the individual sequential phenomena that are clearly resolved in an animal with more protracted development (Sidman, 1961; Young, 1985).

It is because of the longer gestational period and the wealth of information on the structure and organization of the immature and adult retina that we have preferred the cat (gestation 64–65 days) and the ferret (gestation 42 days) over the hamster (gestation 15 days) and the mouse or rat (gestation 19–21 days) for our studies of retinal development.

The magnitude of the changes in the dimensions of the cat eye is illustrated in Fig. 1.1. This figure compares the size of a fetal kitten at E24 to the eye of an adult cat and emphasizes the magnitude of growth of the eye and degree to which the retinal cells produced early in neurogenesis are diluted by those produced subsequently.

## 1.3.  HISTOGENESIS AND MORPHOGENESIS OF THE RETINA

The retina is an outgrowth of the neural tube and initially follows the basic pattern of histogenesis in the vertebrate central nervous system. However, the later phases of retinal development are clearly different from the sequences described in cerebral cortex and cerebellum (Sauer, 1935; Fujita, 1963; Angevine et al., 1970). The wall of the primitive optic vesicle is a pseudostratified columnar epithelium composed of fusiform ventricular cells. These precursor cells extend across the thickness of the retina and are joined by junctional complexes at both the outer limiting membrane (OLM), inner limiting membrane (ILM), and at apposed lateral surface elsewhere in the retina (Sheffield and Fischman, 1970; Hinds and Ruffett, 1971; Fujisawa, 1982; Whiteley and Young, 1986; Zimmerman et al., 1987). Prior to each mitotic division, ventricular cells replicate DNA, release their attachment to the ILM, and round up at the OLM (Sauer, 1935), the equivalent of the

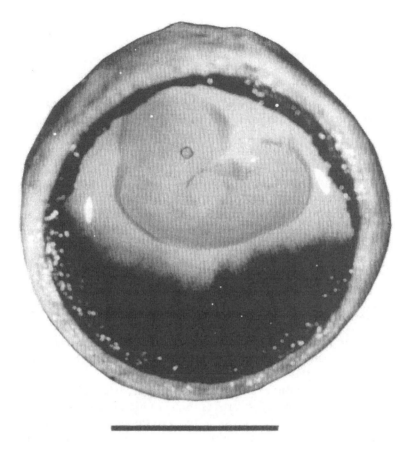

**Figure 1.1.** An E24 cat fetus in the eyecup of an adult cat. The fetal eye is outlined by a thin line of melanin, which is restricted at this age to the peripheral edge of the pigment epithelium. The fetus in its entirety fits within the limits of the dorsal (tapetal) retina, indicating the magnitude of growth that the retina undergoes between the initial period of neurogenesis and maturity. Scale bar = 1 mm.

ventricular zone elsewhere in the developing central nervous system (Angevine *et al.,* 1970). If the products of the mitotic division are destined to be ventricular cells, photoreceptors or Müller glial cells, the connection to the OLM is maintained; otherwise, the daughter cells detach and migrate radially to their appropriate level within the retina (Hinds and Hinds, 1974, 1978, 1983).

Although mitotic figures are seen throughout the neuroblast layer of the fetal retina at these early stages, identifiable differentiating cells are seen only in a limited area or "locus" (Fig. 1.2, E20; Fig. 1.3, E24). Indeed, only the retinal ganglion cells (RGCs), rounded cells with round nuclei found at the inner aspect of the retina within this locus, have migrated out of the neuroblast layer and begun to differentiate. Retinal ganglion cell axons can be found in the optic stalk as early as E19 (Williams *et al.*, 1986). Thus, a "front" of RGC neurogenesis and differentiation is advancing in an expanding central to peripheral pattern of development through the multipotential ventricular cells that comprise the balance of the early optic cup. Few of the ventricular cell divisions result in committed postmitotic cells at this time.

From E20 to E28 the change in the size of the neuroretina reflects cellular addition due to the mitotic activity of ventricular cells throughout the retina. Analysis of serial sections of E20 and E24 eyes demonstrates a dorsal–ventral asymmetry: the tissue of the retinal cup extends principally dorsal to the level of the optic stalk (Figs. 1.2 and 1.3). These serial sections show that

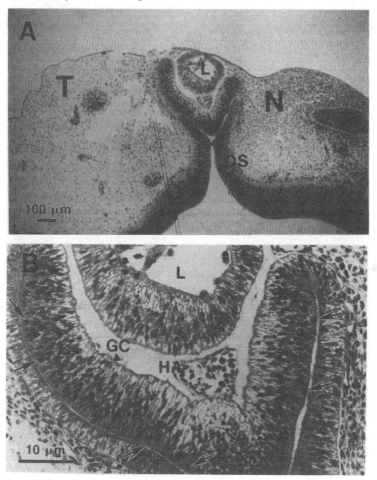

**Figure 1.2.** Retinal morphology of the fetal cat at E20. (A) Low power; note the asymmetry of the regions of retina temporal (T) and nasal (N) to the optic stalk (OS). L, lens. Scale bar = 100 μm. (B) Higher power, detail of region of advanced development. Note migrated round cells (GC, arrowhead), which are first signs of ganglion cell differentiation. Mitotic figures (arrows) are seen in the ventricular zone at the outer limiting membrane. L, lens; HA, hyaloid artery.

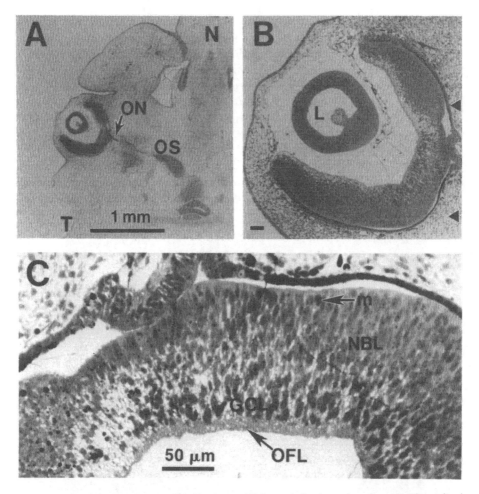

**Figure 1.3.** Retinal morphology of the fetal cat at E24 seen in horizontal section. (A) The retina is asymmetric, having a longer temporal (T) than nasal (N) extent relative to the optic nerve (ON); the optic stalk (OS) is seen in tangential section near its origin from the diencephalon. (B) The asymmetry of the retina and its relationship to the lens (L) is seen more clearly at a higher magnification. Triangular arrowheads at right indicate the area of retina shown in (C). L, lens. (C) Detail of retina. The neuroblast layer (NBL) comprises much of the retina, but a darker row of somata in the proximal retina forms the ganglion cell layer (GCL). The ganglion cell axons collect in the optic fiber layer (OFL). Mitotic figures (m, arrow) can be seen in the ventricular zone. Scale bars: A, 1 mm; B and C, 50 μm.

the retina temporal to the optic stalk is larger than the retina nasal to the stalk. This pattern of early retinal growth is a common feature in vertebrates (human: Mann, 1969; chick: Hilfer, 1983; mouse: Morse and McCann, 1984).

At this stage of development, it is difficult to distinguish with any reliability between premitotic precursor cells and postmitotic cells in early stages of delayed differentiation. However, as noted above, we see an early indication of RGC production, differentiation, and migration in the rounded cells with clear nuclei and dark nucleoli that lie near the inner limiting membrane. These cells first appear in a region superior and temporal to the optic stalk that probably corresponds to the adult *area centralis*. This area is probably equivalent to the developmentally advanced region (DAR) of Robinson (1987).

By E24, in addition to a ganglion cell layer distinct in terms of location, extent, and cellular morphology, the axons of the RGCs have produced an optic fiber layer; optic nerve axons can be traced into the elongating optic stalk (Fig. 1.3). The ganglion cells being produced at this time correspond to the medium sized (beta) ganglion cells labeled in the adult nasal retina by a thymidine injection at E24 (Walsh and Polley, 1985). The lateral dimension of a horizontal section through the retina at this stage has increased by approximately one-third from E20; asymmetry is still marked. Despite the emergence of obvious cell layers, that is, RGC and neuroblast layers, a distinct inner plexiform layer (IPL) is not discernible. This early development of a laminar organization results in an increase in the thickness of the retina with a consequent increase in the length of the ventricular cells whose processes span the thickness of the retina and make up the ILM and OLM.

A horizontal section through the E28 kitten eye no longer shows the obvious asymmetry of the E24 eye (Fig. 1.4). Neurogenesis, evidenced by mitotic figures, is still seen extending across the retina. The optic fiber layer is now seen both temporal and nasal to the optic disk. From E20 to E28 the dendrites of the developing RGCs start to ramify and form the presumptive IPL.

### 1.3.1. Neurogenesis: An Autoradiographic Analysis

We have identified the sequence of cell birth in the cat retina by injecting immature (fetal and neonate) animals with $^3$H-thymidine at various selected stages of development and by subsequent autoradiographic identification of the differentiated and distributed labeled cells (Polley *et al.*, 1982; Zimmerman *et al.*, 1985, 1987; Polley *et al.*, 1985, 1986).

We have previously described the concurrent birth of retinal ganglion cells, A-type horizontal cells, and cone photoreceptors, noting that these latter two classes of cells exhibit delayed morphological differentiation and migration (Polley *et al.*, 1986; Zimmerman *et al.*, 1988). While the retinal ganglion cells differentiate soon after their final mitotic division, the other postmitotic cells produced at this time, the A-type horizontal cell and the cone photoreceptor, are committed but not yet differentiated.

We use the term "cohort" to refer to such dissimilar classes of retinal cells with concurrent periods of neurogenesis as determined by a single application of tritiated thymidine. As in this example, the component cell classes of a cohort may exhibit different rates of differentiation and migration.

Injection of a fetal cat *in utero* at E26 with tritiated thymidine followed by sacrifice at E40 reveals the labeling of medium sized (beta) retinal ganglion cells, as well as morphologically undifferentiated putative cone photorecep-

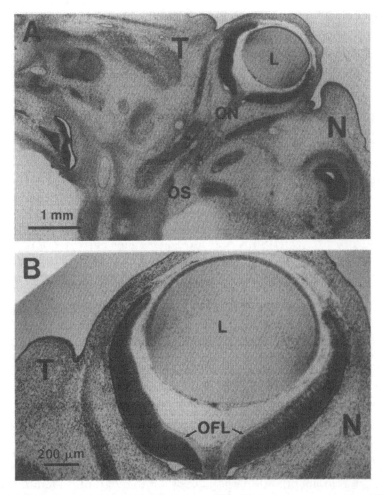

**Figure 1.4.** Retinal morphology of the fetal cat at E28. (A) Low power; note the differences between nasal (N) and temporal (T) regions. While the nasal retina is longer in this plane of section, the most advanced region is still temporal to the optic nerve head. (B) The difference between the nasal and temporal retina can be seen in greater detail. The optic fiber layer (OFL, arrows) is clearly visible.

tors and A-type horizontal cells (Fig. 1.5). Heavily labeled cells are seen in the ganglion cell layer among populations of ganglion cells born at an earlier or later time. The outer neuroblastic layer (NBL) is separated from the ganglion cell layer by a well-developed IPL. Thymidine labeling of morphologically undifferentiated but committed postmitotic cones and A-type horizontal cells provides some early evidence of nonuniformity in cellular composition of the neuroblast layer, yet no OPL is evident. The confirmation of these morphological identifications and the distribution of the components of this cohort in the adult retina are seen in Fig. 1.6.

Production of this cellular cohort continues until E32, when labeling of

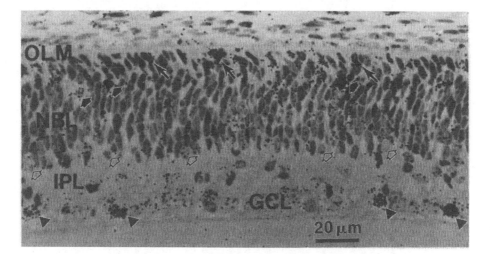

**Figure 1.5.** Autoradiogram of the retina of an E40 fetal cat injected with tritiated thymidine at E26. Three populations of heavily labeled cells can be detected: ganglion cells (arrowheads), A-type horizontal cells (heavy arrows), and cones (thin arrows). Amacrine cells are unlabeled (open arrows). From Zimmerman *et al.* (1988); used with permission of the *Journal of Comparative Neurology.*

A-type horizontal cells ceases and isotope-labeled cells now appear at the inner aspect of the inner nuclear layer. The new population of labeled cells is the amacrine cells; the production of ganglion cells and cones continues (Polley *et al.*, 1982).

During the limited period of RGC neurogenesis (E19–E35), the cells produced early in that time period will develop into the medium sized (X or beta) ganglion cells of the adult. Large (Y or alpha) RGCs are produced during the middle of the period (E25–E28) and small (W or gamma) RGCs develop from later-generated cells (E29–E35). The mechanisms that induce or direct this differential pattern of cell determination are unknown (Walsh *et al.*, 1983; Walsh and Polley, 1985). Ganglion cells are not reported to form recognizably different size classes until about E47–E52 (Stone *et al.*, 1984; Ramoa *et al.*, 1987).

The pattern of neurogenesis is remarkably different in the neonate (Johns *et al.*, 1979). Uptake of tritiated thymidine by premitotic ventricular cells at this postnatal stage in development results in labeling of cells distributed in the ONL and INL (Fig. 1.7). In this P (postnatal day) 9 retina labeled on P2, silver grains are found over neurons in the center of the ONL (bipolar cells, one of which is counterstained by uptake of the fluorescent dye Lucifer Yellow), the inner part of the ONL (rods), and the outer part of the INL (B-type horizontal cells). The pattern of cell generation at this stage is best described as an expanding front of bipolar, rod photoreceptor, and B-type horizontal cell neurogenesis, followed by a trailing edge of diminishing neurogenetic activity. Neurogenesis of these new classes of cells is distributed

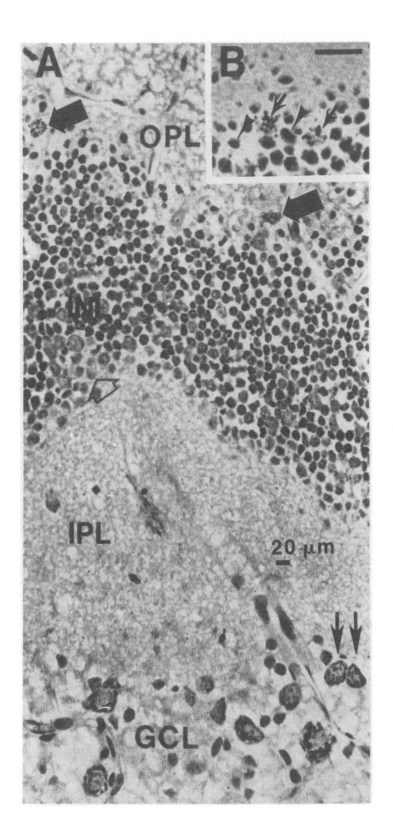

over a wider area of the immature retina; newborn neurons are inserted only into the inner and outer nuclear layers.

At this stage the presence of a central to peripheral gradient of neurogenesis is evidenced by diminished numbers of isotope-labeled cells in more central areas of the retina (Fig. 1.8). The presence of isotope-labeled precursors, along with mitotic figures, in the INL (Robinson et al., 1985, and our unreported observations) suggests that the ventricular cell population is restricted to the INL as the ONL matures. Indeed, the presence of rod photoreceptor nuclei reported proximal to the OPL in immature retina (Young, 1985) may represent incomplete migration from the site where mitosis had occurred. At this time, migration of newly produced rods may be guided by the radial Müller glial cells, as described in a recent study of goldfish retina (Raymond and Rivlin, 1987). The developmental mechanisms that control or direct differentiation and migration of postmitotic cells in the neonate are thus quite different from the condition seen earlier.

This different pattern of isotope labeling of cells born in central and peripheral retina of the neonate gives the impression that fewer ventricular cells have remained mitotically active in central retina than in peripheral retina (Fig. 1.8). The larger numbers of lightly labeled cells in the INL of peripheral retina indicates either that multiple mitotic divisions of the labeled precursors may have occurred in the periphery over the same period of time or that the S phase differs in duration in central and peripheral retina. Changes in cell cycle period during the course of development and in a central to peripheral pattern have been reported (Dütting et al., 1986).

Based on our autoradiographic studies of neurogenesis in the cat, we have constructed a diagram (Fig. 1.9) that illustrates the time course of the birth of individual classes of cells. In this diagram, the labeled ovals represent the time during which the precursor of that cell class incorporated tritiated thymidine while replicating DNA prior to the final mitotic division. As noted above, in some cases, for example, the A-type horizontal cells, the acquisition of the adult cell morphology does not begin immediately at the time of insertion into the neuroretina. The change in background shading from left to right represents the transition from a proliferating population of premitotic, multipotential ventricular cells to the Müller glial cells of the mature retina, which retain a limited proliferative potential (Erickson et al., 1983). The diagram is intentionally vague about the time and rate of the loss of ventricular cells and the initiation of Müller cell generation and maturation. Indeed, a central problem in our understanding of retinal morphogenesis is the rela-

←

**Figure 1.6.** Autoradiogram of the retina of an adult cat that received an injection of tritiated thymidine at E30. (A) Silver grains are found over the nuclei of medium sized ganglion cells (arrows) and A-type horizontal cells (heavy arrows). Note that the largest and smallest ganglion cells are unlabeled. Amacrine cells (open arrows) are not labeled. (B) Cones (arrows) are also labeled by tritiated thymidine injections at this age, but rods (arrowheads) are unlabeled. From Zimmerman et al. (1988); used with permission of the *Journal of Comparative Neurology.*

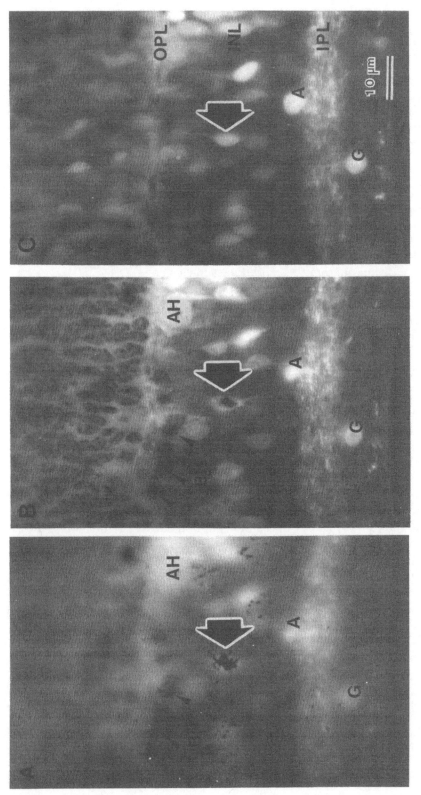

**Figure 1.7.** Neurons labeled by an intravitreal injection of 5 µCi of tritiated thymidine on P2 and stained *in vitro* with Lucifer Yellow on P9. Silver grains are found over the somata of bipolar and B-type horizontal cells, but ganglion cells, amacrine, and A-type horizontal cells are unlabeled. The planes of focus are at the level of the silver grains (A), near the surface of the section (B), and (C) at the level of the stained cell body of a bipolar cell (heavy arrow, center of the field).

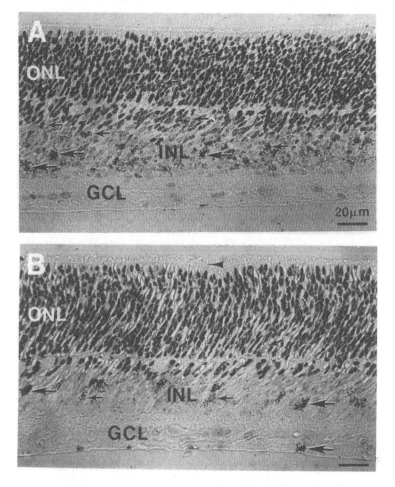

**Figure 1.8.** Autoradiogram of the retina of a neonatal kitten injected intravitreally with [3]H-thymidine on P1 and sacrificed on P4. The region shown in (A) is 2.5 mm peripheral to that shown in (B) and demonstrates the difference in isotope uptake by the cells in a front of neurogenesis. There are fewer and more heavily labeled cells (arrows) seen in the INL of (B) than in the same layer of (A). While cells with similar levels of labeling are found in (A) (arrows), there are numerous lightly labeled cells (thin arrows) in both ONL and INL, suggesting that the isotope was taken up by cells that underwent several subsequent divisions. See text for further discussion.

tionship of Müller cell maturation and the termination of neurogenesis (see Johns *et al.*, 1979).

A vertical line drawn at any point on the abscissa indicates the cohort of cell types being generated at that time. Our use of the term cohort makes no distinctions between the alternatives of single or multiple classes of ventricular cells. Note that the cohort made up of ganglion cells, A-type horizontal cells, and cones is distributed to all three cellular layers of the adult retina, and the last cohort of B-type horizontal cells, bipolar cells, and rods will be added to only the inner and outer nuclear layers. Amacrine cells, perhaps because of

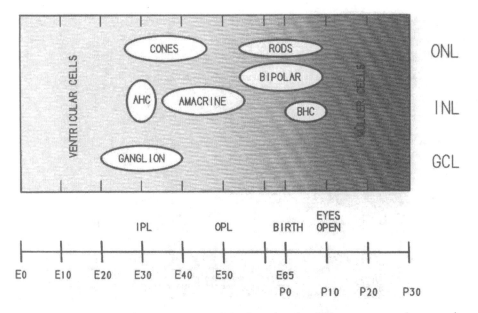

**Figure 1.9.** Summary diagram of neurogenesis in the cat's retina. Ellipses represent the approximate period during which the indicated cell types are produced. The background shading changes gradually from left to right to represent the gradual change from ventricular cells to Müller glial cells as the major radial cell in the retina. See text for further discussion.

their diversity (Kolb *et al.*, 1981), may form a cohort unto themselves. Some of the small isotope-labeled cells in the RGC layer that are generated in the later stages of RGC neurogenesis may well include the numerous "displaced" amacrine cells, long recognized (Ramon y Cajal, 1893) and most recently described by Wong and Hughes (1987) and by Wässle *et al.* (1987). One possible hypothesis of the origin of the displaced amacrine cells has been proposed by Hinds and Hinds (1983), but recent evidence calls this into question (Barnstable *et al.*, 1985). The formation of the adult retina thus involves generation of different cohorts at different times and migration of components of a given cohort to different layers. The mechanisms that control these processes remain unknown.

The progressive differences in the components of each cohort suggest the possibility that either the "capacities" of the ventricular cells or the signals for determination change with time. Because of the increase in the dimensions of the growing retina, the distribution of ventricular cells has been extended and the elements of each new cohort arising from them will be inserted into the cell population generated earlier. Addition by migration of these newly generated cells into the existing cell layers will tend to dilute the concentrations of cells born earlier and perhaps provide an early basis for the generation of mosaic organization (Wässle and Reimann, 1978). Only the ganglion cell layer is not diluted by serial addition but is molded by retinal stretching, cell death, and the insertion of newborn ventricular and Müller cell processes (Lia *et al.*, 1987; see also Chapters 9 and 10 in this volume).

In all instances the initiation of a period of cell production of individual cells is presumed to be controlled by undefined and localized signals for determination (see Zimmerman *et al.*, 1988). However, timing in and of itself does not determine cell type, since several cell types are produced at any given time (see Chapter 3). The same conclusion was reached in a recent study of cell lineage in the *Xenopus* retina, in which all retinal cell types are produced in a single 24 hour period of neurogenesis (Holt *et al.*, 1988).

### 1.3.2. Morphogenesis: A Scanning Electron Microscope Analysis

The changing cellular morphology and cytoarchitecture of the retina are clearly seen in scanning electron micrographs (SEMs) of fetal and neonatal retinas. Scanning electron microscopy enabled us to visualize the development and change of cellular shape and orientation, trace the processes of migrating and differentiating cells, and follow the progressive separation of cellular and plexiform laminae, as well as the appearance and development of the plexiform layers.

An early stage in the evolving laminar organization is seen in Fig. 1.10, a SEM of an E36 ferret retina. The outer neuroblast layer (NBL) consists of fusiform ventricular cells whose processes span the neuroretina. The cells found in the retinal ganglion cell (RGC) layer are among the first cells in the retina to be born and commence differentiation. Following their final mitosis, they migrate away from the OLM toward the ILM and differentiate rapidly; a variety of rounded and oval somata are seen. The cellular layers are separated by the horizontally spreading dendrites elaborated by the differentiated RGCs. These processes, joined later by those of amacrine cells, constitute the IPL at its earliest stage. Individual radial processes (either cylindrical or slightly flattened) of ventricular cells penetrate the IPL and the sparse optic fiber layer and expand to form the ILM. The distal processes of the ventricular cells are joined at the OLM. Smaller, often branched, processes oriented both radially and horizontally in the NBL and IPL probably represent pioneering growing processes of migrating and differentiating cells.

By E54 the laminar organization of the retina is somewhat more complex (Fig. 1.11). The appearance of a population of small more rounded cells at the proximal margin of the INL are amacrine cells. The IPL now contains processes of these amacrine cells as well as the dendrites of the RGCs. Note that the RGCs are separated by the sheetlike radial processes and their expansions at the ILM of what appear to be either ventricular cells or Müller glial cells (Fig. 1.11). The presence of a scattered population of large rounded cells in the distal NBL indicates the level of members of the first cohort, the A-type horizontal cells. The location of these cells is the approximate level of the

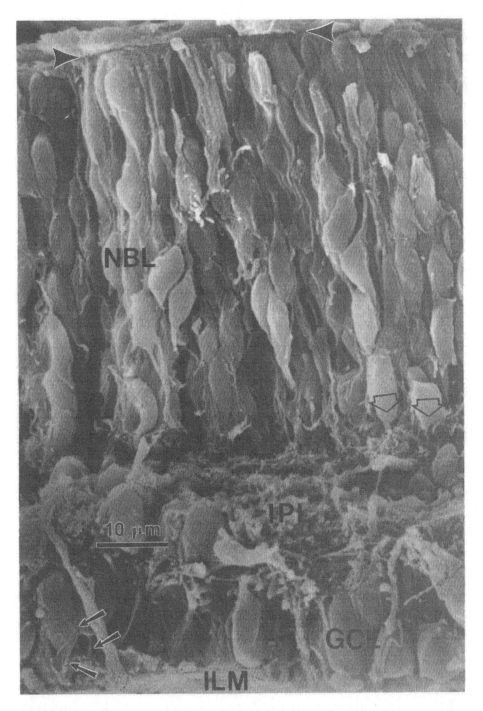

**Figure 1.10.** Scanning electron micrograph of an E36 ferret retina. The retina consists of an outer neuroblast layer (NBL) and an inner layer of retinal ganglion cells (GCL). These cellular layers are separated by an inner plexiform layer (IPL) made up of ganglion cell dendrites, processes of migrating cells, and processes of differentiating cells of the inner NBL, which are most probably amacrine cells (open arrows). The outer limiting membrane, formed at this age by junctions between ventricular cells and postmitotic cones, is indicated by arrowheads. Axons emerging from ganglion cell somata are seen (small arrows).

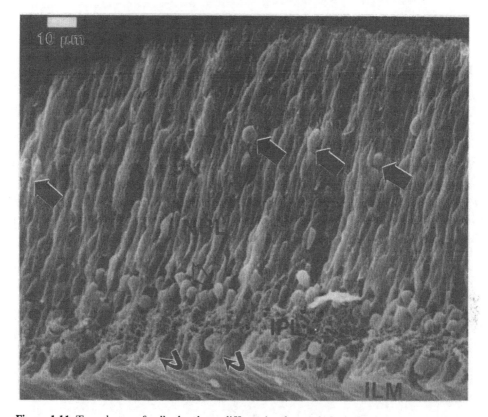

**Figure 1.11.** Two classes of cells that have differentiated morphologically are seen within the neuroblast layer (NBL) in this scanning electron micrograph of the central retina of the cat at E54. The rounded large A-type horizontal cells (heavy arrows) are sparsely distributed at the edge of the distal third of the NBL. Small, rounded amacrine cells, born during a later period of neurogenesis, are found at the proximal edge of the NBL (open arrows). The inner plexiform layer is thicker and more complex. The radial processes of the ventricular cells (curved arrows) form arcadelike spaces distal to the inner limiting membrane.

future OPL. The initial stage of the development of the OPL is evidenced by the horizontal growth of the AHC dendrites that later contact the radially oriented axon terminals of the cones.

All the layers of the adult retina are found in the neonatal retina (Fig. 1.12, P4). As demonstrated by autoradiographic studies discussed above, we know that the cohort now being generated is composed of the bipolar cells, the rod photoreceptors, and the B-type horizontal cells. As the somata of the B-type horizontal and bipolar cells come to lie in the INL, there is an increase in dimensions and complexity of that layer. Differences in cell morphology of INL somata are now more obvious. The organization of the IPL is further complicated by the addition of bipolar cell axon terminals. At this stage, the

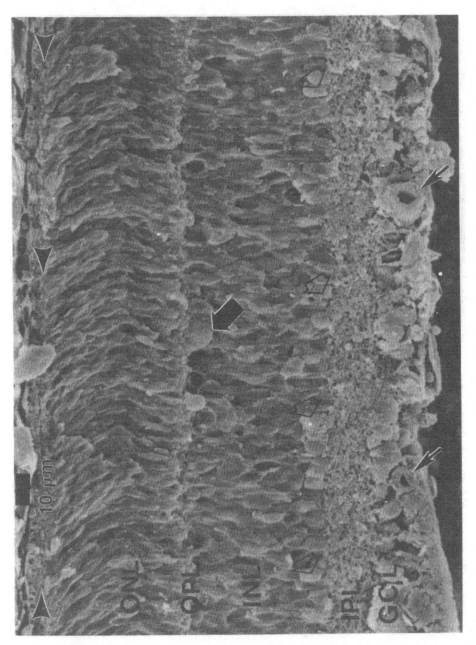

**Figure 1.12.** Scanning electron micrograph of the retina of a P4 cat. The proximal retina is well differentiated, with a thick IPL and continuous GCL. Blood vessels (arrows) are found in the optic fiber layer. The INL is separated from the ONL by a distinct OPL; the soma of a large A-type horizontal cell occupies the center of the field (heavy arrow). The OLM (arrowheads) is closely applied to the pigment epithelium; inner and outer segments of photoreceptor cells are still rudimentary at this stage. Open arrows indicate amacrine cell somata.

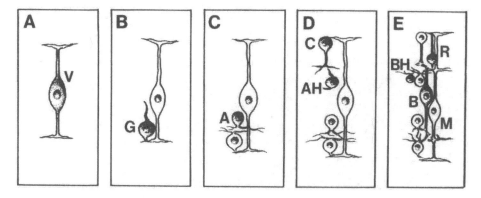

**Figure 1.13.** Diagrammatic summary of the times of appearance of differentiated cells in the fetal retina. The newly differentiated cells added in each phase are indicated by dark stippling. (A) Prior to the onset of neurogenesis (before E19 in the cat), the retina is a pseudostratified columnar epithelium composed of ventricular cells (V). (B) Soon after their final mitosis, the first ganglion cells (G) migrate to their appropriate position in the proximal retina. A trailing process follows the cell body; the axon grows into the nascent optic fiber layer forming distal to the vitreal end feet of the ventricular cells. (C) At about E35, amacrine cells (A) are being produced. These cells migrate immediately to the proximal retina and differentiate in close proximity to the ganglion cells. (D) By E49, differentiation of the A-type horizontal cells (AH) is evidenced by processes that extend radially as well as horizontally from the soma, although the outer plexiform layer is not yet a compact structure. Cone (C) somata are identifiable by their large pale somata near the outer limiting membrane; the long branched process is an exaggeration intended to emphasize that cone and AH processes have yet to make synaptic contacts. (E) Several new classes of differentiated cells are seen in the retina at the end of the first postnatal week. Both rods (R) and cones are now present, but neither type of photoreceptor possesses an outer segment at this time. B-type horizontal cells (BH) and bipolar cells (B) have appeared, and differentiated Müller cells (M) can be detected.

major morphological deficiency in the laminar organization of the retina is the lack of development of the inner and outer segments of the photoreceptors.

In Fig. 1.13 we have attempted to represent diagrammatically some features of development that have become apparent in our studies. In this figure the progression of development is from left to right. In each panel the newly differentiated cells that are morphologically recognizable by light microscopy are stippled. The diagram was based on identification of *in utero* isotope-labeled cells and their distribution in the adult retina.

Figure 1.13A depicts a multipotential premitotic ventricular cell whose radial processes span the neuroretina. The unstippled cell whose radial processes span the neuroretina and whose laterally directed processes extend into inner and outer nuclear layers in Fig. 1.13E is the Müller glial cell. As noted above, this cell is still premitotic but is no longer multipotential. The birthday of the Müller glial cell has yet to be established using tritiated thymidine autoradiography.

The stippled cell added in Fig. 1.13B represents the first cell to differentiate and migrate away from the ventricular zone, the retinal ganglion cell. The other elements of the first cohort, cones and A-type horizontal cells, however, are not morphologically recognizable at the time (cf. Fig. 1.9). They appear as the stippled cells in Fig. 1.13D and only become apparent after the birth and differentiation of the ganglion (Fig. 1.13B) and amacrine (Fig. 1.13C) cells. The members of the last cohort, rods, bipolar cells, and the B-type horizontal cells, are added in Fig. 1.13E.

## 1.4. THEORETICAL CONSIDERATIONS IN RETINAL NEUROGENESIS AND MORPHOGENESIS

### 1.4.1. Differentiation

The initiation of the differentiation of three-dimensional cell morphology must result from gene expression and the production of specific proteins (see Young, 1983, for a discussion in the context of retinal development). However, current knowledge of the molecular mechanisms underlying neuronal differentiation is primitive. One protooncogene, *src*, is thought to be intimately associated with neuronal differentiation (Lynch *et al.*, 1986; Martinez *et al.*, 1987). Both transcription (Vardimon *et al.*, 1986; Grady *et al.*, 1987) and translation (Sorge *et al.*, 1984; Crisanti *et al.*, 1985) of the *src* protooncogene have been reported to be developmentally regulated in the retina. Other specific markers appear coincidentally with regional differentiation (Constantine-Paton *et al.*, 1986) or are restricted to specific cell types, usually after morphological differentiation has been initiated (Barnstable *et al.*, 1985). A special case is the production of enzymes for neurotransmitter synthesis (Schnitzer and Rusoff, 1984).

Although the factors inducing differentiation are not yet identified, it is clear that cell lineage alone does not account for determination and differentiation in the retina (Scholes, 1987; Holt *et al.*, 1988; see Chapter 3 in this volume). It may well be that the first cell type to differentiate directs the subsequent progression of induction and differentiation of other cell types, as in the *Drosophila* retina (Ready *et al.*, 1986; Tomlinson and Ready, 1986).

### 1.4.2. Lamination

Two levels of organization must be considered: the morphological demarcation and growth of separate cellular and synaptic layers, and the subdivision of cellular or synaptic layers in functional units, such as the morphological and functional stratification of the cat inner plexiform layer. Some form of stratified organization has been described for the IPL of several species: cat (Famiglietti and Kolb, 1976; Nelson *et al.*, 1978), fish (Famiglietti *et al.*, 1977; Zimmerman, 1983; Marshak *et al.*, 1984), and macaque monkey (Koontz and Hendrickson, 1987). See Marc (1986) for an excellent review of the stratification of neurotransmitter distribution.

The first obvious step in the development of lamination is the transition from the initially homogeneous pseudostratified neuroepithelium to the appearance of a layer of ganglion cells. The postmitotic RGCs migrate into an area in the proximal retina that is progressively vacated by ventricular cell somata (Fig. 1.3B). The differentiation of the retinal ganglion cells inaugurates the definition of the first two noncellular layers: the optic fiber layer proximal to the GCL, composed of ganglion cell axons, and the inner plexiform layer. At this stage the IPL is first formed by ganglion cell dendrites and the processes of ventricular cells.

The migration of postmitotic differentiated amacrine cells into the proximal edge of the NBL brackets the IPL with another population of maturing neurons (Fig. 1.11). In addition to increasing the intricacy of the IPL by the ingrowth of their processes, they serve to demarcate more sharply the proximal limit of the neuroblast layer. Interactions between amacrine and ganglion cell processes at this early stage may initiate the functional organization of the IPL.

Although cones and axonless (or A-type) horizontal cells (AHC) are produced early on, the initial separation between the outer (ONL) and inner nuclear layers (INL) awaits the appearance of a line of differentiated A-type horizontal cells at approximately the distal third of the NBL, at what will be the distal border of the INL. AHC can be found in this position roughly 2 weeks after their birth, but the development of a recognizable outer plexiform layer occurs still later, as the AHCs produce dendrites and make synaptic contact with cones (Greiner and Weidmann, 1980; Maslim and Stone, 1986; Zimmerman *et al.*, 1988).

The last stage of neurogenesis is the production of bipolar cells, the growth of their dendrites into the OPL, and invasion of the IPL by their axonal growth cones. Only after the development of functional synapses in the OPL and IPL is there a centripetal pathway for the transmission of information about visual input. This phase marks a change in the character of retinal development from neurogenesis to the formation and refinement of synaptic circuitry (McArdle *et al.*, 1977; Morrison, 1982). Subsequently, the plexiform layers become progressively thicker and more complex (Vogel, 1978).

### 1.4.3. Cellular Cohorts

Two basic hypotheses can be put forward to explain the production of different cell types during retinal development, amounting to a restatement of the arguments about "nature versus nurture" or genetic program versus environmental influences dichotomy. In view of our ignorance of the molecular mechanisms involved, these choices are illusory, and the real questions that must be addressed are: What are the factors that act on the genetic programs for proliferation and differentiation, and when do they act?

The idea of diffusible factors regulating neurogenesis is a venerable one; the early literature was reviewed by Ramon y Cajal (1929). Glücksmann (1940) proposed that both cessation of proliferation and differentiation of postmitotic cells might be influenced by unknown factors that are apparently of retinal origin. A more recent form of this hypothesis proposes that the production of specific classes of cells is regulated by cell-specific, density-dependent feedback inhibition by factors produced by differentiated cells (Reh and Tully, 1986; Reh, 1987). This hypothesis was based on experiments in which the specific neurotoxins 6-hydroxydopamine (Negishi *et al.*, 1982; Reh and Tully, 1986) or kainic acid (Reh, 1987) were used to ablate populations of cells in frog or fish retinas (which retain a peripheral proliferative zone). It was suggested that the feedback regulation may affect neuroepithelial cell proliferation as well as neuroblast determination or differentiation (Reh, 1987; see Chapter 3 in this volume).

## 1.5. SUMMARY

Based on our studies, we propose that the cellular architecture and distribution of the adult mammalian retina result from the complex temporal and spatial interaction of neurogenetic and morphogenetic events in the developing retina. The following statements describe some of the events.

The immature retina increases in size by mitotic division and proliferation of ventricular cells that renew themselves during neurogenesis. The ventricular cells also produce several new types of postmitotic cells whose period of neurogenesis, migration, differentiation, and maturation follow specific and sequential programs of development. Recent evidence supports the idea that retinal ventricular cells are multipotential (Turner and Cepko, 1987; Wetts and Fraser, 1988), in contrast to the multiple committed lineages of ventricular cells that are described elsewhere in the central nervous system (Levitt *et al.*, 1981; Rakic, 1982).

More than one cell type may be born during any period of neurogenesis. The differing cell types born at any instant form a cohort. Since the ventricular cells may be multipotential, instructive signals from the local cellular environment would seem to play a prominent role in the commitment of the postmitotic cells to a particular fate (Patterson, 1978; Le Dourin, 1986).

The periods of neurogenesis for members of a cohort overlap but are not necessarily coincident. Thus, "cohort" in our descriptive usage may not always be equivalent to a clone (Scholes, 1976; Turner and Cepko, 1987; Wetts and Fraser, 1988). The pattern of cellular migration and the latency of differentiation of the cells in a cohort may not be similar or coextensive (Zimmerman *et al.*, 1988).

The neurogenesis of specific cell types occurs during limited periods; for example, the retinal ganglion cells are produced from E19 to E35 in the cat.

Newly generated types of cells are added in expanding waves or gradients that are programmed in time of occurrence and position in the retina (e.g., central to peripheral, nasal to temporal) and originate from a "locus" in the immature retina. A pattern of repeated expanding and overlapping gradients of neurogenesis and differentiation from a "locus" results in the insertion of newly formed cells into a continually changing cellular matrix of an enlarging retina.

The diffuse insertion of new ventricular cells in the expanding immature retina dilutes the density of postmitotic cells. Cellular environment into which new postmitotic cells are inserted is different for each successive class of cells.

The end of neurogenesis is due to the loss of the capacity of ventricular cells to generate new types of cells that will be inserted into the developing retinal circuitry. It is from this substrate of cellular neurogenesis that cell death sculpts the organization of the mature retina (see Chapters 7, 9, and 10).

The cellular organization and functional circuitry of the adult mammalian retina develop, at least in part, as a consequence of the phenomena described above. The characteristic distribution and maturation of any individual cell in the retina thus depend on the following factors.

1. The time and location of the final mitotic division of the precursor ventricular cell.
2. The cellular morphological differentiation and the surface extent of retinal development at the time of insertion of the postmitotic cell.
3. The growth and cellular development of the retina after insertion of the postmitotic cell (dilution).
4. Intercellular trophic effects and cell death (refinement).
5. The nonuniform expansion of the growing retina (Mastronarde *et al.*, 1984).

The differing mosaic organization of the rod and cone photoreceptors and the A- and B-type horizontal cells quite possibly derives primarily from their cell birthdays and the extent of retinal development at the time of their insertion into the neuroretina (Fig. 1.9). Both the photoreceptors and the horizontal cells are pairs of cells with separate periods of neurogenesis yet are found in the same cell layer in the adult retina. The high density of cones and medium sized RGCs in the central area may be due to the early common birthdate of these cell types and their insertion into the region of the retina where neurogenesis ceases earliest (Steinberg *et al.*, 1973; Rapaport and Stone, 1984). The differing mosaic pattern of A-type and B-type horizontal cells (Wässle and Boycott, 1978; Wässle *et al.*, 1978) may also be related to their differing cell birthdates (Zimmerman *et al.*, 1988).

ACKNOWLEDGMENTS. We gratefully acknowledge the talented efforts of Linda Warren, who made the original drawings that formed the basis for the summary diagrams. This work was supported in part by NIH grants EY 4593, EY 6163, and BRSG S07RR05477.

## 1.6. REFERENCES

Adler, R., 1986, Developmental predetermination of the structural and molecular polarization of photoreceptor cells, *Dev. Biol.* **117:**520–527.

Angevine, J. B., Bodian, D., Coulombre, A. J., Eddes, M. V., Hamburger, V., Jacobson, M., Lyser, K. M., Prestige, M. C., Sidman, R. L., Varon, S., and Weiss, P., 1970, Embryonic vertebrate central nervous system: Revised terminology, *Anat. Rec.* **166:**257–261.

Barber, A. N., 1955, *Embryology of the Human Eye*, C. V. Mosby, St. Louis.

Barnstable, C. J., Hofstein, R., and Akagawa, K., 1985, A marker of early amacrine cell development in rat retina, *Dev. Brain Res.* **20:**286–290.

Borwein, B., 1985, Scanning electron microscopy in retinal research, *Scan. Electron Microsc.* **I:**279–301.

Constantine-Paton, M., Blum, A. S., Mendez-Otero, R., and Barnstable, C., 1986, A cell surface molecule distributed in a dorsoventral gradient in the perinatal rat retina, *Nature (London)* **324:**459–462.

Crisanti, P., Lorinet, A. M., Calothy, G., and Pessac, B., 1985, Glutamic acid decarboxylase activity is stimulated in quail retina neuronal cells transformed by Rous sarcoma virus and is regulated by pp60$^{v-src}$, *EMBO J.* **4:**1467–1470.

Duke-Elder, S., and Cook, C., 1963, Embryology, in *System of Ophthalmology, Volume III. Normal and Abnormal Development* (S. Duke-Elder, ed.), pp. 11–57 and 81–99, C. V. Mosby, St. Louis.

Dütting, D., Gierer, A., and Hansman, G., 1983, Self-renewal of stem cells and differentiation of nerve cells in the developing chick retina, *Dev. Brain Res.* **10:**21–32.

Erickson, P. A., Fisher, S. K., Anderson, D. H., Stern, W. H., and Borgula, G. A., 1983, Retinal detachment in the cat: The outer nuclear and outer plexiform layers, *Invest. Ophthalmol. Vis. Sci.* **24:**927–942.

Famiglietti, E. V. Jr., Kaneko, A., and Tachibana, M., 1977, Neuronal architecture of on and off pathways in carp retina, *Science* **198:**1267–1269.

Famiglietti, E. V. Jr., and Kolb, H., 1976, Structural basis ON- and OFF-center responses in retinal ganglion cells, *Science* **194:**193–195.

Fujisawa, H., 1982, Formation of gap junctions by stem cells in the developing retina of the clawed frog (*Xenopus laevis*), *Anat. Embryol. (Berlin)* **165:**141–149.

Fujita, S., 1963, The matrix cell and cytogenesis in the developing central nervous system, *J. Comp. Neurol.* **120:**37–42.

Glücksmann, A., 1940, Differentiation of tadpole eye. *Br. J. Ophthalmol.* **24:**153–178.

Grady, E. F., Schwab, M., and Rosenau, W., 1987, Expression of N-myc and c-src during the development of fetal human brain, *Cancer Res.* **47:**2931–2936.

Greiner, J. V., and Weidman, T. A., 1980, Histogenesis of the cat retina. *Exp. Eye Res.* **30:** 439–453.

Hickey, T. L., Whikehart, D. R., Jackson, C. A., Hitchcock, P. F., and Paduzzi, J. D., 1983, Tritiated thymidine experiments in the cat: A description of techniques and experiments to define the time-course of radioactive thymidine availability, *J. Neurosci. Methods* **8:**139–147.

Hilfer, S. R., 1983, Development of the eye of the chick embryo, *Scan. Electron Microsc.* **III:**1353–1369.

Hinds, J. W., and Hinds, P. L., 1974, Early ganglion cell differentiation in the mouse retina: An electron microscopic analysis utilizing serial sections, *Dev. Biol.* **37:**381–416.

Hinds, J. W., and Hinds, P. L., 1978, Early development of amacrine cells in the mouse retina: An electron microscopic, serial section analysis, *J. Comp. Neurol.* **179:**277–300.

Hinds, J. W., and Hinds, P. L., 1983, Development of retinal amacrine cells in the mouse embryo: Evidence for two modes of formation, *J. Comp. Neurol.* **213:**1–23.

Hinds, J. W., and Ruffett, T. L., 1971, Cell proliferation in the neural tube: An electron microscopic and Golgi analysis in the mouse cerebral vesicle, *Z. Zellforsch.* **115:**226–264.

Holt, C. E., Bertsch, T. W., Ellis, H. M., and Harris, W. A., 1988, Cellular determination in the *Xenopus* retina is independent of lineage and birth date, *Neuron* **1:**15–26.

Johns, P., Rusoff, A., and Dubin, M. W., 1979, Postnatal neurogenesis in the kitten retina, *J. Comp. Neurol.* **187:**545–556.

Kolb, H., Nelson, R., and Mariani, A., 1981, Amacrine cells, bipolar cells and ganglion cells of the cat retina: A Golgi study, *Vision Res.* **21**:1081–1114.

Koontz, M. A., and Hendrickson, A. E., 1987, Stratified distribution of synapses in the inner plexiform layer of the primate retina, *J. Comp. Neurol.* **263**:581–592.

Le Douarin, N. M., 1986, Cell line segregation during peripheral nervous system ontogeny, *Science* **231**:1515–1522.

Levitt, P. R., Cooper, M. L., and Rakic, P., 1981, Coexistence of neuronal and glial precursor cells in the cerebral ventricular zone of the fetal monkey: An ultrastructural immunoperoxidase analysis, *J. Neurosci.* **1**:27–39.

Lia, B., Williams, R. W., and Chalupa, L. M., 1987, Formation of retinal ganglion cell topography during prenatal development, *Science* **236**:848–850.

Lynch, S. A., Brugge, J. S., and Levine, J. M., 1986, Induction of altered *c-src* product during neural differentiation of embryonal carcinoma cells, *Science* **234**:873–876.

Mann, I., 1969, *The Development of the Human Eye*, Grune & Stratton, New York.

Marc, R. E., 1986, Neurochemical stratification in the inner plexiform layer of the vertebrate retina, *Vision Res.* **26**:223–238.

Marshak, D., Ariel, M., and Dowling, J. E., 1984, Laminar distribution of retinal ganglion cell inputs in the goldfish, *Invest. Ophthalmol. Vis. Sci. (Suppl.)* **25**:284.

Martinez, R., Mathey-Prevot, B., Bernards, A., and Baltimore, D., 1987, Neuronal pp60[c-src] contains a six-amino acid insertion relative to its non-neuronal counterpart, *Science* **237**:411–415.

Maslim, J., and Stone, J., 1986, Synaptogenesis in the retina of the cat, *Brain Res.* **373**:35–46.

Mastronarde, D. N., Thibeault, M. A., and Dubin, M. W., 1984, Non-uniform postnatal growth of the cat retina, *J. Comp. Neurol.* **228**:598–608.

McArdle, C. B., Dowling, J. E., and Masland, R. H., 1977, Development of outer segments and synapses in the rabbit retina, *J. Comp. Neurol.* **175**:253–274.

Morest, D. K., 1970, The pattern of neurogenesis in the retina of the rat, *Z. Anat. Ent-wicklungsgesch.* **131**:45–67.

Morse, D. E., and McCann, P. S., 1984, Neuroectoderm of the early embryonic rat eye, *Invest. Ophthalmol. Vis. Sci.* **25**:899–907.

Morrison, J. D., 1982, Postnatal development of the area centralis of the kitten retina: An electron microscopic study, *J. Anat.* **135**:255–271.

Negishi, K., Teranishi, T., and Kato, S., 1982, New dopaminergic and indoleamine-accumulating cells in the growth zone of goldfish retinas after neurotoxic destruction, *Science* **216**:747–749.

Nelson, R., Famiglietti, E. V. Jr., and Kolb, H., 1978, Intracellular staining reveals different levels of stratification for on-center and off-center ganglion cells in the cat retina, *J. Neurophysiol.* **41**:472–483.

Patterson, P. H., 1978, Environmental determination of autonomic neurotransmitter functions, *Annu. Rev. Neurosci.* **1**:1–17.

Polley, E. H., Walsh, C., and Hickey, T. L., 1982, Neurogenesis in the inner nuclear layer (INL) and outer nuclear layer (ONL) of the cat retina: A study using ³H-thymidine, *Invest. Ophthalmol. Vis. Sci. (Suppl.)* **22**:114 (Abstract).

Polley, E. H., Zimmerman, R. P., and Fortney, R. L., 1985, Development of the outer plexiform layer (OPL) of the cat retina, *Soc. Neurosci. Abstr.* **11**:14 (Abstract).

Polley, E. H., Zimmerman, R. P., and Fortney, R. L., 1986, Interaction of a temporal sequence of cell birthdays and a spatial gradient of morphological maturation in the mammalian retina, *Invest. Ophthalmol. Vis. Sci. (Suppl.)* **27**:326 (Abstract).

Polyak, S. L., 1941, *The Retina*, University of Chicago Press, Chicago.

Rakic, P., 1982, Organizing principles for development of primate cerebral cortex, in *Organizing Principles of Neural Development* (S. C. Sharma, ed.), pp. 21–48, Plenum Press, New York.

Ramoa, A. S., Campbell, G., and Shatz, C. J., 1987, Transient morphological features of identified ganglion cells in living fetal and neonatal retina, *Science* **237**:522–525.

Ramon y Cajal, S., 1893, *La retine des vertebres*, English translation by D. Maguire and R. W. Rodieck, Appendix I in *The Vertebrate Retina, Principles of Structure and Function* (R. W. Rodieck, ed.), W. H. Freeman, San Francisco, 1973.

Ramon y Cajal, S., 1929, *Studies on Vertebrate Neurogenesis* (translated by L. Guth), C. C. Thomas, Springfield, IL, 1960.

Rapaport, D. H., and Stone, J., 1984, The area centralis of the retina in the cat and other mammals: Focal point for function and development of the visual system, *Neuroscience* **11**:289–301.

Raymond, P. A., and Rivlin, P. K., 1987, Germinal cells in the goldfish retina that produce rod photoreceptors, *Dev. Biol.* **122**:120–138.

Ready, D. F., Tomlinson, A., and Lebovitz, R. M., 1986, Building an ommatidium: Geometry and genes, in *Development of Order in the Visual System* (S. R. Hilfer and J. M. Sheffield, eds.), pp. 97–125, Springer-Verlag, New York.

Reh, T. A., 1987, Cell-specific regulation of neuronal production in the larval frog retina, *J. Neurosci.* **7**:3317–3324.

Reh, T. A., and Tully, T., 1986, Regulation of tyrosine hydroxylase containing amacrine cell number in larval frog retina, *Dev. Biol.* **114**:463–469.

Robinson, S. R., 1987, Ontogeny of the area centralis in the cat, *J. Comp. Neurol.* **255**:50–67.

Robinson, S. R., Rappaport, D. H., and Stone, J., 1985, Cell division in the developing cat retina occurs in two zones, *Dev. Brain Res.* **19**:101–109.

Sauer, F. C., 1935, Mitosis in the neural tube, *J. Comp. Neurol.* **62**:377–405.

Schnitzer, J., and Rusoff, A. C., 1984, Horizontal cells of the mouse retina contain glutamic acid decarboxylase-like immunoreactivity during early developmental stages, *J. Neurosci.* **4**:2948–2955.

Scholes, J., 1976, Neuronal connections and cellular arrangements in the fish retina, in *Neural Principles in Vision* (F. Zettler and R. Weiler, eds.), pp. 63–93, Springer-Verlag, New York.

Scholes, J., 1987, Uncertainties in the retina, *Nature* **328**:114–115.

Sheffield, J. B., and Fischman, D. A., 1970, Intercellular junctions in the developing neural retina of the chick embryo, *Z. Zellforsch.* **104**:405–418.

Sidman, R. L., 1961, Histogenesis of mouse retina studied with thymidine-$^3$H, *The Structure of the Eye* (G. Smelser, ed.), Academic Press, New York.

Sorge, L. K., Levy, B. T., and Maness, P. F., 1984, pp60$^{c\text{-}src}$ is developmentally regulated in the neural retina, *Cell* **36**:249–257.

Steinberg, R. H., Reid, M., and Lacy, P. L., 1973, The distribution of rods and cones in the retina of the cat (*Felis domesticus*), *J. Comp. Neurol.* **148**:229–248.

Stone, J., Maslim, J., and Rapaport, D., 1984, The development of the topographical organisation of the cat's retina, in *Development of Visual Pathways in Mammals* (J. Stone, B. Dreher, and D. H. Rapaport, eds.), pp. 3–21, Alan R. Liss, New York.

Tomlinson, A., and Ready, D. F., 1986, *Sevenless,* a cell-specific homoeotic mutation of the *Drosophila* eye, *Science* **231**:400–402.

Turner, D. L., and Cepko, C. L., 1987, A common progenitor for neurons and glia persists in rat retina late in development, *Nature (London)* **328**:131–136.

Vardimon, L., Fox, L. E., and Moscona, A. A., 1986, Accumulation of *c-src* mRNA is developmentally regulated in embryonic neural retina, *Mol. Cell Biol.* **6**:4109–4111.

Vogel, M., 1978, Postnatal development of the cat's retina, *Adv. Anat. Embryol. Cell Biol.* **54**(4):1–107.

Walsh, C., Polley, E. H., Hickey, T. L., and Guillery, R. W., 1983, Generation of cat retinal ganglion cells in relation to central pathways, *Nature (London)* **302**:611–614.

Wässle, H., and Boycott, B., 1978, Receptor contacts of horizontal cells in the retina of the domestic cat, *Proc. R. Soc. London Sec. B* **203**:247–267.

Wässle, H., and Reimann, H. J., 1978, The mosaic of nerve cells in mammalian retina, *Proc. R. Soc. London Ser. B* **200**:441–461.

Wässle, H., Peichl, L., and Boycott, B., 1978, Topography of horizontal cells in the retina of the domestic cat, *Proc. R. Soc. London Ser. B* **203**:269–291.

Wässle, H., Chun, M. H., and Müller, F., 1987, Amacrine cells of the ganglion cell layer of the cat retina, *J. Comp. Neurol.* **265**:391–408.

Walsh, C., and Polley, E. H., 1985, The topography of ganglion cell production in the cat's retina, *J. Neurosci.* **5**:741–750.

Wetts, R., and Fraser, S. E., 1988, Multipotent precursors can give rise to all major cell types of the frog retina, *Science* **239:**1142–1145.

Whiteley, H. E., and Young, S., 1986, The external limiting membrane in developing normal and dysplastic canine retina, *Tissue Cell* **18:**231–239.

Williams, R. W., Bastiani, M. J., Lia, B., and Chalupa, L. M., 1986, Growth cones, dying axons, and developmental fluctuations in the fiber population of the cat's optic nerve, *J. Comp. Neurol.* **246:**32–69.

Wong, R. O., and Hughes, A., 1987, The morphology, number, and distribution of a large population of confirmed displaced amacrine cells in the adult cat retina, *J. Comp. Neurol.* **255:**159–177.

Young, R. W., 1983, The life history of retinal cells, *Trans. Am. Ophthalmol. Soc.* **81:**193–228.

Young, R. W., 1985, Cell differentiation in the retina of the mouse, *Anat. Rec.* **212:**199–205.

Zimmerman, R. P., 1983, The organization of the ganglion cell dendritic grids in the retina of *Astronotus, Soc. Neurosci. Abstr.* **9:**802 (Abstract).

Zimmerman, R. P., Polley, E. H., and Fortney, R. L., 1985, Stages in the development of the inner plexiform layer of the cat retina, *Soc. Neurosci. Abstr.* **11:**14 (Abstract).

Zimmerman, R. P., Polley, E. H., and Fortney, R. L., 1987, The ultrastructure of the cat's retina during ganglion cell neurogenesis, *Invest. Ophthalmol. Vis. Sci. (Suppl.)* **28:**286 (Abstract).

Zimmerman, R. P., Polley, E. H., and Fortney, R. L., 1988, Cell birthdays and rate of differentiation of ganglion and horizontal cells of the developing cat's retina, *J. Comp. Neurol.* **274:**77–90.

# Retinal Rod Neurogenesis    2

## Russell D. Fernald

> A scientist must also be absolutely like a child. If he sees a thing, he must say that he sees it, whether it was what he thought he was going to see or not. See first, think later, then test. But always see first. Otherwise you will only see what you were expecting.
>
> Douglas Adams, *So Long and Thanks for All the Fish*

## 2.1. INTRODUCTION

The vertebrate eye originally evolved as an underwater visual organ; thus, all vertebrate eyes share a common set of structural properties. This is in striking contrast to the enormous variety of eye types found among invertebrates. Despite the fundamentally similar ocular architecture that exists throughout vertebrate phylogeny, there are still significant differences among eyes. Specifically, fishes, which comprise more than half the extant vertebrate species, have eyes that grow throughout their lifetimes. The advantages of a larger eye are the greater light-capturing ability for deep-sea fish and higher acuity for surface dwellers (for details, see Fernald, 1988).

Retinal growth in fish occurs both via stretching of the existing retinal tissue and addition of new cells at the ciliary margin from a modified epithelial zone (Fernald, 1984). Cells generated at the margin are integrated into existing retina seamlessly, and the complex patterns of photopigment distribution and so on are perfectly matched to the extant retina (Fernald, 1983).

Superficially, such growth would not appear to be a problem since one would assume that it is simply an extension of embryonic growth. However, the cellular consequences of this growth and its regulation have only recently been analyzed. The history of these discoveries is interesting (particularly with perfect hindsight) because they provide new insights about how the selective advantage of having a larger eye was ultimately translated into a developmental program.

---

**Russell D. Fernald** • Institute of Neuroscience, University of Oregon, Eugene, Oregon 97403. Written while R.D.F. was the Hilgard Visiting Professor at Stanford University, Stanford, California 94305.

## 2.2. HISTORICAL REVIEW

It has long been recognized that the eyes of frogs, toads, and fishes grow throughout the lifetimes of the animals. Wunder (1926) proposed that cell division must occur at the margin of the teleost eye, even though he could not find mitotic figures. Some of the earliest experiments were attempts to assess the functional capability of animal eyes of different sizes (ages). So, for example, Birukow (1949) showed that the eyes of larval and adult *Rana temporalis* had the same capabilities in bright light, but not in dim light, where the larvae were not able to see as well. Möller (1950), who first looked at urodeles during growth, found that there was an increase in total number of cells although he noted that the density of ganglion cells and cones decreased whereas the density of rods remained constant (". . ihre ursprüngliche Dichte beibehalten. . .") in the five species he examined. He also found a difference between bright and dim light performance. Möller did not comment on the possible sources of new cells, choosing only to document the changes in absolute number with age. At about the same time, Hans Müller (1952) began working on the development and growth of the guppy retina at the University of Freiburg in the laboratory of Professor H. Lüdtke. Müller was a teacher in the Gymnasium (H. Wässle, personal communication) and was studying for an advanced degree. His choice of Lüdtke undoubtedly influenced both his choice of topic and ultimately his conclusions. Earlier, Lüdtke had published several studies about the backswimmers (*Notonecta glauca*), most notably on the embryonic and postembryonic development of the eye (Lüdtke, 1940). His conclusion was that the eye grew from moult to moult by the concentric addition of cells at its margin.

Müller arduously counted ganglion cells, inner nuclear layer cells, rods, and cones in four size classes of guppy (*Lebistes reticulatus*). In this analysis, he found that the absolute number of all cell types increased (Müller, 1952, Table 14, p. 298), but that the *relative* increase in number of rods was greater (Müller, 1952, Table 16). When presented as the ratio of other cell types to 10 ganglion cells (Müller, 1952, Fig. 24), as the fish increased fourfold in length, the ratio of inner nuclear layer cells increased by about 12%, the ratio of cones remained constant, and the ratio of rods to ganglion cells increased by 250%! Although Müller did not express the values as percentages, he was nonetheless impressed by the fact, as also found by Möller, that the rod photoreceptor density remained approximately constant with growth. Since he observed mitotic figures only in the marginal zone, Müller concluded that the margin was the only source of new rod cells which must somehow then move into the center of the retina.

The improbability of this can be seen from the data provided by Müller himself. As the guppy grows from 7 to 21 mm in 50 days, 293,000 new rods are added, or about 5860 per day. On average, each would have to move about 325 $\mu$m radially, and more importantly, presumably every cell, not just the new ones, would have to move every other day! Since rods consist of long outer segments, approximately 50 $\mu$m in length, connected to a cell body located 10–15 μm toward the center of the eye, where connections with other retinal cells occur, and since rods are distributed among the cones that need not move to account for the data, this migration is very difficult to contemplate indeed. This essentially continuous lateral migration of such a large fraction of the rods would seem to cause too great a disruption of retinal function in an animal dependent on vision. Nonetheless, Müller concluded that his results were parallel to those of his mentor, Lüdtke, on the backswimmer, with cells added only from the marginal neurogenic zone.

Other investigators subsequently confirmed Müller's central observation, that rod density remains constant with eye growth, while the density of all other cell types drops. But the issue of just *how* this was achieved remained a matter of debate until quite recently. Since cell addition at the retinal margin could account for the increase in total cell number as the eye grows, the question is: Where do the extra rods come from? Müller states that the rods must slide through regions, since their density remains constant over the whole retina (Müller, 1952, p. 315).

In the 35 years since Müller's analysis, a handful of investigators have worked on the same problem. Vilter and Lewin (1954) studied the retina of an adult deep-sea teleost (*Bathylagus bendicti* = *B. euryops*) which has a pure-rod retina with several layers of banks of rods in the receptor layer. They reported mitoses in both the outer and inner nuclear layers, suggesting that they probably produced rods needed to maintain the resolving power of the retina. Much later, also in an all-rod retina, Munk and Jørgensen (1983) reported finding mitoses in the outer nuclear layer in two deep-sea teleosts.

Baburina (1955) found that rods appeared coincident with mitoses in the ONL and that there was a second layer of photoreceptor nuclei added vitread to the cone nuclei. Lyall suggested two hypotheses based on an analysis of changes in the cone arrangements during growth. First, that migration of undifferentiated cells into the outer nuclear layer is "probably responsible for the increase in the number of rods" (Lyall, 1957a, p. 109), and second, commenting on the loss of single cones in the trout, that "the transmutation of these cones into rods seems the most probable explanation" (Lyall, 1957b, p. 197). Ahlbert (1973, 1976), studying perch, salmon, and trout, found mitoses primarily in the outer nuclear layer, but some as well in the inner nuclear layer and the ganglion cell layers. He concluded that "it seems plausible that the cells resulting from mitoses in the outer nuclear layer differentiate into rods as these are known to increase in number throughout the retina during growth" (Ahlbert, 1976, p. 33). Blaxter and Jones (1967) suggested that new cells might be recruited from the INL in the herring. As Blaxter (1975) noted, however, the source of the new rods in the retina was not at all clear.

None of these investigators used unambiguous markers for cell division, relying instead entirely on the appearance of mitotic figures in their histological preparations. Hollyfield (1972) used ³H-thymidine, a precursor of DNA, to label cells that had had a mitotic division between its injection and sacrifice. Although interested primarily in the histogenesis of the retina, he also injected animals later in life and stated that "labeled muclei were only found near the margin of the retina" (Hollyfield, 1972, p. 378).

Johns and Easter (1977) undertook a quantitative analysis of retinal growth in the goldfish and reported results comparable to those of Müller (1952) in the guppy. They found that rod density remained constant with growth, while the density of all other cell types decreased. In her further analyses of this discrepancy, Johns (1976, 1977) argued that the *only* site of retinal neurogenesis was the retinal margin and that the new rods needed to maintain rod density arose there, just as Müller had argued 25 years earlier. Johns elaborated the original hypothesis of Müller, suggesting that the rods do not "move" but rather "resist" retinal expansion of the other layers and thus maintain their density. Johns' retinal shearing hypothesis stated specifically that "rods are relatively more dense in the central retina of big fish *not* because rods have been added there, but because non-rod cells have left, pulled peripherally in the retinal expansion" (Johns, 1977, p. 356). Although she stated that "some of the rods are pulled peripherally by the retinal expansion, others remain behind" (Johns, 1977, p. 356), *any* relative movement would, of course, require that rods, whether some or all, disconnect and then reconnect frequently. As in the analysis of Müller's data, the number of photoreceptors and distances involved are quite large.

By 1976, autoradiographic evidence of cell division in the outer nuclear layer of the fish retina had been published but not recognized as the source of newly added rods. Scholes (1976) has photomicrographs of such labeled cells and Meyer (1978, p. 102) notes "many scattered labeled cells were found centrally in this layer [the outer nuclear layer] for another several hundred micrometers and some fewer cells were observed throughout the central retina." In Johns' paper (1977, Fig. 4), four labeled cells in the outer nuclear layer are shown, although identified as "displacement of labeled rods" to be consistent with her "retinal shearing hypothesis."

## 2.3. PRESENT VIEWS OF ROD NEUROGENESIS

Presently, I am working on the highly visual cichlid fish, *Haplochromis burtoni*, which grows so rapidly that the eye size can double within weeks. It maintains a constant rod density, and one can predict that every rod would have to move about every few hours during the lifetime of the animal. In collaboration with Johns, I repeated her experiments on the cichlid and from the very first animal found numerous labeled rod nuclei (see Fig. 2.1) distributed throughout the outer nuclear layer (Fernald and Johns, 1980a,b). Repeating the experiments on goldfish confirmed that there too cell division was common throughout the outer nuclear layer (Johns and Fernald, 1981). Sandy and Blaxter (1980) had also been looking at teleost retinal development, examining the sequence of development in larval herring and sole. They found that rods appear late in retinal development and, based on $^3$H-thymidine autoradiography, postulated that rods arise from scattered

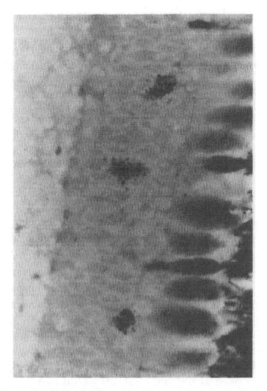

**Figure 2.1.** Autoradiograph of a retinal section (3 μm, sclera upward) of an adult *H. burtoni* injected 24 hr previously with ³H-thymidine. Three labeled nuclei in the outer nuclear layer are clearly visible. Labeled nuclei are found only in the outer nuclear layer and these nuclei are cytologically indistinguishable from the surrounding nuclei of rods. From Fernald (1988).

"basal" cells in the outer nuclear layer. These basal cells they postulated came originally from the margin of the eye (Sandy and Blaxter, 1980). It now seems clear that the rods needed to maintain constant rod density come from cell divisions in the ONL. The ultimate origin and lineage of these precursor cells is not yet clear.

### 2.3.1. Discovery of the Secondary Marginal Zone and Its Regulatory Function

Thus, in the teleost retina, new retinal tissue is being added all around the iris margin at the ora serrata. In addition, the existing retina is being stretched, accommodating the intercalation of newly generated rods exclusively. These two processes preserve visual sensitivity by maintaining constant rod density while adjusting the angular spacing of cones and hence visual acuity to increasing eye size (see Fernald, 1985, and this volume for details). As a consequence of these essentially competing processes, there is a critical regulatory problem at the margin of the eye (Fernald and Scholes,

1985, 1988). Since the density of rods is constant at all sizes of fish, while the density of all other cell types decreases, the *ratio* of the number of rods to the number of other cells is increasing. For example, at hatching, *H. burtoni* has approximately 1 rod/cone, whereas at 6 months the ratio may be 4–6 rods/cone. In the central retina, rod addition is distributed and is responsible for incrementing the rod/cone ratio throughout life. At the margin of the eye, however, the germinal zone is obliged to produce an increasing ratio of rods to other retinal neurons, depending on the size of the fish.

How does the germinal zone "know" the current ratio of rods to cones, particularly since the individual growth rates can vary enormously, depending on rearing conditions (Fernald, 1983)? We have recently discovered a second exclusive rod germinal zone near the margin of the eye that achieves this regulatory feat (Fernald and Scholes, 1985, 1988).

Examining the outer nuclear layer (ONL) near the retinal margin in fish of different species and widely different sizes, we found that there is a single sheet of cells with the rods arranged in a sparse square lattice among the cones, which are arranged in their typical array. Nearer the center of the retina, however, rod nuclei become much more numerous, packed confluently in a separate layer vitread to the cone nuclei, more than one deep. As seen in Fig. 2.2, the density of rods increases abruptly from approximately 2.5 $\times 10^4$ to $9 \times 10^4/mm^2$, a value typical in *H. burtoni* of all sizes (Fernald, 1988). The increase begins about 100 μm from the margin. The density toward the center of other retinal neurons declines somewhat in this region of the retina but is uniform elsewhere in the retina at any given age.

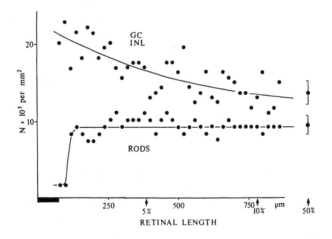

**Figure 2.2.** Retinal cell density plotted against retinal length (expressed in μm and as %), with *black bar* signifying marginal retinal blastema. The density of rod nuclei (RODS) increases 80–100 μm inside the margin, contrasting with inner nuclear layer (INL) and ganglion cell density (GC), which falls in peripheral 10% of retinal length but thereafter remains constant throughout like the rod density. Values are corrected cell counts from photomicrographs using 20 μm sampling bins over first 12% of retinal length and then 100 μm bin at the center of the retina (retinal length = 50%), where standard deviations are also shown. From Fernald (1988).

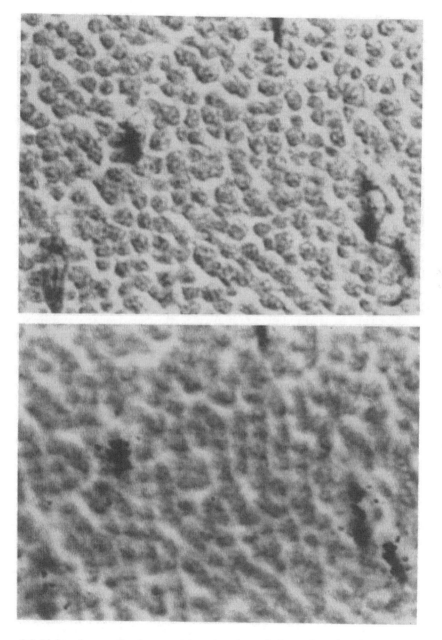

**Figure 2.3.** Light micrograph of tangential section through the retina of *H. burtoni*, showing pleomorphic cells (top), which have been labeled by injection of [3]H-thymidine (bottom). From Fernald (1988).

Corresponding spatially with the sharp increase in rod density, we found numerous extra cell profiles concentrated in a narrow annulus in the ONL, supernumerary to the rods and cones (see Fig. 2.3). We have called these *pleomorphic cells* (Fernald and Scholes, 1985) because of their variable shapes, which include occasional mitotic figures. After intermediate survival times (4 days), pleomorphic cells were comparatively weakly labeled, whereas heavy clusters of grains were seen over differentiated rod nuclei. After a long survival time (10 days), only rod nuclei were labeled.

These findings suggest a novel solution to the problem of generating the required *variable* ratio of rods to cones at the margin of the eye during the life of the animal. The pleomorphs operate independently of the primary retinal germinal zone, continually moving outward through the ONL to colonize new tissue made at the margin during growth in the adult to produce rods. Moreover, this same zone appears to populate the newly generated retina with the precursors that are responsible for inserting rods later as the tissue stretches (Fernald and Scholes, 1985, 1988). This mechanism is shown schematically in Fig. 2.4.

An important feature of the second, exclusive rod germinal zone is that the stem cells (pleomorphs) themselves appear to *increase* in number, maintaining a constant density as the eye grows, whereas the primary germinal zone appears to *decrease* in density with growth. This crucial difference allows the pleomorphs to accumulate a numerical advantage over the marginal stem

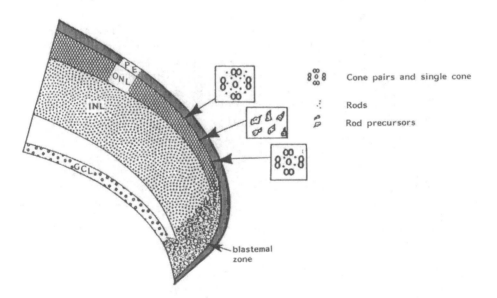

**Figure 2.4.** Schematic summary of the retinal edge as reconstructed from adult *H. burtoni* retinal tissue. The cone array as seen peripheral to and central to the zone of rod precursors shows a massive increase in rod photoreceptors central from this region.

cells so that the secondary rod production in new retinal tissue can produce enough rods to match the existing retina.

## 2.4. EMBRYONIC ORIGIN OF RODS

Since the margin of the adult eye is a spatial recapitulation of developmental time, the discovery of an exclusive zone of rod neurogenesis at the margin of the adult eye suggested that we might find a period of exclusive rod neurogenesis during embryonic time. To examine this, we have begun a study of embryonic retinal neurogenesis.

We have used $^3$H-thymidine injections to map the position of dividing cells during retinal embryogenesis (Fernald and Shelton, 1986). Initially, to produce eggs at various developmental stages, we relied on the fertilizations that occurred in our breeding colonies. We discovered, however, a high level of variability in developmental stages within single broods. Recently, we have been successful with *in vitro* fertilization so that we can now produce animals of known age exactly.

We anesthetized embryos using iced tank water and injected them with 2–3 $\mu$Ci $^3$H-thymidine directly into the body cavity since injections into the yolk resulted in longer availability of $^3$H-thymidine. Following survival times ranging from 1.5 hr to 10 days, animals were sacrificed and immersed in Bouin's fixative. Following conventional procedures, they were embedded in plastic (JB-4, Polysciences) and sectioned at 4 $\mu$m. To reveal the location of thymidine-labeled cell nuclei, slides were coated with emulsion (NTB-2, Kodak) and developed using conventional procedures. The tissue was examined under the light microscope. The distribution of labeled cells was mapped using the Lambert azimuthal projection, which preserves area when a hemisphere is projected onto a plane.

The outpocketing of the diencephalon into optic vesicles occurs at about 24 hr, followed by formation of the lens placode and optic cup. By day 3 (72 hr), rapid cell division is occurring in the future neural retina and by day 4 (96 hr), there is evidence of considerable organization in the inner nuclear layer. By this time, cell division is occurring throughout the eyecup, with a clear vitread–sclerad gradient of development extending from the ganglion cell layer to the photoreceptor layer as is common in most developing vertebrate retinas (e.g., Carter-Dawson and LaVail, 1979). Figure 2.5 summarizes the developmental data for retinal growth from typical cases.

By the fourth day of development, approximately half the eyecup contains dividing cells and the remainder differentiated, layered retina. By the fifth day after fertilization, the ratio of dividing to differentiated cells has decreased and a significant amount of cell division appears in the vicinity of the temporal pole of the retina, near the site of the first cell divisions in retinal tissue. By day 7, this second phase of cell divisions covers much more of the differentiated retinal tissue and has begun to take the shape of a ring. By day 14, the distribution of the late labeled cells is a ring close to the margin of the eye.

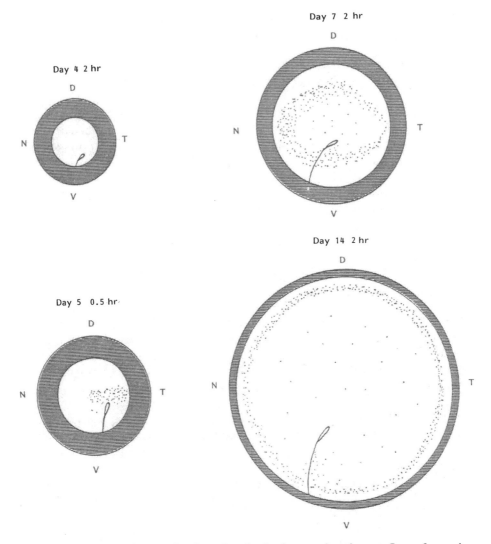

**Figure 2.5.** In each case, the curved embryonic retina has been projected onto a flat surface and its outer edge is represented by a circle. Retinal orientation is given by D = dorsal, V = ventral, N = nasal, and T = temporal. Shaded areas are where mitotic activity is occurring as revealed by $^3$H-thymidine label. Clear areas are where retinal layering is distinct and cellular differentiation is proceeding. Individual dots in the central areas are individual cell divisions which occur in the ONL. The line represents the embryonic fissure and the small loop the optic nerve.

To identify the progeny of these labeled cells, we examined the retina with electron microscopy and we looked at labeled progeny in animals that survived for long periods of time. For electron microscopy, we reembedded outer nuclear layer tissue from the central retina, from the center of the mitotically active zone, and from the peripheral retina and sectioned it for electron microscopy. We found that central tissue had differentiated rod and cone photoreceptors. The marginal tissue had an undifferentiated population of cells. In the midst of the area with cell divisions, we found differentiated cone photoreceptors with incompletely assembled outer segments and no evidence of rods. In tissue from animals that survived for a longer time, we found only rod nuclei labeled in the central retina.

This late phase of rod addition in the embryonic tissue appears to be the origin of the second germinal zone described above (Fernald and Scholes, 1985, 1988). This suggests that the adult retinal edge contains an exact map of development, where developmental time is mapped onto distance along the retinal edge.

ACKNOWLEDGMENTS. This work was supported by the NIH (EY 05051) and the Medical Research Foundation of Oregon. I would like to thank Dr. J. Scholes and Ms. L. Shelton for scientific help and Ms. J. Kinnan for unstinting clerical help.

## 2.5. REFERENCES

Ahlbert, I. B., 1973, Ontogeny of double cones in the retina of perch fry (*Perca fluviatillis*, Teleostei), *Acta Zool. (Stockholm)* **54**:241–254.

Ahlbert, I. B., 1976, Organization of the cone cells in the retinae of salmon (*Salmo salar*) and trout (*Salmo trutta trutta*) in relation to their feeding habits, *Acta Zool. (Stockholm)* **57**:13–35.

Baburina, E. A., 1955, The eye of the retina in the Caspian shad, *Dokl. Akad. Nauk. S.S.S.R.* **100**:1167–1170.

Birukow, G., 1949, Die Entwicklung des Tages- und Dämmerungssehens im Auge des Grasfrosches (*Rana temporaria*), *Z. Vergl. Physiol.* **31**:322–347.

Blaxter, J. H. S., 1975, The eyes of larval fish, in *Vision in Fishes* (M. A. Ali, ed.), pp. 427–443, Plenum Press, New York.

Blaxter, J. H. S., and Jones, M. P., 1967, The development of the retina and retinomotor responses in the herring, *J. Mar. Biol. U.K.* **47**:677–697.

Carter-Dawson, L. D., and LaVeil, M. M., 1979, Rods and cones in the mouse retina. II. Autoradiographic analysis of cell generation using tritiated thymidine, *J. Comp. Neurol.* **188**:263–272.

Fernald, R. D., 1983, Neural basis of visual pattern recognition in fish, in *Advances in Vertebrate Neuroethology* (J.-P. Ewert, R. R. Capranica, and D. J. Ingle, eds.), pp. 569–580, Plenum Press, New York.

Fernald, R. D., 1984, Vision and behavior in an African cichlid fish, *Am. Sci.* **72**(1):58–65.

Fernald, R. D., 1985, Growth of the teleost eye: Novel solutions to complex constraints, *Env. Biol. Fish* **13**:113–123.

Fernald, R. D., 1988, Aquatic adaptations in fish eyes, in *Sensory Biology of Aquatic Animals* (J. Atema, R. R. Fay, A. N. Popper, and W. N. Tavolga, eds.), pp. 435–465, Springer-Verlag, New York.

Fernald, R. D., and Johns, P., 1980a, Retinal structure and growth in the cichlid fish, *Invest. Ophthal, Vis. Sci. (Suppl.)* **69**.

Fernald, R. D., and Johns, P., 1980b, Retinal specialization and growth in the cichlid fish, *H. burtoni, Am. Zool.* **20**:943.

Fernald, R. D., and Scholes, J., 1985, A zone of exclusive rod neurogenesis in the teleost retina, *Soc. Neurosci. Abstr.* **11**:810.

Fernald, R. D., and Scholes, J., 1988, Retinal neurogenesis in teleosts: A second germinal zone, in press.

Fernald, R. D., and Shelton, L., 1986, Zone of exclusive rod neurogenesis in teleost retina arises late in embryogenesis, *Soc. Neurosci. Abstr.* **11**:389.

Hollyfield, J. G., 1972, Histogenesis of the retina in the killifish *Fundulus heteroclitus, J. Comp. Neurol.* **144**(3):373–380.

Johns, P. R., 1976, Synaptic connections must change in the adult goldfish retina, *Neurosci. Abstr.* **6**:826.

Johns, P. R., 1977, Growth of the adult goldfish eye. III. Source of the new retinal cells, *J. Comp. Neurol.* **176**(3):343–357.

Johns, P. R., and Easter, S. S., 1977, Growth of the adult goldfish eye. II. Increase in retinal cell number, *J. Comp. Neurol.* **176**(3):331–341.

Johns, P. R., and Fernald, R. D., 1981, Genesis of rods in teleost fish retina, *Nature (London* **293**:141–142.

Lüdtke, H., 1940, Die embryonale und postembryonale Entwicklung des Auges bei *Notonecta glauca, Z. Morphol. Oekol. Tiere* **37**(1):1–37.

Lyall, A. H., 1957a, The growth of the trout retina, *Q. J. Microsc. Sci.* **98**:101–110.

Lyall, A. H., 1957b, Cone arrangements in teleost retinae, *Q. J. Microsc. Sci.* **98**:189–201.

Meyer, R. L., 1978, Evidence from thymidine labelling for continuing growth of retina and tectum in juvenile goldfish, *Exp. Neurol.* **59**:99–111.

Möller, A., 1950, Die Struktur des Auges bei Urodelen verschiedener Körpergrösse, *Zool. Jahrb. Abt. Zool. Physiol. Tiere* **62**(2):138–182.

Müller, H., 1952, Bau und Wachstum der Netzhaut des Guppy, *Zool. Jahrb.* **63**:275–324.

Munk, O., and Jørgensen, J. M., 1983, Mitoses in the retina of two deep-sea teleosts, *Vidensk. Medd. Dan. Naturhist. Foren. Khobenhavn* **144**:75–81.

Sandy, J. M., and Blaxter, J. H. S., 1980, A study of retinal development in larval herring and sole, *J. Ma . Biol. Assoc. U.K.* **60**:59–71.

Scholes, J. H., 976, Neuronal connections and cellular arrangement in the fish retina, in *Neural Principles i Vision* (F. Zettler and R. Weiler, eds.), pp. 63–93, Springer-Verlag, New York.

Vilter, V., and Lewin, L., 1954, Existence et répartition des mitoses dans la rétine d'un poisson abyssal, *Ba hylagus benedicti, C. R. Soc. Biol.* **148**:1771–1775.

Wunder, W., 1 )26, Über den Bau der Netzhaut von Süsswasser fischen, die in grosser Tiefe leben, *Z. V rl. Physiol.* **4**(1):22–36.

# The Regulation of Neuronal Production during Retinal Neurogenesis

# 3

Thomas A. Reh

## 3.1. INTRODUCTION

The vertebrate central nervous system is composed of a large number of different types of neurons that can be distinguished on the basis of morphology, biochemistry, and electrophysiology. During the development of the nervous system, these various kinds of neurons are produced in precise ratios with respect to one another, resulting in the appropriate numbers of the different cell types necessary for the functioning of the adult nervous system. How these different cell types are generated in the correct numbers and ratios is a central problem of developmental neurobiology. While there is some information concerning the factors involved in the origin of cellular diversity in the peripheral nervous system (Patterson, 1978; Le Douarin, 1986; Rohrer *et al.*, 1986) and work in neural crest has shown that a variety of environmental factors can influence the particular phenotype the neural crest cell ultimately achieves, little is known about the mechanisms that give rise to the even greater cellular diversity in the central nervous system. The question of how the precise ratios of neurons found in the adult CNS arise during development is the subject of this chapter. While cell death is likely to play a very important role in regulating final neuronal numbers in many areas of the CNS (Cowan, 1973), this aspect of neural development is reviewed in other chapters in this volume (Chapters 7, 9, and 10) and therefore will only be considered briefly in this chapter.

---

**Thomas A. Reh** • Department of Medical Physiology, University of Calgary, Calgary, Alberta, Canada T2N 4N1.

## 3.2. POTENTIAL MECHANISMS REGULATING CELL NUMBER AND TYPE

### 3.2.1. Are Neurons Produced in the Appropriate Ratios as a Result of Strict Lineal Relationships and Highly Deterministic Cell Division Patterns?

Cell number in most vertebrate tissues is thought to be regulated at the level of cell proliferation and differentiation by several factors, including growth factors, mitotic inhibitors, and various extracellular matrix components. Such factors have also been implicated in the production of neurons by the neural crest during development (Weston, 1986; Erickson and Turley, 1987; Perris *et al.*, 1988). However, little evidence exists for the presence or action of similar types of factors to regulate the production of the various neuronal cell types in the vertebrate CNS. In fact, it is possible that CNS neurons are produced in the appropriate ratios as a result of strict lineal relationships and highly deterministic patterns of cell division. Evidence from certain invertebrate systems, like *C. elegans* and the leech (Stent *et al.*, 1982; but see also Sternberg, 1988), indicates that just such a rigid, deterministic mode of cell production can give rise to the various types of neurons in a simple nervous system.

Several chimeric studies in mice have also argued that the numbers of particular cells in the CNS arise by a highly deterministic pattern of cell divisions from a few founder cells (Wetts and Herrup, 1982a,b,c; Herrup, 1986). From Purkinje cell counts in wild-type/lurcher mutant cerebella, Herrup (1986) further proposed that the number of neurons in the CNS is achieved as "an autonomous property of the lineage itself and hence, presumably, of the progenitor cell." This conclusion was based on the fact that although there was extensive mixing of the cells of the two genotypes, the final number of Purkinje cells conformed to values expected from contributions of autonomous progenitors of each genotype. However, due to the indirect nature of this type of analysis, other interpretations are possible (Mead *et al.*, 1987). While the presence of developmental clones of a strain-specific size provides clear evidence for a *genetic* contribution to neuronal numbers, this contribution may not be based on lineage at the level of the Purkinje cell precursor, but rather could arise from the local interactions among all the cells within a region of the cerebellar germinal zone of the same genotype. Subsequent cell mixing during the stages of migration, differentiation, and death would then obscure these relationships.

More direct evidence for a lineage-based control of cell number comes from *in vitro* studies of muscle stem cells. Quinn *et al.* (1985) found that the final number of cells in chick muscle cell clones *in vitro* was an integral exponent of 2, suggesting that a "division-counting mechanism" and symmetrical division patterns could regulate cell number in this system. However, more recently, *in vitro* time-lapse cinematographic studies of the same cells demonstrated many exceptions to this simple division pattern (Konigsberg and

Pfister, 1986). Moreover, *in vivo* studies of muscle lineages in zebrafish embryos, by clonal analysis of dye-filled blastomeres, found no evidence for symmetric muscle cell lineages; muscle cell progenitors could generate two, three, five, or six cells, in the twelve examples examined (Kimmel and Warga, 1986, 1987). Recently, similar studies of the lineage relationships of the vertebrate CNS at this level of detail have been undertaken. Both dye injections of progenitor cells and retrovirus-mediated gene transfer techniques indicate that a strict deterministic model of cell division patterns is not applicable for CNS progenitors. For example, clusters of labeled cells (presumably clones) found in rat retina following retroviral infection on the first few postnatal days consist of several different combinations of the various retinal cell types still being generated at those ages (Price *et al.*, 1987; Turner and Cepko, 1987). Also, in zebrafish, the final division of a CNS progenitor derived from a particular blastomere might be a motoneuron and an interneuron, two interneurons, or two motoneurons in different animals (C. D. Kimmel, personal communication). These results indicate that the final phenotype, and consequently the final number of cells of any particular phenotype, is not strictly based on that cells' lineage, and they suggest that the types of neurons produced in the retina depend on interactions with the neighboring cells and the microenvironment (see also Hinds and Hinds, 1978, 1979). This conclusion is hardly surprising in light of previous experimental embryological studies of inductive interactions in CNS development, in which particular regions of the embryonic nervous system can be induced to change their commitment or regenerate adjacent areas (Detwiler, 1945, 1946; Harrison, 1947; see Cowan and Finger, 1982, for a review). Such studies have demonstrated the importance of *position* in the specification of phenotype. On the other hand, this is not to say that lineage plays no role in the specification of ultimate phenotype, since lineage will normally determine the position of the progeny. In addition, there is substantial evidence that the germinal cells of the various regions of the CNS are stably determined to differentiate into the appropriate neuronal phenotypes according to their position along the anterior–posterior axis as early as neural tube stage (Holtfreter and Hamburger, 1955; Model, 1978).

### 3.2.2. Are All Neuronal Cell Types Overproduced, with the Appropriate Ratios Established as a Result of a Later Phase of Cell Death?

Cell death and other regressive events are important aspects of CNS development. Cell death is widespread in the nervous system and is particularly apparent in—though not confined to—those neurons that project axons over relatively long distances, where it appears to be of major importance in the quantitative matching of projection neurons with their postsynaptic targets (Oppenheim, 1981). In the retina, ganglion cell loss during normal development can be as great as 80% of the cells initially produced (Lia *et al.*, 1987). On the other hand, there is much less cell loss typically observed in local circuit neurons, and consequently it has received relatively little attention.

Nevertheless, recent studies in rat have reported that a wave of cell death occurs in the retina involving all retinal cell types (Spira *et al.*, 1984; Young, 1984). Might this not be used as a mechanism for eliminating unnecessary retinal elements, paring the appropriate cellular ratios out of an initially undefined population? While this is clearly a potentially powerful mechanism for the establishment of the correct numbers of the various types of retinal neurons, there is some evidence to suggest such a simple matching process is probably not operating in developing retina. If a pre- and postsynaptic interdependence existed among the cells of the retina, removal of one or another cell type would result in the transneuronal degeneration of those cells in contact with the ablated type of neuron. This experiment has recently been carried out by Beazley *et al.* (1987), in which the ganglion cell death that normally occurs in the rat was amplified by a crush of the optic nerve. Such a treatment eliminated all the retinal ganglion cells and yet had no effect on the subsequent rate, magnitude, or distribution of pycnotic nuclei present in the retina (see also Chapter 7 in this volume). Also, the photoreceptor degeneration that naturally occurs in the PRD mutant mouse has no dramatic transneuronal degenerative effects on the other retinal cells (Blanks and Bok, 1977). These negative results can perhaps be explained by the ability of local circuit neurons to make synapses with cells other than their normal target cell types; that is, a sustaining collateral, albeit inappropriate, might provide the necessary trophic support. However, even given this explanation, it is clear that trophic interdependency and cell death would not provide an effective mechanism for quantitatively matching cell populations whenever an alternative target cell population is available.

### 3.2.3. Selective Neurotoxin Lesions in Developing Retina Demonstrate That Neuronal Production Can Be Regulated by the Microenvironment

The frog retina has proved to be a very useful preparation for studying the regulation of central neurogenesis. The neural retina of the frog, *Rana pipiens*, like the retinas of other vertebrate classes, develops with a distinct central to peripheral gradient of cell proliferation and differentiation (Hollyfield, 1968; Beach and Jacobson, 1979; Reh and Constantine-Paton, 1983). However, this development is prolonged in the frog, with new neurons added at the extreme periphery of the retina for the entire larval period of several months. At metamorphosis, this new cell addition slows considerably, but retinal neurogenesis continues in adult frogs. Thus, it is possible experimentally to manipulate the composition of the differentiated retina and examine the effects on the production of new neurons by the proliferative zone.

There are several possible ways in which cell production can be regulated in vertebrate tissues, including the central nervous system. If a single type of germinal neuroepithelial cell can give rise to all types of retinal neurons, as appears to be true for rat, this multipotent cell might be stimulated to generate the different cell types required, depending on the prevailing extracellular conditions. For example, a balance between two factors secreted by two different cell types might be necessary to keep a multipotent precursor generating new cells of each type; when the factors are no longer balanced due to the overproduction of one or the other of the two cell types, the precursor cell might be directed to produce more of the type that is under-represented. This model would then require some degree of regulative interaction among the dividing and differentiating cells of the retina. Another possible model entails no regulation of cell division in the germinal neuroepithelial cells, but rather the production of new cells could be controlled at the level of differentiation. Germinal cells produce undetermined, undifferentiated postmitotic cells that differentiate into particular types of retinal neurons and glia, depending on their local environment. This model also requires cellular interactions to regulate the types and numbers of CNS cells that are produced, but now the interactions occur among differentiating cells: cell division is not regulated.

To determine whether the production of new neurons is regulated by previously differentiated neurons, we made selective neurotoxic lesions of the larval frog retina and examined the effects on new cell production (Reh and Tully, 1986). In the first series of experiments, the neurotoxin 6-hydroxy-dopamine (6-OHDA) was used to destroy the dopamine-containing amacrine cells in the retina (Reh and Tully, 1986). By assaying dopamine and serotonin with HPLC and substance P by a radioimmunoassay, we found that a particular intraocular dose of the toxin would induce selective degeneration of only this one class of catecholamine-containing cells in the retina. In addition, immunohistochemical analysis, using antibodies to serotonin, tyrosine hydroxylase, and substance P demonstrated that, of the three, only the tyrosine hydroxylase immunoreactive (THIR) amacrine cells were affected by the toxin. In addition, estimates of the total number of retinal ganglion cells from counts of every tenth (10 $\mu$m) section in treated and uninjected retinas were similar to normal retinas of similarly staged animals. Thus, it appears that 6-OHDA is selective in its effect of destroying only the dopamine-containing THIR amacrine cells of the tadpole retina, consistent with its use as a selective neurotoxin in other systems.

Over the next 6 weeks, the 6-OHDA lesioned retinas were examined for newly produced THIR amacrine cells by immunohistochemistry. No new cells were ever observed in central retinal regions at any time after the 6-OHDA injections that we examined; however, within 1 week, the first new THIR cells appeared adjacent to the germinal neuroepithelial zone at the peripheral retinal margin. By 3 weeks following the toxin treatment, this area contained numerous THIR cells (Fig. 3.1C), with a density more than twice that of normal retina (Table 3.1). Thus, it appeared that although the 6-OHDA destroyed the differentiated THIR amacrine cells, the toxin did not destroy the stem cells for these neurons at the germinal zone. Moreover, THIR amacrine cells were generated at a higher rate than normal for up to 3 weeks after the neurotoxin administration. Finally, this up-regulation of THIR amacrine cell production was very specific: substance-P-containing and serotonin-containing amacrine cells did not show a similar increase in their density

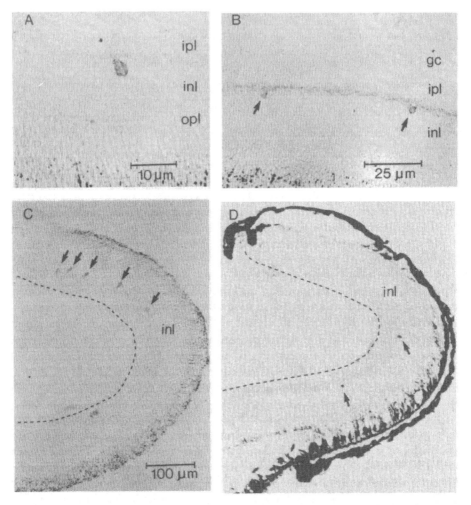

**Figure 3.1.** Ten micrometer paraffin sections stained with (A,B,C) tyrosine hydroxylase or (D) serotonin antisera. The sections shown in (A) and (B) are high- and low-power (respectively) micrographs of THIR amacrine cells of control retinas, while (C) and (D) show THIR and 5-THIR cells (arrows) in retinal sections from an animal that had received an intraocular injection of 6-OHDA 3 weeks prior to fixation. Note the increased density of THIR cells in the peripheral retina in (C) as compared with (D). Magnification is the same for (C) and (D). GC, ganglion cell layer; IPL, inner plexiform layer; INL, inner nuclear layer, OPL, outer plexiform layer. This figure reprinted with permission of Academic Press.

at the retinal margin, and the overall level of $^3$H-thymidine incorporation in the germinal zone was not significantly increased following 6-OHDA treatment (Table 3.1).

To determine whether the cell specific up-regulation of neuronal production that was observed for the THIR amacrine cells following 6-OHDA treatment was a general phenomenon of the germinal neuroepithelial cells in the larval *Rana* retina, selective neurotoxin lesions of other types of neurons

**Table 3.1. Immunoreactive Cells in 6-OHDA-Treated Retinas**

| | Cells/mm$^2$ in most peripheral tenth retina[a] | | | $^3$H-thymidine-labeled cells[b] |
|---|---|---|---|---|
| | THIR | 5-HTIR | SPIR | |
| Injected eye | 296 ± 33[c] | 116 ± 23 | 107 ± 20 | 57 ± 6 |
| | (n = 8) | (n = 6) | (n = 4) | (n = 4) |
| Uninjected eye | 143 ± 18 | 104 ± 35 | 98 ± 14 | 52 ± 9 |
| | (n = 8) | (n = 6) | (n = 4) | (n = 4) |
| Normal | 138 ± 27 | 121 ± 31 | 113 ± 28 | — |
| | (n = 15) | (n = 6) | (n = 3) | |

[a]Cell density 3 weeks after unilateral intraocular injection of 6-OHDA.
[b]Number of labeled cells in retinas from animals sacrificed 3 days after a 1 μCi/g body weight intraperitoneal injection of $^3$H-thymidine (New England Nuclear, spec. act. 6.7 Ci/mmol), 3 weeks following unilateral intraocular injection of 6-OHDA.
[c]Significantly different from uninjected control eye ($p < 0.002$) using nonparametric randomization test.
*Source:* This table reprinted with permission of Academic Press.

in the differentiated regions of retinas were made using kainic acid (KA) (Reh, 1987a,b). Kainic acid causes degeneration in the inner nuclear layer and ganglion cell layer of several vertebrate classes, including chicken and mammals. This laminar specificity of the neurotoxic actions of KA is thought to represent the sensitivity of neurons to the intense and prolonged depolarization that results from the activation of the kainate/glutamate receptor on certain types of retinal neurons (Erlich and Morgan, 1980; Hampton *et al.*, 1981; Ingham and Morgan, 1983). Ganglion cells, photoreceptors, and ON bipolar cells appear to be generally insensitive to the toxin, while OFF bipolar cells, horizontal cells, and most classes of amacrine cells are destroyed at relatively low concentrations of KA. In the *Rana* tadpole, intraocular injections of KA resulted in a 52% decline in the cell density of the inner nuclear layer, a 37% decline in the cell density of the ganglion cell layer, and no significant change in the density of cells in the outer nuclear layer (Table 3.2). This particular laminar distribution of KA sensitivity is similar to that observed in other species, with the loss of cells in the ganglion cell layer most likely due to destruction of displaced amacrine cells (Stelzner and Strauss, 1986; Beazley *et al.*, 1987).

In the 3 weeks following the KA damage to the developing retina, we injected the tadpoles with $^3$H-thymidine to detect any changes in the numbers and laminar distribution of the newly produced cells. Within the first week after the KA injection, there was a decline in the number of $^3$H-thymidine labeled cells in the KA-treated eye as compared with the control retina. The number of mitotic figures in the germinal zone decreased in a corresponding manner. However, in animals that received the $^3$H-thymidine injections more than 1 week after the KA lesion, the number of labeled neurons as well as the number of mitotic figures in the germinal neuroepithelial cells at the margin of the retina was significantly increased over that found in the contralateral control retina. Thus, the damage to the differentiated retina results in an increase in neuronal production at the level of the mitotic germinal neu-

**Table 3.2. Percentage of Cells in Retinal Layers following Kainic Acid Treatment**

|  | RGC | INL | ONL |
|---|---|---|---|
| Degenerating cells (%) in layers, 1 day after KA (n = 4) | 10.3 ± 4.5 | 23.3 ± 8.7 | <1 |
| Change (%) in cell density 2–6 weeks after KA (n = 16) | −37.6 ± 3.8 | −52.3 ± 4.2 | −5.8 ± 2.9 |
| Change (%) in ³H-thymidine-labeled cells/layer: KA/control (< 1 week) (n = 5) | −2.1 ± 19 | −9 ± 12 | −10 ± 20 |

*Note:* Asterisks signify level of statistical significance using the nonparametric randomization test: $*p < 0.01$; $**p < 0.002$.
*Source:* This table reprinted with permission of Society for Neuroscience.

roepithelial cells, either by the release of mitogenic factors by the remaining cells or by the elimination of mitotic inhibitors normally produced by the differentiated cells.

To determine whether particular retinal cell types were selectively increased in their production after the KA treatment, the laminar distribution of the newly produced cells was quantified in those cases that received ³H-thymidine more than 1 week after the toxin. As Table 3.2 demonstrates, the lamina that shows the greatest increase in the number of labeled cells over the contralateral control retina is the inner nuclear layer, while there was only a small increase over the control in the ganglion cell layer and no significant increase above control values in the number of labeled cells in the outer nuclear layer. Thus, there is selective increase in the production of the types of neurons that were destroyed by the toxin. This result indicates that regulation of neuronal production observed for the THIR amacrine cells following 6-OHDA lesions is not confined to this cell type, but rather appears to be a general property of retinal histogenesis.

Similar studies of fish retinal histogenesis also support the hypothesis that the types of neurons produced by germinal cells of CNS can be regulated by the microenvironment. Using a similar paradigm of neurotoxic lesions, Negishi *et al.* (1982, 1985, 1987) have observed a similar up-regulation in the production of catecholaminergic amacrine cells. In addition, a specialized progenitor cell for new rod production in fish appears to be regulated by the density of neighboring differentiated rods. Such a density-dependent regulatory mechanism appears to be responsible for keeping rod density constant as the eye of the fish expands (see Chapter 2 in this volume). This natural experiment also lends support to the following model we have proposed to explain the results of the experimental lesions on retinal histogenesis (Reh and Tully, 1986; Reh, 1987a,b). Germinal neuroepithelial cells are normally exposed to a complex microenvironment composed of mitotically active cells, postmitotic migrating immature neurons, and differentiating neurons (Hinds and Hinds, 1978, 1979). A particular type of differentiating cell produces a signal that inhibits the multipotent germinal neuroepithelial cell from generating new cells of that type. If the density of that particular cell type is reduced below a certain level, the germinal neuroepithelial cells are stimulated to produce more of that type. This then provides a way by which the production of the various types of neurons found in the CNS can be regulated by the differentiated cell density.

## 3.3. WHAT MOLECULAR FACTORS REGULATE THE PROLIFERATION OR DIFFERENTIATION OF GERMINAL NEUROEPITHELIAL CELLS IN THE DEVELOPING CNS?

### 3.3.1. Peptide Mitogens

Nonneuronal cells of the CNS have been shown to be regulated in their proliferation by a number of characterized peptide mitogens. Epidermal growth factor increases thymidine uptake and cell number in cultured astrocytes (Heldin *et al.*, 1981; Leutz and Schachner, 1981; Simson *et al.*, 1982; Saneto and de Villis, 1985). Fibroblast growth factor (Saneto and de Villis, 1985) and platelet-derived growth factor (Ek *et al.*, 1982; Besnard *et al.*, 1987) also stimulate proliferation of astrocytes and a human glial cell line, respectively, and a large number of uncharacterized mitogens have been purified to varying degrees from whole brain or pituitary that are mitotically active on glia as well as other tissues. Studies by Giulian and colleagues (Giulian and Lachman, 1985; Giulian and Young, 1986; Giulian *et al.*, 1986) have identified several brain-derived "glial-promoting" factors (GPFs) that stimulate mitosis in either astrocytes ($GPF_2$ and $GPF_4$) or oligodendrocytes ($GPF_1$ and $GPF_3$). In addition, plasminogen activator, bFGF, and interleukin I have been shown to be mitogens for astrocytes, while interleukin II (Benveniste and Merrill, 1986) and bFGF stimulate oligodendroglial proliferation. However, at the present time, only one of these peptide mitogens, PDGF, has been tested, without effect, for activity on germinal neuroepithelial cells (Besnard *et al.*, 1987). Indeed, while the number of well-characterized mitogens and mitotic inhibitors for other vertebrate cells and tissues has increased substantially over the last few years, this area of CNS cell biology has received remarkably little attention.

One of the main reasons why so little is known about neuronal proliferation is that there is no suitable culture system in which germinal neuroepithelial cells remain mitotically active for prolonged periods. Explant cultures of the developing nervous system, in a number of studies, have been shown to support some degree of mitotic activity in neuronal precursors (Lyser, 1968, 1977; Anderson and Waxman, 1985a,b); however, these culture systems have many limitations for further characterization of neuronal precursors, since the reproducibility and penetration of reagents is less than that of a dissociated cell culture system.

Unfortunately, neurogenesis is very limited in dissociated cell cultures.

The most well-characterized preparation is that of Sensenbrenner *et al.* (1971, 1973, 1980a,b), in which 6 day chick embryo brain hemispheres are mechanically dissociated and plated onto collagen substrates. These cultures are reported to consist primarily of neuroblasts, proliferating for the first 2–3 days *in vitro*. Thereafter, glial proliferation predominates. Similar results have been reported by other groups that have studied neurogenesis in culture: neurogenesis does not continue for more than a few days (Juurlink and Federhoff, 1982; Kriegstein and Dichter).

Despite this lack of a suitable culture system, some attempts have been made to identify germinal neuroepithelial mitogens. Using the 6 day chick embryo brain-dissociated culture system described above, and assaying for $^3$H-thymidine uptake during the first 2–3 days *in vitro* when "neuroblast" proliferation was predominant, Sensenbrenner and colleagues were able to show that chloroform–methanol extracts of bovine brain could stimulate cell proliferation (Sensenbrenner *et al.*, 1980b; Barakat and Sensenbrenner, 1981). When this material was further purified, it was found that the active factors in these extracts are most likely nucleotides, specifically purines (Barakat *et al.*, 1983, 1984). While purines have been shown to be rate limiting for other chick and amphibian primary cell cultures (Sooy and Metzger-Freed, 1970), no evidence yet supports the notion that nucleotides regulate cell proliferation *in vivo*. Therefore, the status of purines as *specific* mitogens for germinal neuroepithelial cells is not yet clear. The effects of the above-mentioned peptide mitogens on this culture system are only beginning to be examined; at the present time, only PDGF has been reported to have been tested in this system and, as mentioned above, has been found to have no effect.

An alternative model for studies of neurogenesis can be found in transformed neuronal cell lines. Neuroblastoma, a neural-crest-derived tumor cell line, has been a popular model for studying the factors that regulate neuronal proliferation and differentiation (Schubert *et al.*, 1973; Haffke and Seeds, 1975; Prasad, 1975). A good deal of work by Sato, Bottenstein (Bottenstein, 1985), and colleagues has gone into defining a serum-free medium for various neuronal cell lines. These studies indicate that the basic requirements of neuronal cell lines are similar to other cell types; insulin, transferrin, and selenium are required for proliferation in defined medium. In addition, neuroblastoma cells require polyamines and progesterone for maximum cell growth. Two observations suggest these latter two compounds may be important regulatory molecules involved in germinal neuroepithelial proliferation and differentiation. First, putrescine and progesterone are necessary for survival, if not proliferation, of many primary neuronal cultures in defined media. Second, polyamines are present in high concentrations in the developing CNS (Bottenstein, 1985). Most of the various peptide mitogens have also not yet been tested on neuroblastoma cells; however, one recent study has found that insulin-like growth factors I and II can show mitogenic activity on the human SH-SY-5Y neuroblastoma line (Mattsson *et al.*, 1986). Taken together, the work on the various neuronal cell lines has contributed to our understanding of the basic media requirements for primary neuronal cultures and has pointed to possible growth regulators for germinal neuroepithelial cells; unfortunately, since transformed cells are clearly aberrant in their growth regulatory processes, and there is a lack of a well-defined primary culture of CNS germinal cells, it is not possible to ensure that neuroblastoma is an appropriate model for central neurogenesis.

### 3.3.2. Differentiating Factors or Mitotic Inhibitors

As an alternative to peptide mitogens, perhaps germinal neuroepithelial cells are regulated by mitotic inhibitors or factors that induce differentiation. Several agents have been described that promote some of the differentiated features of neurons, for example, nerve growth factor (Levi-Montalcini, 1982). Also, studies of neuroblastoma indicate that mitotically active cells do possess receptors for many of the components active on mature, differentiated cells. In addition, neuroblastoma can be stimulated to differentiate by factors that chronically increase cAMP levels (Prasad and Hsie, 1971; Schubert et al., 1973; Prasad, 1975; Breakefield, 1976). Therefore, it is possible that mitotic inhibitors or differentiating factors could act on the germinal cells to suppress their proliferation by triggering their differentiation.

The most well-characterized mitotic inhibitor in other tissues is transforming growth factor β (TGF-β). This 25 kDa protein has inhibitory effects on proliferation of many different tissues (for review, see Sporn et al., 1986; Massaqué, 1987) including hepatocytes, embryo fibroblasts, T and B lymphocytes, keratinocytes, and bronchial epithelial cells. While its antiproliferative actions are largely thought due to antagonism of the mitogenic effects of other peptide growth factors, in many cases TGF-β also has specific effects on cell differentiation. For example, TGF-β inhibits the mitotic stimulation of kidney epithelial cells by insulin and hydrocortisone, yet it does not inhibit the increased protein synthesis caused by these hormones. Although TGF-β has diverse cell-type-specific modulatory actions on various different cell lines, its antiproliferative effects predominate, and it has been proposed that it acts as a "feedback control on excessive clonal expansion" (Sporn et al., 1986). Such mitotic inhibitors could be active in the developing CNS, and the experiments on *Rana* retina using neurotoxic lesions are certainly consistent with their presence; however, at present no similar antiproliferative peptides have been shown to act in the CNS.

Alternatively, Lauder and her colleagues (Lauder and Krebs, 1976; Lauder et al., 1980, 1981; Lauder, 1985) have suggested that certain neurotransmitters may have effects on the production of neurons in the developing CNS. At the present time, however, the evidence is largely indirect. For example, neurotransmitters, such as dopamine, serotonin, norepinephrine, and acetylcholine, have been detected as early as neural tube formation during embryogenesis, suggesting these molecules have roles other than communication between neurons. Also, immunohistochemical studies of catecholamine- and 5-HT-containing neurons have demonstrated that these cells are among the first to differentiate in the CNS and send processes to many regions of the brain that still contain mitotically active germinal neuroepithelial zones. Such a close spatial correlation led Lauder to propose that serotonin or dopamine acts as a "differentiation signal" for the proliferative cells of the CNS. However, attempts to test this hypothesis by treating pregnant female rats with monoamine antagonists or agonists have yielded conflicting results. Lauder *et al.* (1981) found that *p*-chlorophenylalanine (pCPA), which inhibits 5-HT synthesis, results in an increase in cell proliferation in the developing neural tube, while other studies (Lewis *et al.*, 1977; Patel *et al.*, 1980; Patel, 1985) using a range of monoamine synthesis inhibitors (including pCPA) have found that these drugs cause a 50% decrease in $^3$H-thymidine labeling in the developing brain. The most obvious problem with these studies is that they have been carried out *in vivo;* drugs administered to pregnant rats will clearly have effects on the physiology of the mothers, which could secondarily affect fetal development. This concern is highlighted by the fact that in one study (Lauder *et al.*, 1980) saline-injected controls exhibited some of the changes in the pattern of $^3$H-thymidine incorporation that were observed with the drug.

### 3.3.3. Other Factors

Another way in which neuronal production is regulated is via hormonal factors. Evidence for regulation of neurogenesis by hormonal factors has been found for teleosts (see Chapter 2 in this volume) and frogs. In frog retina, the mitogenic effects of thyroxine were first described by Kaltenbach (Kaltenbach and Hobbs, 1972). Beach and Jacobson (1979) further demonstrated a differential effect of thyroxine on neuronal production in the dorsal and ventral retina of *Xenopus* (see Chapter 6 in this volume). It is possible that the mitogenic effect of thyroxine is a general phenomenon since thyroid functioning influences brain development in mammals as well. However, the mitogenic effects of thyroxine could be indirect; for example, lens epithelial mitotic regulation occurs through a cascade that is *initiated* by thyroxine but has as its final result the release of somatomedin C, the actual mitogen for lens epithelium (Rothstein *et al.*, 1981).

Chronic depolarization has also been reported to be mitogenic for neurons. Cone and Cone (1976, 1978) and Stillwell *et al.* (1973) claimed that fully differentiated, postmitotic neurons *in vitro* could be induced to divide by chronic application of ouabain and related agents. Unfortunately, these studies lacked adequate markers, leaving open the possibility that these researchers were not describing mitosis in neurons in their cultures, but rather the cells they studied were a type of glial cell. In light of the large number of studies showing that differentiated neurons *in vitro* are terminally postmitotic, it would be very surprising if the cells described by Cone and Cone were dividing neurons. In any case, these studies should clearly be followed up using cell-type-specific markers.

### 3.4. GERMINAL NEUROEPITHELIAL CELLS IN THE PROLIFERATIVE ZONE OF THE RETINA CONTINUE TO UNDERGO MITOSIS *IN VITRO* FOR UP TO 3 WEEKS

In order to study in detail the mechanisms involved in regulating the proliferation of germinal neuroepithelial cells, we began to develop methods to keep them mitotically active for prolonged periods *in vitro*. In frogs and fishes, development of the nervous system proceeds much more slowly than in homeothermic vertebrates. As mentioned above, in the frog retina a specialized zone of germinal neuroepithelial cells is preserved at the retinal margin throughout the animal's life (Straznicky and Gaze, 1972; Hollyfield, 1968; Reh and Constantine-Paton, 1983); therefore, these cells might provide a source for a germinal neuroepithelial cell line. To this end we have developed a new slice culture technique. The germinal neuroepithelium is sliced parallel to the long axis of the cells, thereby preserving the overall organization of the proliferative zone, as well as the cellular contacts with the extracellular matrix at their basal surface.

The slice cultures are made as follows. Midlarval staged *Rana pipiens* tadpoles are anesthetized, the eyes removed, and the retinas, with RPE still attached, were embedded in 4% low melting point agarose and sectioned at 100–200 μm. Retinal slices are cultured in media consisting of 50% L-15, 1–10% fetal bovine serum, glucose, and antibiotics, or alternatively in a serum-free medium consisting of 50% L-15, insulin, selenium, transferrin, and hypoxanthine (Reh and Radke, 1988).

Immediately after the retinal slices are made, the laminar pattern of the retina can be clearly discerned. The retinal ganglion cell layer (RGC), inner nuclear layer (INL), outer nuclear layer (ONL), as well as the RPE and vitreal vascular membrane, all retain their normal appearance (Fig. 3.2a). At the peripheral retinal margin, the zone of proliferating germinal neuroepithelial cells is also well preserved and time-lapse recordings show cells in mitosis in this marginal zone (T. Reh, unpublished observations).

Electrophysiological recordings can be made from cells in the retinal slices, both in the differentiated zones and the germinal region. Stable resting potentials of up to 60 mV are routinely encountered and spontaneous action potentials can be recorded following intracellular penetrations of retinal ganglion cells. Cells in the differentiated retina have the typical morphological appearance of retinal neurons described previously by many methods; Fig. 3.2d shows an example of a dye-filled amacrine cell from central retina. Lucifer injections of single germinal neuroepithelial cells at the marginal zone always resulted in the labeling of a group of tightly coupled cells (Fig. 3.2c).

When slices are incubated in media containing $^3$H-thymidine, the mitotically active germinal neuroepithelial cells are well labeled. Figure 3.2b shows an autoradiogram of a 4 μm plastic section from a slice cultured for 3 days, to show the degree of labeling at the marginal zone. In addition, labeled cells are also now present in the differentiated laminated areas of the retina. While the level of mitotic activity remains high in these slice cultures for at least the next

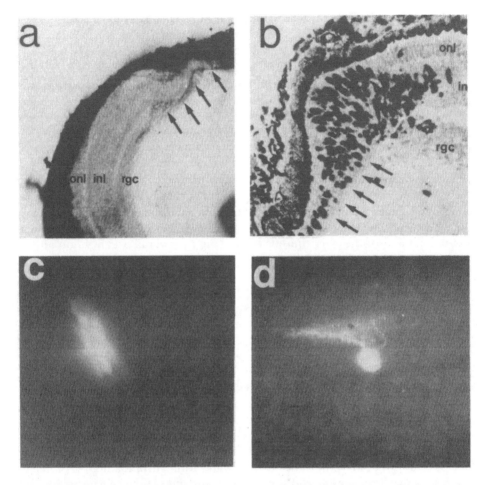

**Figure 3.2.** Slice cultures of the *Rana* tadpole retina. In (a) a freshly cut 100 μm slice can be seen at low power; the retinal ganglion cell (rgc) layer, inner nuclear layer (inl), and outer nuclear layer (onl) can all be identified in the differential central retina, while the arrows point to the peripheral germinal zone. The autoradiogram of a similar slice shown in (b) demonstrates that these slice cultures retain the ability to incorporate ³H-thymidine at normal levels, and some labeled cells are found in the differentiated retina layers 3 days after labeling. The photomicrographs in (c) and (d) show cells injected with Lucifer Yellow in the germinal zone and differentiated retina, respectively.

3 weeks, the lamination of differentiated retina is progressively obscured, and it becomes increasingly difficult to discern the retinal lamina. Throughout this period the germinal cells at the marginal zone retain their pseudostratified columnar epithelial morphology and continue to incorporate ³H-thymidine.

Within the first week *in vitro*, the ³H-thymidine labeling in the retinal slices is almost exclusively in the cells of the germinal zone. A few labeled cells are occasionally encountered in the RPE and the vitreal vascular membrane, but $^3$H-thymidine uptake was rarely observed in central retinal regions, either by glial cells or by any other cell type in the central retina. Even with continuous $^3$H-thymidine exposure for 3 days, no labeled cells were ever found in the outer nuclear layer, indicating that the frog tadpole does not possess mitotically active rod precursor cells like those found in teleosts (see Chapter 2 in this volume).

Neuron-specific monoclonal antibodies were used to identify the types of new cells that were generated by the germinal neuroepithelial cells in culture. Antibodies to neurofilament proteins normally only stain the ganglion cells in the retina of the rat. We have found a similar staining pattern in *Rana* and have used a monoclonal antibody to the 200 kDa subunit of neurofilament to determine if new ganglion cells are generated *in vitro*. Also, we have produced two monoclonal antibodies specific to *Rana* neurons. One (2D3) recognizes an epitope present on a molecule in *Rana* similar to NCAM (Nagy and Reh, 1987), while initial characterization of the other (2C6) indicates that it is likely to be directed against a cell surface glycolipid. Neither of these antibodies stain glia (GFAP+) in retinal cultures. Using these neuron-specific monoclonal antibodies, we have been able to determine that most of the new cells generated in these slice cultures are neurons (Reh, 1987b; Reh and Radke, 1988).

## 3.5. WHAT FACTORS CAN MODIFY GERMINAL NEUROEPITHELIAL CELL PROLIFERATION *IN VITRO?*

### 3.5.1. Peptide Mitogens and Growth Factors

We now have a well-characterized *in vitro* system with which to screen potential mitogens or mitotic inhibitors for germinal neuroepithelial cells. The first factor we examined was nerve growth factor (NGF). This particular molecule is well known for its actions on neuronal differentiation and survival, both *in vivo* and *in vitro* (Levi-Montalcini, 1982). Although its actions have been most extensively studied in neural-crest-derived neurons, recent work in CNS (Large *et al.*, 1986), including amphibian and fish retina (Turner and Delaney, 1979; Turner *et al.*, 1981, 1982), indicates NGF has activity on certain central neuronal populations as well. We were initially interested in NGF as a differentiating agent or mitotic inhibitor, since its most thoroughly characterized actions involve differentiated functions of neurons; however, a recent report has demonstrated that NGF has mitogenic activity on chromaffin cells (Lillien and Claude, 1985). Therefore, we had reason to believe that NGF could be active on frog retinal germinal neuroepithelial cells, but *a priori* we could not predict whether this activity would be mitogenic or inhibitory.

The results of the initial experiments we have carried out with NGF are shown in Fig. 3.3a. After a 2 day incubation with the 7S form of NGF, there is a dose-dependent decline in $^3$H-thymidine uptake to approximately 60% of

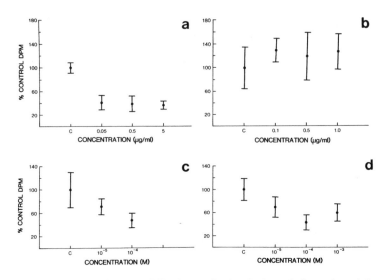

**Figure 3.3.** TCA precipitated counts following 24 hr incubation of slice cultured frog retinal germinal neuroepithelial cells with $^3$H-thymidine (20 Ci/mmol; 1 µCi/µl). Error bars represent one standard deviation from the mean of at least three cultures. (a) Nerve growth factor, 7S; (b) epidermal growth factor; (c) serotonin; (d) dopamine. In all cases, the slices were cultured in serum-free medium consisting of 50% L-15 with insulin, transferrin, selenium, hypoxanthine, antibiotics, and glucose added. In the experiments shown in (c) and (d) ascorbate was also added to the cultures.

the control levels. This decline in proliferation could reflect a slowing of the cell cycle of the germinal cells or an induction of their differentiation, as a direct action of NGF. At this point we cannot determine which of these alternatives is the correct interpretation; however, it is clear that the proliferation of the germinal neuroepithelial cells in the larval frog retina can be regulated by NGF, and therefore these cells are likely to possess receptors for this molecule. Since NGF is widespread in the developing CNS in chick and mouse embryos, it is possible that germinal neuroepithelial cell differentiation is regulated by NGF in these species as well.

To test for the specificity of this effect, we have also examined other previously characterized mitogens for activity on germinal neuroepithelial cells. Epidermal growth factor (EGF), as noted above, has been shown to be a mitogen for rat brain astrocytes, as well as chick neural crest cells (Erickson and Turley, 1987). Since EGF is also a molecule that is highly conserved in vertebrates, we tested it for mitogenic activity in the frog retinal germinal cell cultures. As Fig. 3.3b shows, we found no significant changes in $^3$H-thymidine incorporation when EGF was added to the cultures for either 1 or 2 days prior to the addition of the $^3$H-thymidine. Similar results were obtained with basic fibroblast growth factor; no significant increase in $^3$H-thymidine incorporation was observed, when serum-starved cells were incubated in concentrations of 0.1–100 µg/ml bFGF.

### 3.5.2. Monoamine Neurotransmitters and Germinal Cell Proliferation

As described above, certain neurotransmitters, that is, the monoamines, have also been implicated in the regulation of proliferation of germinal neuroepithelial cells. Since most of the evidence for such a regulatory role for dopamine and serotonin comes from *in vivo* studies, we attempted to verify that these compounds could influence $^3$H-thymidine incorporation in our culture system. We found that when either serotonin or dopamine were added to the cultures, $^3$H-thymidine incorporation was significantly decreased. As can be seen from Fig. 3.3c,d both monoamines suppress DNA synthesis in tadpole retinal germinal neuroepithelial cells in a dose-dependent manner. This effect is likely to be specific to these neurotransmitters, since incubation in 100 µM kainate, a concentration sufficient to destroy much of the differentiated INL, has no effect on the number of germinal cells labeled with $^3$H-thymidine (Reh, 1987a). Although these results are far from a demonstration that monoamines play a regulatory role during normal neurogenesis, they do imply that at least some mitotically active, neural precursors possess receptors for monoamines, and their proliferation can be influenced by these substances.

### 3.5.3. External Basement Membrane Is Also Necessary for the Proliferation of Germinal Neuroepithelial Cells

The CNS is surrounded by a basement membrane from the very earliest stages of its development (Svoboda and O'Shea, 1987; Tuckett and Morris-Kay, 1986). Germinal neuroepithelial cells span the proliferative zone, such that one process makes contact with this basement membrane, while the other end of the cell remains attached to the ventricular surface. One of the steps typically employed in preparing dissociated cell culture is the enzymatic digestion of this basement membrane. To determine if germinal neuroepithelial cells require contact with this basement membrane to continue proliferating, we treated the cultures with 200 units/ml highly purified collagenase to remove this matrix. In most cases, the majority of the basement membrane was removed, as determined in some cases by immunohistochemical staining for laminin, a major component of this basement membrane (see Reh *et al.*, 1987), and in other cases, by toluidine blue staining.

The results of these experiments are shown in Table 3.3. If the cultures are labeled 1 day after the removal of the external basement membrane, germinal cells still incorporate the $^3$H-thymidine at nearly normal rates. There are no obvious differences in most other aspects of their morphology; however, the cells do appear less elongated than in control cultures. By 3 days, the collagenase-treated cultures have a significantly reduced level of $^3$H-thymidine uptake. In addition, this zone is beginning to differentiate to give a laminate pattern of cell distribution. By 5 days after removal of the basement membrane, germinal neuroepithelial proliferation has declined considerably

**Table 3.3. Decline of Germinal Neuroepithelial Proliferation after Collagenase Removal of Basement Membrane in Vitro[a]**

|                    | 1 day         | 3 day        | 5 day         |
|--------------------|---------------|--------------|---------------|
| Control            | 1250 ($n = 4$) | 841 ($n = 4$) | 1916 ($n = 3$) |
| EDTA               | 1660 ($n = 4$) | 1067 ($n = 4$) | 1231 ($n = 4$) |
| EDTA/collagenase   | 1050 ($n = 4$) | 545 ($n = 4$) | 632 ($n = 7$) |

[a]Control cultures were 200 μm retinal slices from *Rana pipiens* tadpoles, while 5 min EDTA treatment facilitated RPE removal and EDTA/collagenase treatment (5 min/ 10 min) resulted in cultures free from RPE and the majority of the basal basement membrane. Cultures were labeled with [3]H-thymidine for 24 hr, fixed, embedded in JB-4 plastic resin, sectioned, and processed for autoradiography. Cells with over 5 grains were counted as labeled.

overall, and the zone has more fully differentiated (Table 3.3). The cells appear to be intact and no signs of degeneration were observed in this region following the enzyme treatment. The fact that these cells do not die, but rather progress from a proliferative state to a differentiated one, indicates that the removal of the basal basement membrane exerts a specific effect and does not simply destroy the germinal cells.

It has previously been reported that the retinal pigmented epithelial cells also have an effect on the organization and [3]H-thymidine uptake of the germinal neuroepithelial zone (Vollmer and Layer, 1986). We also found that in those cases where the RPE was not completely removed during the collagenase treatment, the remaining RPE cells maintained a close association with the cells of the germinal zone; however, when the RPE was removed without the concomitant removal of the basement membrane (by brief EDTA incubation) the levels of [3]H-thymidine incorporation were not different from the control values (Table 3.3).

These results suggest that germinal neuroepithelial cells must retain their contacts with the basal basement membrane, and the loss of this contact may be a necessary step in the terminal differentiation of neurons in the CNS. Such an interpretation is consistent with earlier observations, using serial section electron microscopy, which led Hinds and Hinds (1979) to suggest that one of the first steps in the differentiation of germinal neuroepithelial cells into neurons is the failure to reestablish contact with the inner limiting membrane following cytokinesis.

These experiments indicate that the basal basement membrane is important for continued germinal cell proliferation, but the mechanisms for this interaction can only be speculated on. Epithelial cells are typically polarized, with a basement membrane found only at their basal surface, and it has been shown in several cases that this polarity is necessary for the normal functioning of the epithelium. In addition, culture of many different types of epithelial cells is profoundly influenced by the type of basement membrane on which they are plated, both in terms of their differentiated state and their rate of proliferation. We suggest that germinal neuroepithelial cells may be regulated by similar mechanisms, and that certain components of the extracellular matrix may be essential for regulating the production and differentiation of neurons.

## 3.6. SUMMARY AND CONCLUSIONS

The results of the experiments carried out on the developing *Rana* retina have revealed several aspects of neurogenesis that are likely to be relevant to the CNS as a whole. First, the *in vivo* experiments, using selective neurotoxins, indicate that germinal neuroepithelial cells can be regulated by intrinsic factors, both in their overall level of neuronal production and in the types of neurons that will be produced. In addition, it is likely that some level of cellular interaction via these intrinsic factors is necessary for the creation of the appropriate ratios of neuronal cell types during neurogenesis, since neither strictly deterministic lineage-based mechanisms nor subsequent elimination of unnecessary elements by cell death will adequately account for the data currently available. Second, the *in vitro* experiments demonstrate that germinal neuroepithelial cells can be regulated in their overall rate of proliferation, primarily in an inhibitory direction, by previously characterized peptide growth factor NGF and also by the neurotransmitters dopamine and serotonin. Since both of these classes of compounds have been shown to act via similar second messenger systems in many different cell types, it is likely that germinal neuroepithelial proliferation is ultimately controlled at this level. Third, the results of the experiments entailing removal of the basement membrane from the germinal neuroepithelial cells are consistent with recent studies in other vertebrate tissues that demonstrate a role for the ECM in regulating cell division by modulating the cell's response to other growth factors. In addition, it appears that the polarized morphology of central germinal zones may be necessary for maintaining their proliferative potential.

Normal histological studies with both light and electron microscopy indicate that the common feature distinguishing mitotically active germinal neuroepithelial cells from the differentiating neurons in the retina is the presence of a process that extends to the vitreal surface. Although studies of ganglion cell development also show a near simultaneous loss of the ventricular process in most cells (Hinds and Hinds, 1974; but see Morrest, 1970), the developing photoreceptors never lose their ventricular contact during their differentiation (Hinds and Hinds, 1979). In addition, the only differentiated cells that retain their contacts with the vitreal surface are the Müller cells, and they retain their mitotic activity in adult animals. Therefore, it is possible that the failure to reestablish contact with this surface after cytokinesis is the first step toward terminating neuronal differentiation of a germinal neuroepithelial cell. The results reported in this chapter are also consistent with this possibility; removal of the basement membrane which these processes normally contact induces the premature differentiation of all the germinal cells in the tadpole marginal zone.

These conclusions suggest the following model. As more cells are produced in the developing retina, a certain percentage of germinal neuroepithelial cells will be unable to reestablish contact with the basement membrane at any given time and go on to differentiate. What they differentiate into will be determined by "when" (developmentally) they lose contact with the basement membrane. This developmental time is translated into the cellular phenotype by the balance of factors that are present in the microenvironment. For example, perhaps the germinal neuroepithelial cells will differentiate into either ganglion cells or cones until the amount of some RGC or cone inhibitory factor is too high; at this time, any newly differentiating germinal cells will differentiate into amacrine cells or horizontal cells. These cell types will then be made until such time as factors they produce during differentiation inhibit the differentiation of new cells of these types and the germinal cells now are restricted to differentiating into bipolars, rods, and Müller cells. While this is very speculative, the model is only intended to organize the various results in the area of neurogenesis in the CNS into a coherent scheme. Although the molecular details of the cellular interactions necessary to generate the enormous complexity of the CNS remain obscure, some of the classes of interactions likely to be involved can now be specified.

## 3.7. REFERENCES

Anderson, M. J., and Waxman, S. G., 1985a, Neurogenesis in adult vertebrate spinal cord in situ and in vitro: A new model system, *Ann. N.Y. Acad. Sci.* **xx:**213–233.

Anderson, M. J., and Waxman, S. G., 1985b, Neurogenesis in tissue cultures of adult teleost spinal cord, *Dev. Brain Res.* **20:**203–212.

Barakat, I., and Sensenbrenner, M., 1981, Brain extracts that promote the proliferation of neuroblasts from chick embryo in culture, *Dev. Brain Res.* **1:**355–368.

Barakat, I., Sensenbrenner, M., and Labourdette, G., 1983, Purines stimulate chick neuroblast proliferation in culture, *Neurosci. Lett.* **41:**325–330.

Barakat, I., Sensenbrenner, M., and Labourdette, G., 1984, In bovine extracts, RNAs are active factors that stimulate the proliferation of chick neuroblasts in culture, *J. Neurosci. Res.* **11:**117–129.

Beach, D. H., and Jacobson, M., 1979, Patterns of cell proliferation in the retina of the clawed frog during development, *J. Comp. Neurol.* **183:**603–614.

Beazley, L. D., Perry, V. H., Baker, B., and Darby, J. E., 1987, An investigation into the role of ganglion cells in the regulation of division and death of other retinal cells, *Dev. Brain Res.* **33:**169–184.

Benveniste, E. N., and Merril, J. E., 1986, Stimulation of oligodendroglial proliferation and maturation by interleukin 2, *Nature (London)* **321:**610–613.

Besnard, F., Perraud, F., Sensenbrenner, M., and Labourdette, G., 1987, Platelet derived growth factor is a mitogen for glial but not for neuronal rat brain cells *in vitro, Neurosci. Lett.* **73:**287–292.

Blanks, J. C., and Bok, D., 1977, An autoradiographic analysis of postnatal cell proliferation in the normal and degenerative mouse retina, *J. Comp. Neurol.* **174:**317–328.

Bottenstein, J. E., 1985, Growth and differentiation of neural cells in defined media, in *Cell Culture in the Neurosciences* (J. E. Bottenstein and G. Sato, eds.), pp. 3–44, Plenum Press, New York.

Breakefield, X. O., 1976, Neurotransmitter metabolism in murine neuroblastoma, *Life Sci.* **18:**267–278.

Cone, C. D., and Cone, C. M., 1976, Induction of mitosis in mature neurons in central nervous system by sustained depolarization, *Science* **192:**155–158.

Cone, C. D., and Cone, C. M. 1978, Evidence of normal mitosis with complete cytokinesis in central nervous system neurons during sustained depolarization with guavaine, *Exp. Neurol.* **60:**41–55.

Cowan, W. M., 1973, Neuronal death as a regulative mechanism in the control of cell number in the nervous system, in *Development and Aging in the Nervous System* (M. Rockstein, ed.), pp. 19–41, Academic Press, New York.

Cowan, W. M., and Finger, T. E., 1982, Regeneration and regulation in the developing central nervous system with special reference to the reconstitution of the optic tectum of the chick following removal of the mesencephalic alar plate, in *Neuronal Development* (N. C. Spitzer, ed.), pp. 377–412, Plenum Press, New York.

Detwiler, S. R., 1945, The results of unilateral and bilateral extirpation of the forebrain of *Amblystoma*, *J. Exp. Zool.* **100:**103–115.

Detwiler, S. R., 1946, Midbrain regeneration in *Amblystoma*, *Anat. Rec.* **94:**229–241.

Ek, B., Westermark, B., Wateson, A., and Heldin, C. H., 1982, Stimulation of tyrosine-specific phosphorylation by platelet derived growth factor, *Nature (London)* **295:**419–420.

Erickson, C. A., and Turley, E. A., 1987, The effects of epidermal growth factor on neural crest cells in tissue culture, *Exp. Cell Res.* **169:**267–279.

Erlich, O., and Morgan, I. G., 1980, Kainic acid lesions chick retina amacrine cells, *Neurosci. Lett.* **17:**43–48.

Giulian, D., and Lahman, L. B., 1985, Interleukin I stimulation of astroglial proliferation after brain injury, *Science* **228:**497–499.

Giulian, D., and Young, D. G., 1986, Brain peptides and glial growth. Identification of cells that secrete glial promoting factors, *J. Cell Biol.* **102:**812–820.

Giulian, D., Allen, R. L., Baker, T. J., and Tomozawa, Y., 1986, Brain peptides and glial growth, *J. Cell Biol.* **102:**803–811.

Haffke, S. C., and Seeds, N., 1975, Neuroblastoma: The *E. coli* of neurobiology, *Life Sci.* **16:**1649–1658.

Hampton, C. K., Garcia, C., and Redburn, D. A., 1981, Localization of kainic acid sensitive cells in the mammalian retina, *J. Neurosci. Res.* **6:**99–111.

Harrison, R. G., 1947, Wound healing and reconstitution of the central nervous system of the amphibian embryo after removal of parts of the neural plate, *J. Exp. Zool.* **106:**27.

Heldin, C. H., Westermark, B., and Wasteson, A., 1981, Specific receptors for platelet-derived growth factor on cells derived from connective tissue and glia, *Proc. Natl. Acad. Sci. U.S.A.* **78:**3664–3668.

Herrup, K., 1986, Cell lineage relationships in the development of the mammalian CNS: Role of cell lineage in control of cerebellar Purkinje cell number, *Dev. Biol.* **115:**148–154.

Herrup, K., Diglio, T. J., and Letsou, A., 1984, Cell lineage relationships in the development of the mammalian CNS: The facial nerve nucleus, *Dev. Biol.* **103:**329–336.

Hinds, J. W., and Hinds, P. L., 1974, Early ganglion cell differentiation in the mouse retina: An electron microscopic analysis utilizing serial sections, *Dev. Biol.* **37:**381–416.

Hinds, J. W., and Hinds, P. L., 1978, Early development of amacrine cells in the mouse retina: An electron microscopic, serial section analysis, *J. Comp. Neurol.* **179:**277–300.

Hinds, J. W., and Hinds, P. L. 1979, Differentiation of photoreceptors and horizontal cell in the embryonic mouse retina: An electron microscopic serial section analysis, *J. Comp. Neurol.* **187:**495–512.

Hollyfield, J. G., 1968, Differential addition of cells to the retina in *Rana pipiens* tadpoles, *Dev. Biol.* **18:**163–179.

Holtfreter, J., and Hamburger, V., 1955, Amphibians, in *Analysis of Development* (B. Willier, P. Weiss, and V. Hamburger, eds.), pp. 230–296, W. B. Saunders, Philadelphia.

Ingham, C. A., and Morgan, I. G., 1983, Dose-dependent effects of intravitreal kainic acid on specific cell types in chicken retina, *Neuroscience* **9:**151–181.

Juurlink, B., and Federhoff, S., 1982, The development of mouse spinal cord in tissue culture. II. Development of neuronal precursor cells, *In Vitro* **18:**179–182.

Kaltenbach, J. C., and Hobbs, A. W., 1972, Local action of thyroxine on amphibian meta-morphosis. V. Cell Division in the eye of the anuran larvae affected by thyroxine-cholesterol implants, *J. Exp. Zool.* **179**:157–165.

Kimmel, C. D., and Warga, R. M., 1986, Tissue specific cell lineages originate in the gastrula of the zebrafish, *Science* **231**:365–368.

Kimmel, C. B., and Warga, R. M., 1987, Cell lineages generating axiomuscle in the zebrafish embryo, *Nature (London)* **327**:234–237.

Konigsberg, I. R., and Pfister, K. K., 1986, Replicative and differentiative behavior in daughter pairs of myogenic stem cells. *Exp. Cell Res.* **167**:63–74.

Kriegstein, A., and Dichter, M. A., 1984, Neuron generation in dissociated cell cultures from fetal rat cerebral cortex, *Brain Res.* **295**:184–189.

Large, T. H., Bodary, S. C., Clegg, D. O., Westcamp, G., Otten, U., and Reichardt, L. F., 1986, Nerve growth factor gene expression in the developing rat brain, *Science* **234**:352–355.

Lauder, J. M., 1985, Roles for neurotransmitters in development: Possible interaction with drugs during the fetal and neonatal periods. Prevention of physical and mental congenital defects, part C: *Basic and Medical Science, Education, and Future Strategies*, pp. 375–380, Alan R. Liss, New York.

Lauder, J. M., and Krebs, H., 1976, Effects of *p*-chlorophenolaline on time of neuronal origin during embryogenesis in the rat, *Brain Res.* **107**:638–644.

Lauder, J. M., Wallace, J. A., Krebs, H., and Petrusz, P., 1980, Serotonin as a timing mechanism in neuroembyrogenesis, in *Progress in Psychoneuroendocrinology* (E. Bambrillo, G. Recogni, and D. Dewead, eds.), pp. 539–556, Elsevier/North-Holland Biomedical Press, New York.

Lauder, J. M., Paul, Y. Z., and Krebs, H., 1981, Maternal influences on tryptophan hydroxylase activity in embryonic rat brain, *Dev. Neurol.* **4**:291–295.

Le Douarin, N. M., 1986, Cell line segregation during peripheral nervous system ontogeny, *Science* **231**:1515–1522.

Leutz, A., and Schachner, M., 1981, Epidermal growth factor stimulates DNA synthesis of astro-cytes in primary cerebellar cultures, *Cell Tissue Res.* **220**:393–404.

Levi-Montalcini, R., 1982, Developmental neurobiology and the natural history of nerve growth factor, *Annu. Rev. Neurosci.* **5**:341–362.

Lewis, P. D., Patel, A. J., Bendek, G., and Balazs, R., 1977, Effect of reserpine on cell proliferation in the developing rat brain: A quantitative histological study, *Brain Res.* **129**:299–308.

Lia, B. A., Williams, R. W., and Chalupa, L. M., 1987, Formation of retinal ganglion cell topography during pre-natal development, *Science* **236**:848–851.

Lillien, L. E., and Claude, P., 1985, Nerve growth factor is a mitogen for cultured chromaffin cells, *Nature (London)* **317**:632–634.

Lyser, K. M., 1968, Early differentiation of motor neuroblasts in the chick embryo as studied by electron microscopy, *Dev. Biol.* **17**:117–142.

Lyser, K. M., 1977, Differentiation of cells in organ culture, in *Cell Tissue and Organ Cultures in Neurobiology* (S. Fedoroff and L. Hertz, eds.), pp. 121–140, Academic Press, New York.

Massagué, J., 1987, The TGF-β family of growth and differentiation factors, *Cell* **49**:437–438.

Mattsson, M. E. K., Engberg, G., Ruussala, A.-I., Hall, K., and Pahlman, S., 1986, Mitogenic response of human SH-SY5Y neuroblastoma cells to insulin-like growth factor I and II is dependent on the stage of differentiation, *J. Cell Biol.* **102**:1949–1954.

Mead, R., Schmidt, G. H., and Ponder, B. A. J. 1987, Calculating numbers of tissue progenitor cells using chimeric animals, *Dev. Biol.* **121**:273–276.

Model, P. G., 1978, Regulation of the Mauthner cell following unilateral rotation of the prospective hindbrain in axolotl neurulae (Ambystoma mexicanum), *Brain Res.* **153**:135–143.

Morrest, D. K., 1970, The pattern of neurogenesis in the retina of the rat, *Z. Anat. Entwicklungs-gesch.* **131**:45–67.

Nagy, T., and Reh, T., 1987, Nervous system specific antibodies in frog, *Soc. Neurosci. Abstr.* **13**:385–412.

Negishi, K., Teranishi, T., and Kato, S., 1982, New dopaminergic and indoleamine-accumulating cells in the growth zone of goldfish retinas after neurotoxic destruction, *Science* **216**:747–749.

Negishi, K., Teranishi, T., and Kato, S., 1985, Growth rate of a peripheral annulus defined by neurotoxic destruction in the goldfish retina, *Dev. Brain Res.* **20:**291–295.

Negishi, K., Teranishi, T., Kato, S., and Nakamura, Y., 1987, Paradoxical induction of dopaminergic cells following intravitreal injection of high doses of 6-hydroxydopamine in juvenile carp retina, *Dev. Brain Res.* **33:**67–79.

Oppenheim, R. W., 1981, Neuronal cell death and some related regressive phenomena during neurogenesis: A selective historical review and progress report, in *Studies in Developmental Neurobiology* (W. M. Cowan, ed.), pp. 74–98, Oxford University Press, New York.

Patel, A. J., 1985, Neurotrophic drugs and brain development: Effects on cell replication in vivo and in vitro. Prevention of physical and mental congenital defects, part C: *Basic and Medical Science, Education, and Future Strategies,* pp. 301–305, Alan R. Liss, New York.

Patel, A. J., Vertes, Z. S., Lewis, P. D., and Lie, M., 1980, Effect of chlorpromazine on self-proliferation in the developing rat brain. A combined biochemical and morphological study, *Brain Res.* **202:**414–428.

Patterson, P. H., 1978, Environmental determination of autonomic neurotransmitter functions, *Annu. Rev. Neurosci.* **1:**1–17.

Perris, R., von Boxberg, Y., and Löfberg, J., 1988, Local environmental matrices determine region-specific phenotypes in neural crest cells, *Science* **241:**86–88.

Prasad, K. N., 1975, Differentiation of neuroblastoma in culture, *Biol. Rev.* **2:**129–165.

Prasad, K. N., and Hsie, A. W., 1971, Morphological differentiation of mouse neuroblastoma cells induced *in vitro* by db-cAMP, *Nature (London)* **233:**141–142.

Price, J., Turner, D., and Cepko, C. 1987, Lineage analysis in the vertebrate nervous system by retrovirus-mediated gene transfer, *Proc. Natl. Acad. Sci. U.S.A.* **84:**156–160.

Quinn, L. S., Holtzer, H., and Nimeroff, M., 1985, Generation of chick skeletal muscle cells in groups of 16 from stem cells, *Nature (London)* **313:**692–694.

Reh, T. A., 1987a, Cell-specific regulation of neuronal production in the larval frog retina, *J. Neurosci.* **7:**3317–3324.

Reh, T. A., 1987b, A role for the extracellular matrix in CNS neurogenesis, *Soc. Neurosci. Abstr.* **13:**55.

Reh, T. A., and Constantine-Paton, M., 1983, Qualitative and quantitative measures of plasticity during the normal development of the *Rana pipiens* retinotectal projection, *Dev. Brain Res.* **10:**187–200.

Reh, T. A., Nagy, T., and Gretton, H., 1987, Retinal pigmented epithelial cells induced to transdifferentiate to neurons by laminin, *Nature (London)* **330:**68–71.

Reh, T. A., and Radke, K., 1988, A role for the extracellular matrix in retinal neurogenesis *in vitro, Dev. Biol.,* **129:**283–293.

Reh, T. A., and Tully, T., 1986, Regulation of tyrosine hydroxylase containing amacrine cell number in larval frog retina, *Dev. Biol.* **114:**463–469.

Rohrer, H., Acheson, A. L., Thibault, J., and Thoenen, H., 1986, Developmental potential of quail dorsal root ganglion cells analyzed *in vitro* and *in vivo. J. Neurosci.* **6:**2616–2624.

Rothstein, H., Worgul, B., and Weinsieder, A., 1981, Regulation of lens morphogenesis and cataract pathogenesis by pituitary-dependent, insulin-like mitogens, in *Cellular Communication During Ocular Development* (J. B. Sheffield and S. R. Hilfer eds.), pp. 111–144, Springer-Verlag, Berlin.

Saneto, R. P., and de Villis, J., 1985, Hormonal regulation of the proliferation and differentiation of astrocytes and oligodendrocytes in primary culture, in *Cell Culture in the Neurosciences* (J. E. Bottenstein and G. Sato, eds.), pp. 125–159, Plenum Press, New York.

Schubert, D., Harris, A. J., Heinemann, S., Kidokoro, Y., Partrick, J., Steinbach, J. H., 1973, Induced differentiation of a neuroblastoma, in *Tissue Culture of the Nervous System* (G. Sato, ed.), p. 55, Plenum Press, New York.

Sensenbrenner, M., Booher, J., and Mandel, P., 1971, Cultivation and growth of dissociated neurones from chick embryo cerebral cortex in the presence of different substrates, *Z. Zellforsch. Mirosk. Anat.* **117:**559–569.

Sensenbrenner, M., Booher, J., and Mandel, P., 1973, Histochemical study of dissociated nerve cells from embryonic chick cerebral hemispheres in flask cultures, *Experientia* **29:**699–701.

Sensenbrenner, M., Labourdette, G., Delaunoy, J. P., Pettmann, B., Devilliers, G., Moonen, G., and Bock, E., 1980a, Morphological and biochemical differentiation of glial cells in primary culture, in *Tissue Culture in Neurobiology* (E. Giacobini, A. Vernadakis, and A. Shahar, eds.), pp. 385–395, Raven Press, New York.

Sensenbrenner, M., Wittendorp, E., Barakat, I., and Rechenmann, R. V., 1980b, Autoradiographic study of proliferating brain cells in culture, *Dev. Biol.* **75**:268–277.

Sensenbrenner, M., Barakat, I., and Labourdette, G., 1982, Proliferation and maturation of neuronal cells from the central nervous system in culture, in *Neurotransmitter Interaction and Compartmentation* (H. F. Bradford, ed.), pp. 497–514, Plenum Press, New York.

Simpson, D. L., Morrison, R., de Vellis, J., and Herschman, H. R., 1982, Epidermal growth factor binding and mitogenic activity on purified populations of cells from the central nervous system, *J. Neurochem. Res.* **8**:453–462.

Sooy, L. E., and Metzger-Freed, L., 1970, A serum macromolecule-supplemented medium for frog cell lines, *Exp. Cell Res.* **60**:482–485.

Spira, A. W., Hudy, S., and Hannah, R. S., 1984, Ectopic photoreceptor cells and cell death in the developing rat retina, *Anat. Embryol.* **169**:293–301.

Sporn, M. B., Roberts, A. B., Wakefield, L. M., and Assoian, R. K., 1986, Transforming growth factor—β: Biological function and chemical structure, *Science* **233**:532–534.

Stelzner, D. J., and Strauss, J. A., 1986, A quantitative analysis of frog optic nerve regeneration: Is retrograde ganglion cell death or collateral axonal loss related to selective reinnervation? *J. Comp. Neurol.* **245**:83–106.

Stent, G. S., Weisblat, D. A., Blair, S. S., and Zackson, S. L., 1982, Cell lineage in the development of the leech nervous system, in *Neuronal Development* (N. C. Spitzer, ed.), pp. 1–25, Plenum Press, New York.

Sternberg, P. W., 1988, Control of cell fate within equivalence groups in *C. elegans*, *Trends Neurosci.* **11**:259–264.

Stillwell, E. F., Cone, C. M., and Cone, C. D., 1973, Stimulation of DNA synthesis in CNS neurons by sustained depolarization, *Nature (London)* **246**:110–111.

Straznicky, C., and Gaze, R. M., 1972, The growth of the retina in *Xenopus laeris:* An autoradiographic study, *J. Embryol. Exp. Morph.* **26**:67–79.

Svoboda, K. K. H., and O'Shea, K. S. 1987, An analysis of cell shape and the neuroepithelial basal lamina during optic vesicle formation in the mouse embryo, *Development* **100**:185–200.

Tuckett, S., and Morris-Kay, G. M., 1986, The distribution of fibronectin, laminin and entactin in the neurulating rat embryo studied by indirect immuno-fluorescence, *J. Embryol. Exp. Morphol.* **94**:95–112.

Turner, D. L., and Cepko, C., 1987, A common progenitor for neurons and glia persists in rat retina late in development, *Nature* **328**:131–136.

Turner, J. E., and Delaney, R. K., 1979, Retinal ganglion cell response to axotomy and nerve growth factor antiserum in the regenerating visual system of the newt: An ultrastructural morphometric analysis, *Brain Res.* **177**:35–47.

Turner, J. E., Delaney, R. K., and Johnson, J. E., 1981, Retinal ganglion cell response to axotomy and nerve growth factor antiserum treatment in the regenerating visual system of the goldfish: An in vivo and in vitro analysis, *Brain Res.* **204**:283–294.

Turner, J. E., Schwabb, M. E., and Thoenen, H., 1982, Nerve growth factor stimulates neurite outgrowth from goldfish retinal explants: The influence of a prior lesion, *Dev. Brain Res.* **4**:59–66.

Vollmer, G., and Layer, P. G., 1986, An in vitro model of proliferation and differentiation of the chick retina: Coaggregates of retinal and pigment epithelial cells, *J. Neurosci.* **6**(7):1885–1896.

Weston, J. A., 1986, Phenotypic diversification in neural crest derived cells: The time and stability of commitment during early development. *Curr. Topics Dev. Biol.* **20**:195–210.

Wetts, R., and Herrup, K., 1982a, Interaction of granule, Purkinje, and olivary neurons in lurcher chimeric mice. I. Qualitative studies, *J. Embryol. Exp. Morphol.* **68**:87–98.

Wetts, R., and Herrup, K., 1982b, Cerebellar Purkinje cells are descended from a small number of progenitors committed during early development: Quantitative analysis of lurcher chimeric mice, *Brain Res.* **250:**358–362.

Wetts, R., and Herrup, K., 1982c, Interaction of granule, Purkinje, and olivary neurons in lurcher chimeric mice. II. Granule cell death, *J. Neurosci.* **2:**1494–1498.

Young, R. W., 1984, Cell death during differentiation of the retina in the mouse, *J. Comp. Neurol.* **229:**362–373.

# Development of the Visual System in Hypopigmented Mutants

# 4

Maree J. Webster, Ursula C. Drager, and Jerry Silver

## 4.1. INTRODUCTION

The ganglion cells of most vertebrate retinas are distributed nonuniformly across the retina. Three major retinal specializations occur within the ganglion cell layer: the visual streak, the area or fovea centralis, and the nasotemporal division. Understanding the mechanisms that influence the development of these specializations has become a central theme in visual and developmental neuroscience (see also Chapters 6, 9, and 10 in this volume).

The visual streak is a horizontally oriented region of high ganglion cell density found in some species, which allows scanning of the horizon without the need for eye movements as are necessary for scanning with an area centralis (Rowe and Stone, 1976). The fovea centralis of primates and the area centralis found in many other mammals are considered homologous, the difference being greater specialization of structure in the fovea. Both are regions of high ganglion cell and cone density subserving high-resolution vision and in many species binocular fixation (Rodieck, 1973). The nasotemporal division is the transition between regions of retina containing contralaterally projecting cells (in nasal retina) and those containing the ipsilaterally projecting cells in temporal retina. The overlap of the visual fields seen by the two eyes is larger in species with frontally placed eyes, thus allowing for extensive binocular vision, and smaller in species with more laterally placed eyes (Lund, 1978). In most mammalian species the area centralis or its

**Maree J. Webster and Jerry Silver** • Neuroscience Program, Department of Developmental Genetics, School of Medicine, Case Western Reserve University, Cleveland, Ohio 44106. **Ursula C. Drager** • Department of Neurobiology, Harvard Medical School, Boston, Massachusetts 02115.

homologue is located at the intersection of the visual streak and the nasotemporal division (Stone, 1983). Rodents, however, are an exception with an area of peak ganglion cell density located in superior temporal retina at the temporal edge of the visual streak (Fukuda, 1977; Keens, 1981) and the nasotemporal division located separately in inferior temporal retina (Cowey and Franzini, 1979; Cowey and Perry, 1979; Jeffrey *et al.*, 1981; Keens, 1981). Subsequent observations by Dreher *et al.* (1985) have described a region of large ganglion cell concentration at the nasotemporal division of the rat retina; thus suggesting that this large cell region, rather than the region of peak ganglion cell density, is homologous to the area centralis of other species, because of its location on the nasotemporal division and its involvement in binocular vision.

As an approach to the mechanisms responsible for the nonuniform distribution of ganglion cells and in particular for the laterality of retinofugal projections, one can either compare species that differ with respect to retinal specializations, such as cat and rodents, or one can compare patterns of axon development in normal animals with those in mutants which exhibit abnormal projections. The albino mutation, which decreases ocular melanin, has been shown to cause a larger than normal proportion of optic axons to cross at the optic chiasm, thereby reducing the ipsilateral projection and shifting the nasotemporal division into temporal retina. This anomaly has been described in all mammalian species studied, which include the rat (Lund, 1965; Creel and Giolli, 1976; Wise and Lund, 1976), mouse (Drager, 1974; LaVail *et al.*, 1978; Drager and Olsen, 1980), rabbit (Giolli and Guthrie, 1969), mink (Sanderson *et al.*, 1974; Guillery *et al.*, 1979), cat (Creel *et al.*, 1982), monkey (Gross and Hickey, 1980; Guillery *et al.*, 1984), and human (Creel *et al.*, 1978). The same abnormality is present in the Siamese cat, which carries an incomplete (temperature-sensitive) mutation at the albino locus (Guillery, 1969; Guillery and Kaas, 1971; Guillery *et al.*, 1974; Shatz, 1977; Stone *et al.*, 1978; Cooper and Pettigrew, 1979; Murakami *et al.*, 1982). Moreover, the crossing defect is not simply keyed to a mutation at the albino locus. Similar defects are seen as a consequence of mutations in several unrelated genes, which cause a reduction in ocular melanin pigment (Sanderson *et al.*, 1974; LaVail *et al.*, 1978), and the severity of the optic nerve crossing defect correlates roughly with the degree of hypopigmentation in the retinal pigment epithelium. These observations point to melanin pigment itself, rather than an unknown pleiotropic effect of the albino locus, as serving a role in the establishment of normal retinofugal projections. In addition to the defect in the nasotemporal division, hypopigmentation mutants show a defect in another retinal specialization: the area or fovea centralis. Albino primates lack a proper fovea (Naumann *et al.*, 1976; Fulton *et al.*, 1978) and, as in the case of the albino green monkey, have abnormally large ganglion cells in the central retinal areas (Guillery *et al.*, 1984). Similarly, Siamese cats have only a rudimentary area centralis, with low ganglion cell density and increased ganglion cell size (Stone *et al.*, 1978). An abnormal area centralis homologue has also been demonstrated in the albino rabbit (Choudhury, 1981). Thus, these studies show that a congenital reduction in melanin pigment in the eye, independent of coat color, is associated with a number of abnormalities in retinal organization and fiber projections, which in turn alter the organization of the dorsal lateral geniculate nucleus and visual cortex (for reviews see Murakami *et al.*, 1982; Stone, 1983).

The abnormal retinal projections have been observed very early in development: they were found in the optic tract of Siamese cats (Shatz and Kliot, 1982; Kliot and Shatz, 1985) and albino ferrets (Cucchiaro and Guillery, 1984) before the first axons enter their targets. In albino rats the onset of the abnormality is not as certain: whereas initially the earliest ipsilateral projection was found to be normal, and the diminished projection was thought to emerge secondarily following elimination of axons after they have reached their targets (Land and Lund, 1979; Land *et al.*, 1981; Maxwell and Land, 1981), recent results are more in accord with those in cat and ferret (Bunt *et al.*, 1983): the ipsilateral projection appears somewhat abnormal at the earliest stages of development in the albino rat in that it is significantly reduced in size and delayed in outgrowth as compared to pigmented rats. Thus, it seems likely that the initial chiasmatic choice of optic axons is abnormal in *all* albino mammals, and the albino abnormality may even be expressed more peripherally in the visual pathway.

The exact cellular localization of the melanin defect that causes the retinal abnormalities and the chiasmatic misrouting has yet to be defined. Since melanin is found in the pigment epithelium of the retina and the dorsal cells of the optic stalk, a number of studies have attempted to correlate abnormalities in either of these structures to the abnormalities of retinal projections associated with genetic states of hypopigmentation.

## 4.2. THE ROLE OF RETINAL PIGMENT EPITHELIUM IN PRODUCING ABNORMALITIES IN THE VISUAL SYSTEM OF HYPOPIGMENTED ANIMALS

Shatz and Kliot (1982) report a reduction in the levels of pigment present within cells of the retinal pigment epithelium (RPE) of Siamese cats throughout fetal development and suggest that actual pigment levels may control the laterality of retinal ganglion cell axons. Consistent with this suggestion is the finding that within true albino cats, which have no retinal pigment at all, the abnormalities in the visual system are even more pronounced than in Siamese cats (Creel *et al.*, 1982). In addition, a recent study by Guillery and colleagues (1986) using albino/pigmented mosaic mice indicates that a small patch of RPE cells at the transition from the eye to the optic stalk may be essential for producing the albino abnormalities.

The cells of the RPE become pigmented at the time when the first ganglion cells leave the mitotic cycle (Drager, 1985a). During early stages of eye development, the RPE cells are coupled via gap junctions (Fujisawa *et al.*, 1976; Hayes, 1976) or desmosomal junctions (Kuwabara and Weidman, 1974) with the developing neural retina, and these junctions disappear with the onset of retinal differentiation. Recently, Drager (1985b) demonstrated that melanin pigment is a very effective calcium chelator, which provides heavily pigmented tissues, such as the RPE, with enormous calcium buffering capacity. No comparable calcium binding capacity is seen in the RPE of albino animals (see Fig. 4.6). Since intercellular junctions are known to be affected by calcium (Garrod, 1986), it is possible that the presence of a powerful calcium buffer in RPE cells affects their junctional communication with cells of the neural retina and through this the timing of neuronal differentiation. Indeed, the timing of retinal events such as outer plexiform layer (OPL) development and cessation of cytogenesis appear abnormal in both the albino rat (Webster and Rowe, 1985, 1987) and Siamese cat (Webster, 1985).

### 4.2.1. Developmental Abnormalities in the Retina of the Albino Rat and Siamese Cat

In pigmented rats (Webster and Rowe, 1985, 1988) and normally pigmented cats (Rapaport and Stone, 1983), both the reduction of mitotic activity at the ventricular surface and the formation of the OPL occur first at a location superior and temporal to the optic disc. This site corresponds to the future area of peak ganglion cell density in the adult. In addition, in the pigmented rat (Webster and Rowe, 1985, 1988) and normal cat (Rapaport and Stone, 1983) both processes begin at the same time and remain spatially and temporally coincident until they are complete. However, in albino rats and Siamese cats the two processes are not always spatially and temporally coincident. While both processes commence in a similar region in pigmented and albino rat, that is, in the region superior and temporal to the optic disc, in the albino animal cessation of cytogenesis can proceed without OPL formation. Similarly, in the Siamese cat at E54 (Fig. 4.1), there is a notable decline in mitotic activity but no indication of OPL formation. It was initially proposed (Rapaport *et al.*, 1984) that OPL formation caused the end of cytogenesis by blocking the migratory movements characteristic of the cell cycle. While OPL formation, when it is present, may contribute to this process, it is clear that some other mechanism can stop cytogenesis, since, as in the albino, OPL formation is long delayed. An alternative explanation for the cessation of cytogenesis may be that each cell is programmed to divide a given number of times, before it starts to differentiate. In this case, the OPL would simply form as a consequence of the dendritic differentiation in horizontal cells, rather than serving primarily in signaling the cessation of cytogenesis. Indeed, the strict radial migration of cells from the ventricular surface to their final position has led to the suggestion that the sequence of developmental events may be caused or influenced by some intrinsic mechanism residing in the precursor cells at the ventricular surface (Rakic, 1981; Rapaport and Stone, 1982) or perhaps via communication with the pigment epithelium as in the case of the retina.

Furthermore, in albino rats and Siamese cats the time of onset as well as the rate of both processes is more variable than in pigmented rats and normal cats. For example, in the pigmented rat a thin seam of OPL first appears in a

**Figure 4.1.** Map of the distribution of mitotic figures at the ventricular surface from an E54 Siamese cat retina. One section every 120 μm was traced with a motor-driven microscope linked to an Apple IIe computer. Along each section, the edge of the OPL was marked, together with optic disc and every mitotic figure along the ventricular surface. A zone of decreasing mitotic activity has appeared over the central retina. The OPL has not formed.

relatively localized retinal region at postnatal day 4 (P4) and gradually spreads peripherally to completion by P9. In the albino rat, the OPL often makes a more sudden appearance over a wide region of retina (Webster and Rowe, 1985, 1988). Similarly, in the Siamese cat the development of the OPL can be delayed considerably as compared to the normal cat; no OPL is visible at E54 (see Fig. 4.1) and still at E58 (Webster, 1985), whereas in the normal cat the OPL appears at E50 and by E60 has spread across 6% of the retinal area (Rapaport and Stone, 1982).

### 4.2.1.1. Possible Effect of Melanin Pigment on Retinal Ganglion Cell Topography

The defect in OPL maturation in albinos suggests that melanin pigment, in addition to its action on the chiasmatic pathways, has a direct effect on retinal development, and the phenomenon may be related to the rudimentary area centralis in albinos (Stone *et al.*, 1978). For example, the presence of melanin in the pigment epithelium may affect the first ganglion cells leaving the cell cycle and thereby influence the spatial distribution of cell proliferation. This hypothesis is supported by observations on the albino rat and Siamese cat retina, which indicate the timing of developmental events such as cessation of cytogenesis and OPL formation are both delayed and mismatched in these species. Such consequences may cause further disruption in the sequence of maturation and thereby result in anomalous organization in the retina and within the visual system.

Murakami *et al.* (1982) propose a similar idea to account for the development of retinal topography and in particular the area centralis in normal and Siamese cats. They propose that the final gradient of ganglion cells is largely dependent on the spatiotemporal gradients of ganglion cell cytogenesis, synaptogenesis, and cell death. Anything that alters this spatiotemporal gradient will also alter the ganglion cell gradient. Since it has been shown in both Siamese cats (Stone *et al.*, 1978) and albino primates (Guillery *et al.*, 1984) that the gradient of ganglion cell density is reduced, this may have arisen from a disruption in the timing sequence of developmental events, possibly influenced by events resulting from the lack of melanin in the pigment epithelium. The spatiotemporal gradients in cell birth (Kliot and Shatz, 1982; Walsh *et al.*, 1983; Walsh and Polley, 1985; Webster, 1985; see Chapter 1 in this volume), cell death (Sengelaub and Finlay, 1982; Sengelaub *et al.*, 1986; Dunlop *et al.*, 1987; Provis, 1987; however also see Dreher *et al.*, 1984; Stone and Rapaport, 1986; see also Chapters 9 and 10 in this volume), and synaptogenesis in target nuclei (Hendrickson and Rakic, 1977; Shatz, 1983) of normally pigmented animals support the timing hypothesis. Thus, the development of a *normal* retinal topography would depend on a sequence of events, tightly orchestrated both spatially and temporally, commencing with the initial point of contact between neural ectoderm and surface ectoderm during eyecup invagination as Robinson (1987) suggested. Any disruption in the timing of one event may lead to a cascade of mismatched events resulting in the abnormalities which are seen in the adult hypopigmented retinas.

This hypothesis implies a mechanism that is not very specific and hence allows for much individual variation within the system. Indeed, considerable variation is evident in the development of both albino rat and Siamese cat retinas (Webster, 1985; Webster and Rowe, 1985, 1988) and probably reflects the significant variability in the extent of retinal and chiasmatic abnormalities observed in other adult albinos (Guillery and Kaas, 1971; Guillery *et al.*, 1971, 1979; Sanderson *et al.*, 1974; Shatz, 1977; LaVail *et al.*, 1978; Stone *et al.*, 1978; Cooper and Pettigrew, 1979).

### 4.2.1.2. Possible Effect of Melanin Pigment on Retinal Projections

Although the nasotemporal division appears early in ontogeny, the factors influencing its development seem to be separate from those forming the density distribution of ganglion cells (Webster and Rowe, 1988). For example, studies on the developing rat retina (Beazley *et al.*, 1987; Webster and Rowe, 1988) indicate that the early maturation of a restricted region of retina at the future site of peak ganglion cell density is more likely to be involved in establishing the ganglion cell gradients in that region than in the nasotemporal division which is located in the inferior temporal retina. However, while the abnormal patterns of projection present very early in the gestation of hypopigmented animals (Shatz and Kliot, 1982; Bunt *et al.*, 1983; Cucchiaro and Guillery, 1984; Kliot and Shatz, 1985) are likely to be a direct consequence of chiasmatic misrouting, they could also be a consequence of abnormal genesis of cells. Indeed, Drager (1985a) has shown that the first ganglion cells born in the normally pigmented mouse retina later come to be located in two distinct areas, one in central retina, probably corresponding to the future area of peak ganglion cell density, and the other in inferior temporal retina corresponding to the region of ipsilaterally projecting cells. If, as Webster and Rowe (1987) suggest, the normal arrangement of retinal ganglion cell axons in the optic nerve depends on the sequence or precise timing of their outgrowth from the retina, then the altered spatiotemporal pattern of retinal maturation could be related to the abnormal ganglion cell projection in hypopigmented animals. It seems possible that the absence of melanin in the pigment epithelium may result in a general metabolic abnormality in the cells which may influence cytogenesis in the neural retina. As a consequence, cytogenesis may be affected to a variable degree both spatially and temporally, resulting in a cascade of abnormally timed events in the developing visual system.

## 4.3. THE OPTIC STALK'S ROLE IN PRODUCING ABNORMALITIES IN THE VISUAL SYSTEM OF HYPOPIGMENTED ANIMALS

Examination of the optic stalk in normal and Siamese cats has revealed an abnormal pattern of development in the Siamese which may be related to the subsequent formation of an abnormal retinal projection (Webster *et al.*, 1986). Because the abnormalities associated with albinism are more extensive in Siamese cats than in previously studied species, a careful investigation of the morphological differentiation of the optic stalk and of the guidance and/or misguidance of optic axons in this structure was undertaken in both normally pigmented and Siamese cats. The embryogenesis of the eye is briefly reviewed here, before we describe the morphogenesis of the optic stalk in normal and Siamese cats.

### 4.3.1. Embryogenesis of the Optic Stalk

The development of the eye in the mammalian embryo begins with an outpocketing of the neuroectoderm of the diencephalon, which forms the optic vesicle (Mann, 1969; Silver and Hughes, 1973). The cavity of the optic vesicle is continuous with the third ventricle of the diencephalon via the optic stalk. The tip of the optic vesicle closely apposes the surface ectoderm, which thickens to become the lens placode. The vesicle then invaginates to form the optic cup, the outer layer transforms into the pigment epithelium, and the inner layer into the neural retina. The optic stalk is a tubular structure that can be divided into dorsal and ventral tiers. Along the distal part of the ventral tier lies a groove, the optic fissure, which forms by a process of invagination. The fissure eventually disappears, leaving only a small opening that is occupied by the hyaloid artery (Silver and Robb, 1979). The lumen of the stalk (before axons have entered it) is continuous with the third cerebral ventricle and the space between the neural retina and the pigment epithelium of the eye. The dorsal tier of the stalk is continuous with the pigment epithelium while the ventral cells are continuous with the neural retina. The ganglion cell axons grow toward the optic disc, along channels produced by processes between the surrounding neuroepithelial cells (Silver and Robb, 1979; Silver and Sidman, 1980) to enter the optic stalk through the optic disc. Once in the stalk, the optic axons travel a stereotypic route as they infiltrate between the neuroepithelial cell processes along the marginal wall of the optic stalk.

### 4.3.2. Pigment in the Optic Stalk

A number of authors have examined the potential role of a small population of transiently pigmented cells in the developing optic stalk in determining the path taken by the retinal axons as they first exit from the eye. Normally, the distodorsal stalk cells, that is, those nearest the eye, contain pigment but the proximodorsal cells and all cells in the ventral tier do not. Because within the stalk optic axons grow only among the pigment-free cells, Silver and Sapiro (1981) suggested that pigmented cells may somehow prevent optic axons from growing in their vicinity. Albino animals have no pigment in the distodorsal stalk cells, and axons invade the dorsal area ectopically almost immediately after they leave the eye. This misrouting of axons disrupts their normal neighborhood relations in the stalk and may lead directly to the misrouting in the optic chiasm (Oberdorfer *et al.*, 1981; Silver and Sapiro, 1981). In an alternative hypothesis, Strongin and Guillery (1981) believed that the stalk melanin lyses, and they suggested that this lytic process proceeds in a spatial and temporal pattern that allows for the breakdown products to have an effect on the outgrowing fibers. They did *not* observe any relation of axons with pigmented stalk cells or any changes in the trajectory of retinal axons in the optic stalk of the albino.

### 4.3.3. Morphogenesis of the Normally Pigmented Cat Optic Stalk

In the fetal cat on embryonic day 20 (E20), before axons enter the optic stalk, the only clear morphological distinction between Siamese and normal cat is the distribution of pigment in the stalk. Pigment is found in the dorsal stalk cells of the normal cat for 200 µm from the optic disc, but not in the Siamese optic stalk (Fig. 4.2).

By E23 axons invade the extracellular spaces in the ventral optic stalk. Concurrent with the initial stages of axonal exit from the retina, the dorsal and ventral cell groups in the *normal* stalk are partitioned into completely separate components via a two-stage process: (1) desmosomal junctions form between the two cell groups obliterating the stalk lumen (Fig. 4.3) and (2) the formation of desmosomes is followed by complete separation of the stalk's dorsal and ventral tiers by invagination of the basal lamina which cleaves the stalk along the plane of the old lumen (Fig. 4.3). The ventral tier fills with axons while the dorsal tier is shed gradually. Pioneering axons invade the ventral margin of the stalk symmetrically and later (E24–E25) move relatively more dorsally but still remain within the ventral tier, below the old lumen, while dorsal pigment cells are eliminated. Additional, observations in the mouse suggest that dorsal stalk cells may be removed slowly, one at a time, by basal lamina invagination, through similar mechanisms as in the cat (Fig. 4.4).

Previous studies describing optic stalk morphogenesis have not appreciated the loss of dorsal stalk cells by basal lamina invagination. However, all authors have described the displacement of the lumen dorsally as axons invade the ventral stalk cells. The ventral region expands as axons enter, while early on at least, the dorsal region remains one cell layer thick (Assheton,

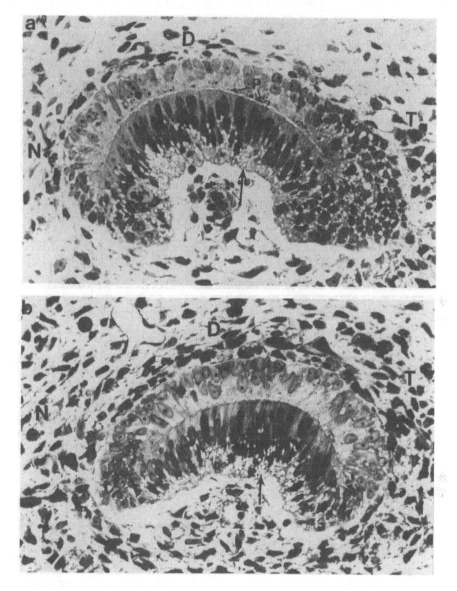

**Figure 4.2.** Optic stalk of normal (a) and Siamese (b) cat (60 μm from optic disc) are similar at E21. Note less-dense cytoplasm of dorsal tier as compared to ventral tier. Extracellular spaces are apparent at the vitreal surface of the ventral tier (arrows). Pigment granules (P) are present in the dorsal tier of the normal cat (a) but absent in the Siamese (b). D, dorsal; N, nasal; T, temporal. 40×.

1892; Robinson, 1896; Kuwabara, 1975; Rager, 1980a,b; Navascues *et al.*, 1985; Horsburgh and Sefton, 1986). A number of theories have been proposed to account for the eventual loss of the lumen. These include obliteration of the lumen, either by proliferation of the apical cells to fill the lumen (Kuwabara, 1975) or by merging and fusion of the two sides (Robinson, 1896;

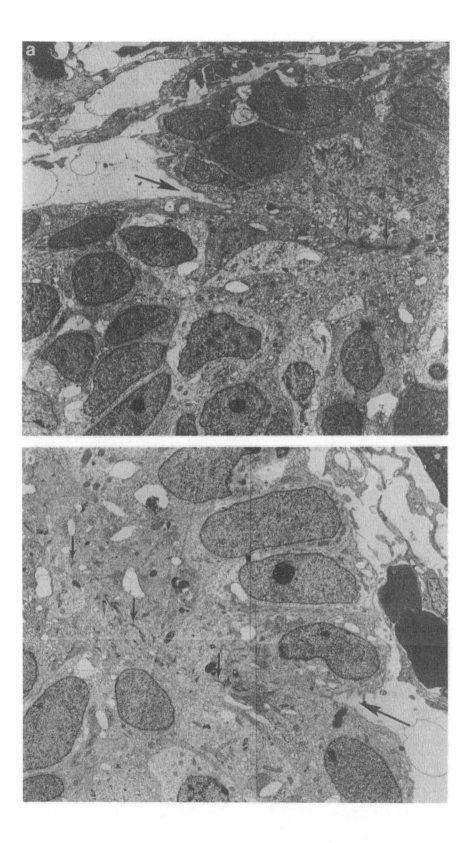

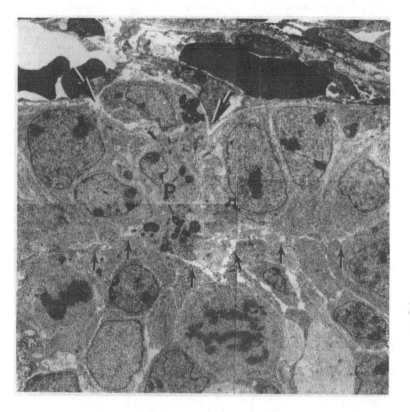

**Figure 4.4.** Electron micrograph of an E14 pigmented mouse optic stalk illustrating the invagination of basal lamina (large arrows) around a dorsal stalk cell. Melanosomes containing pigment (P) are present in the dorsal stalk cells. The stalk lumen is indicated by small arrows. 3000×.

Horsburgh and Sefton, 1986), or alternatively by degeneration of the dorsal stalk cells (Rager, 1980a,b; Navascues *et al.*, 1985). Careful examination of the cat optic stalk found no evidence to support any of these theories (Webster *et al.*, 1986, 1988). However, because the time period for stalk transformation is so short (2–3 days in cat), previous studies (e.g., Robinson, 1896; Williams *et al.*, 1986) may have overlooked the sloughing process and basal lamina invagination.

### 4.3.4. Morphogenesis of the Siamese Optic Stalk

In contrast to the morphological differentiation of the normal cat optic stalk is the differentiation of the stalk in Siamese cats. Although the retinal

**Figure 4.3.** Electron micrographs of E24 normal cat optic stalk illustrating the invagination of the basal lamina (large arrows) at the level of the lumen, which is indicated by junctions (small arrows). 12,000×.

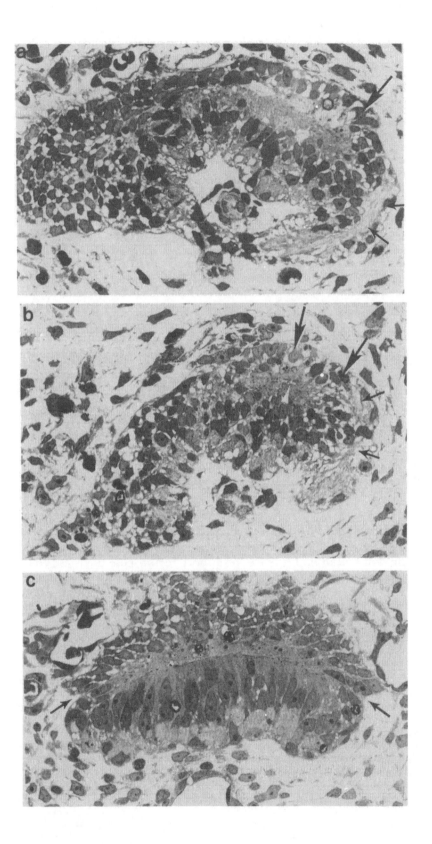

pigment epithelium of the Siamese cat embryo contains pigment (Shatz and Kliot, 1982), no pigment could be detected in the Siamese optic stalk (Fig. 4.2). In the Siamese, dorsal stalk cells are not sloughed off properly and are therefore incorporated ectopically into the nerve (Fig. 4.5a,b). Basal lamina invagination is irregular, and while much of the lumen is eventually oblite-rated, it is not by junction formation and subsequent basal lamina invagina-tion as in normal animals, but rather by coalescence of the dorsal and ventral tiers. Evidence for this coalescence is the persistence within the core of the Siamese stalk of remnants of desmosomal junctions and cilia, hallmarks of the old luminal surface. Axons do not fill the Siamese stalk symmetrically but enter the region of ectopic cells in dorsal stalk, which in turn disrupts normal fiber topography (Fig. 4.5a,b).

The abnormal morphogenesis of the developing Siamese optic stalk is likely to reflect the absence of pigment from the dorsal stalk cells. As in the RPE, the melanin pigment in the stalk chelates large amounts of calcium (see Fig. 4.6). Because desmosomal junctions are known to be affected by calcium (Garrod, 1986), the enormous calcium buffering capacity of melanin may in some way contribute to the stability of junctions between the dorsal and ven-tral tiers and to the sloughing-off process. It seems particularly important to reiterate that while the Siamese cat does have pigment in the retinal pigment epithelium, there is very little if any in the dorsal stalk cells, suggesting that levels of pigment in the stalk may play a key role in the albino visual malfor-mations.

In the Siamese cat it appears that axons originating from the region of retina temporal to the optic fissure are those that invade the dorsal tier of ectopic cells before they have separated. The altered position of optic axons in the albino stalk may provide an explanation for the chiasmatic misrouting of optic axons in the albino mutant. For example, in the optic stalk of the nor-mally pigmented cat, axons arising from ventrotemporal and dorsotemporal retina are positioned laterally in the optic stalk and project ipsilaterally at the chiasm (e.g., see Silver, 1984; Naito, 1986). They are separate from the con-tralateral group which are positioned more centrally and nasally in the stalk. Because many dorsal stalk cells remain ectopic in the Siamese cat, the tem-poral axons that move into this area fasciculate with those in the central region above and adjacent to the normal contralaterally projecting fiber con-tingent. Since the ectopic fibers come to lie above and adjacent to the normal contralaterally projecting fiber contingent, they may be shunted contralateral-ly with them.

**Figure 4.5.** (a) and (b) Transverse sections through the optic stalk of an E24 Siamese cat showing persistence of lumen (arrowheads), dorsal cells (large arrows, b), and ectopic axons laterally among these cells (small arrows, a, b). (c) Transverse section through the optic stalk of a E24 normal cat at similar location to (b). Note the symmetry of basal lamina invagination (arrows) at the level of the lumen and the symmetry of axon invasion in ventral stalk. 40×.

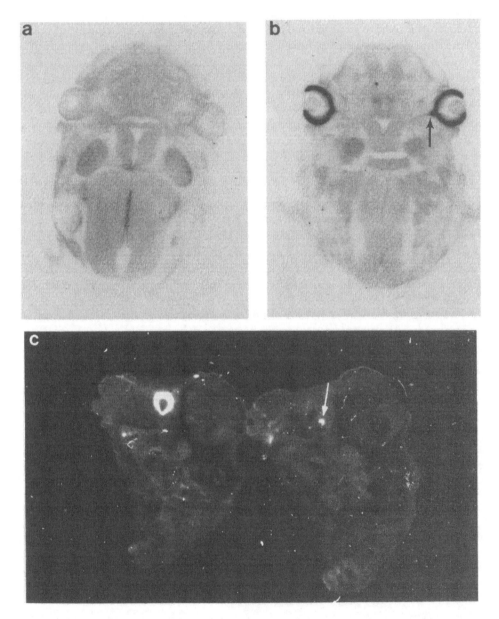

**Figure 4.6.** Autoradiographs of horizontal sections through E14 (a) albino and (b) pigmented mouse embryos. Sections were incubated in radioactive calcium and exposed for 40 min. Note heavy labeling over the RPE, extending into the stalk (arrow) of the pigmented mouse. No labeling is apparent in the albino. (c) Sagittal section through E14 mouse head showing label over the dorsal stalk cells (arrow).

## 4.4. CONCLUSION

The studies outlined in this chapter have described a number of developmental abnormalities at the cellular and subcellular level which occur in the visual system of hypopigmented animals and which are likely to contribute to the abnormalities of the adult system. Although a reduction of melanin synthesis and content in both the pigment epithelial and stalk cells appears to be directly related to the abnormal patterns of development in the retina and the stalk, the exact cellular mechanisms responsible for the developmental abnormalities remain to be determined.

ACKNOWLEDGMENTS. The authors would like to thank Catherine Doller for expert technical assistance. The Siamese cat retina was taken from a study done by MJW in partial fulfillment of a Ph.D. degree in the laboratory of Professor Jon Stone.

## 4.5. REFERENCES

Assheton, R., 1892, On the development of the optic nerve of vertebrates, and the choroidal fissure of embryonic life, *Q. J. Microsc.* **34:**85–104.

Beazley, L. D., Perry, V. H., Baker, B., and Darby, J. E., 1987, An investigation into the role of ganglion cells in the regulation of cell division and death of other retinal cells, *Dev. Brain Res.* **33:**169–184.

Bunt, S., Lund, R. D., and Land, P. W., 1983, Prenatal development of the optic projection in albino and hooded rats, *Dev. Brain Res.* **6:**149–168.

Choudhury, B. P., 1981, Ganglion cell distribution in the albino rabbit's retina, *Exp. Neurol.* **72:**638–644.

Cooper, M. L., and Pettigrew, J. D., 1979, The retinothalamic pathways in Siamese cats, *J. Comp. Neurol.* **187:**313–348.

Cowey, A., and Franzini, C., 1979, The retinal organization of uncrossed optic nerve fibers in rats and their role in visual discrimination, *Exp. Brain Res.* **35:**445–455.

Cowey, A., and Perry, V. H., 1979, The projection of temporal retina in rats, studied by retrograde transport of horseradish peroxidase, *Exp. Brain Res.* **35:**457–464.

Creel, D., and Giolli, R. A., 1976, Retinogeniculate projections in albino and ocularly hypopigmented rats, *J. Comp. Neurol.* **166:**445–456.

Creel, D., O'Donell, F. E., and Witkop, C. J., 1978, Visual system anomalies in human ocular albinos, *Science* **201:**931–933.

Creel, D., Hendrickson, A. E., and Levinthal, A. G., 1982, Retinal projections in tyrosinase-negative albino cats, *J. Neurosci.* **2**(7):907–911.

Cucchiaro, J., and Guillery, R. W., 1984, The development of the retinogeniculate pathways in the normal and albino ferrets, *Proc. R. Soc. London Ser. B* **223:**141–164.

Drager, U. C., 1974, Autoradiography of tritiated proline and sucrose transported transneuronally from the eye to the visual cortex in pigmented and albino mice, *Brain Res.* **82:**284–292.

Drager, U. C., 1985a, Birth dates of retinal ganglion cells giving rise to the crossed and uncrossed optic projections in the mouse, *Proc. R. Soc. London Ser. B* **224:**57–77.

Drager, U. C., 1985b, Calcium binding in pigmented and albino eyes, *Proc. Natl. Acad. Sci. U.S.A.* **82:**6716–6720.

Drager, U. C., and Olsen, J. F., 1980, Origin of crossed and uncrossed retinal projections in pigmented and albino mice, *J. Comp. Neurol.* **191**:383–412.

Dreher, B., Potts, R. A., Ni, S. Y. K., and Bennett, M. R., 1984, The development of heterogeneities in distribution and soma sizes of rat retinal ganglion cells, in *Development of visual pathways in mammals* (J. Stone, B. Dreher, and D. H. Rapaport, eds.), pp. 39–57, Alan R. Liss, New York.

Dreher, B., Sefton, A. J., Ni, S. Y. K., and Nisbett, G., 1985, The morphology, number, distribution and central projections of Class 1 retinal ganglion cells in albino and hooded rats, *Brain Behav. Evol.* **26**:10–48.

Dunlop, S. A., Longley, W. A., and Beazley, L. D., 1987, Development of the area centralis and visual streak in the grey kangaroo *Macropus fuliginosus, Vision Res.* **27**(2):151–164.

Fujisawa, H., Morioka, H., Watanabe, H., and Nakamura, H., 1976, A decay in gap junctions associated with cell differentiation of neural retina in chick embryonic development, *J. Cell Sci.* **22**:585–596.

Fukuda, Y., 1977, A three-group classification of rat retinal ganglion cells: Histological and physiological studies, *Brain Res.* **119**:327–344.

Fulton, A. B., Albert, D. M., and Craft, J. L., 1978, Human albinism. Light and electron microscopy study, *Arch. Ophthalmol.* **96**:305–310.

Garrod, P. R., 1986, Desmosomes, cell adhesion molecules and the adhesive properties of cells in tissues, *J. Cell Sci. (Suppl.)* **4**:221–237.

Giolli, R. A., and Guthrie, M. D., 1969, The primary optic projections in the rabbit. An experimental degeneration study, *J. Comp. Neurol.* **136**:99–126.

Gross, K. J., and Hickey, T. L., 1980, Abnormal laminar patterns in the lateral geniculate nucleus of an albino monkey, *Brain Res.* **190**:231–237.

Guillery, R. W., 1969, An abnormal retino-geniculate projection in Siamese cats, *Brain Res.* **14**:739–741.

Guillery, R. W., and Kaas, J. H., 1971, A study of normal and cogenitally abnormal retinogeniculate projections in cats, *J. Comp. Neurol.* **143**:73–100.

Guillery, R. W., Amorn, C. S., and Eighmy, B. B., 1971, Mutants with abnormal visual pathways: An explanation of anomalous geniculate laminae, *Science* **174**:831–832.

Guillery, R. W., Casagrande, V. A., and Oberdorfer, M. D., 1974, Congenitally abnormal vision in Siamese cats, *Nature (London)* **252**:195–199.

Guillery, R. W., Oberdorfer, M. D., and Murphy, E. H., 1979, Abnormal retinogeniculate and geniculocortical pathway in several genetically distinct color phases of the mink (*Mustela vison*), *J. Comp. Neurol.* **185**:623–656.

Guillery, R. W., Hickey, T. L., Kaas, J. H., Felleman, D. J., Dedruyn, E. J., and Sparks, D. L., 1984, Abnormal central visual pathways in the brain of an albino green monkey (*Cercopithecus aethiops*), *J. Comp. Neurol.* **226**:165–183.

Guillery, R. W., Jeffery, G., and Cattanach, B. M., 1986, The number of uncrossed retinofugal fibers in mice that show albino mosaicism: High variability indicates that a small cell group determines the albino abnormality, *Soc. Neurosci. Abstr.* **12**:122.

Hayes, B. P., 1976, The distribution of intercellular gap junctions in the developing retina and pigment epithelium of *Xenopus laevis, Anat. Embryol.* **150**:99–111.

Hendrickson, A. E., and Rakic, P., 1977, Histogenesis and synaptogenesis in the dorsal lateral geniculate of the fetal monkey brain, *Anat. Rec.* **187**:602.

Horsburgh, G. M., and Sefton, A. J., 1986, The early development of the optic nerve and chiasm in embryonic rat, *J. Comp. Neurol.* **243**:547–560.

Jeffrey, G., Cowey, A., and Kuypers, H. G. J. M., 1981, Bifurcating retinal ganglion cell axons in the rat, demonstrated by retrograde double labelling, *Exp. Brain Res.* **44**:34–40.

Keens, J., 1981, Aspects of retinal projections in rats, *Proc. Aust. Physiol. Pharmacol. Soc.* **12**:163P.

Kliot, M., and Shatz, C. J., 1982, Genesis of different retinal ganglion cell types in the cat, *Soc. Neurosci. Abstr.* **8**:815.

Kliot, M., and Shatz, C. J., 1985, Abnormal development of the retinogeniculate projection in Siamese cats, *J. Neurosci.* **5**(10):2641–2653.

Kuwabara, T., 1975, Development of the optic nerve of the rat, *Invest. Opthalmol.* **14**(10):732–745.

Kuwabara, T., and Weidman, T. A., 1974, Development of the prenatal rat retina, *Invest. Ophthalmol.* **3:**725–739.

Land, P. W., and Lund, R. D., 1979, Development of the rat's uncrossed retinotectal pathway and its relation to plasticity studies, *Science* **205:**698–700.

Land, P. W., Hargrove, K., Eldridge, J., and Lund, R. D., 1981, Differential reduction in the number of ipsilaterally projecting ganglion cells during the development of retinofugal projections in the albino and pigmented rats, *Soc. Neurosci. Abstr.* **7:**141.

LaVail, J. H., Nixon, R. A., and Sidman, R. L., 1978, Genetic control of retinal ganglion cell projections, *J. Comp. Neurol.* **182:**399–422.

Lund, R. D., 1965, Uncrossed visual pathways of hooded and albino rats, *Science* **149:**1506–1507.

Lund, R. D., 1978, *Development and Plasticity of the Brain: An Introduction*, Oxford University Press, New York.

Mann, I., 1969, *Development of the Human Eye*, pp. 22–28, Grune & Stratton, New York.

Maxwell, P. E., and Land, P. W., 1981, Development of retinogeniculate projections in albino and pigmented rats, *Anat. Rec.* **199:**165A.

Murakami, D., Sesma, M. A., and Rowe, M. H., 1982, Characteristics of nasal and temporal retina in Siamese and normally pigmented cats, *Brain. Behav. Evol.* **21:**67–113.

Naito, J., 1986, Course of retinogeniculate projection fibers in the cat optic nerve, *J. Comp. Neurol.* **251:**376–387.

Naumann, G. O. H., Lerche, W., and Schroeder, W., 1976, Foveola-Aplasie bei Tyrosinase-positivem oculocutanen Albinismus, *Von Graefe's Arch. Ophthalmol.* **200:**39–50.

Navascues, J., Rodriguez-Gallardo, L., Martin-Partido, G., and, Alvarez, I. S., 1985, Proliferation of glial precursors during early development of the chick optic nerve, *Anat. Embryol.* **172:**365–373.

Oberdorfer, M., Miller, N., and Silver, J., 1981, Distribution of axons in albino and pigmented embryonic optic stalks, *Invest. Ophthalmol. Vis. Sci.* **20:**174.

Provis, J. M., 1987, Patterns of cell death in the ganglion cell layer of the human retina, *J. Comp. Neurol.* **259:**237–246.

Rager, G. H., 1980a, Development of the retinotectal projection in the chicken, *Adv. Anat. Embryol. Cell Biol.* **63:**41–62.

Rager, G. H., 1980b, Die ontogenese der retinotopen projektion Beobachtung und Reflexion, *Naturewissenschaften* **67:**280–287.

Rakic, P., 1981, Neuronal-glia interactions during brain development, *Trends Neurosci.* **4:**184.

Rapaport, D. H., and Stone, J., 1982, The site of commencement of maturation in mammalian retina: Observations in the cat, *Dev. Brain Res.* **5:**273–279.

Rapaport, D. H., and Stone, J., 1983, The topography of cytogenesis in the developing retina of the cat, *J. Neurosci.* **3:**1824–1838.

Rapaport, D. H., Robinson, S. R., and Stone, J., 1984, Cell movement and birth in the developing cat retina, in *Development of Visual Pathways in Mammals* (J. Stone, B. Dreher, and D. H. Rapaport, eds.), pp. 23–38, Alan R. Liss, New York.

Robinson, A., 1896, On the formation and structure of the optic nerve, and its relation to the optic stalk, *J. Anat. Physiol. (London)* **30:**319–333.

Robinson, S. R., 1987, Ontogeny of the area centralis in the cat, *J. Comp. Neurol.* **255:**50–67.

Rodieck, R. W., 1973, *The Vertebrate Retina*, W. H. Freeman, San Francisco.

Rowe, M. H., and Stone, J., 1976, Properties of ganglion cells in the visual streak of the cat's retina, *J. Comp. Neurol.* **169:**99–126.

Sanderson, K. E., Guillery, R. W., and Shackelford, R. M., 1974, Congenitally abnormal visual pathways in mink (*Mustela vison*) with reduced retinal pigment, *J. Comp. Neurol.* **154:**225–248.

Sengelaub, D. R., and Finlay, B. L., 1982, Cell death in the mammalian visual system during normal development: I. Retinal ganglion cells, *J. Comp. Neurol.* **204:**311–317.

Sengelaub, D. R., Dolan, R. P., and Finlay, B. L., 1986, Cell generation, death, and retinal growth in the development of the hamster retinal ganglion cell layer, *J. Comp. Neurol.* **246:**527–543.

Shatz, C. J., 1977, A comparison of visual pathways in Boston and Midwestern Siamese cats, *J. Comp. Neurol.* **171:**205–228.

Shatz, C. J., 1983, The prenatal development of the cat's retinogeniculate pathway, *J. Neurosci.* **3**:482–498.

Shatz, C. J., and Kliot, M., 1982, Prenatal misrouting of the retinogeniculate pathway in Siamese cats, *Nature (London)* **300**(9):525–529.

Silver, J., 1984, Studies on factors that govern directionality of axonal growth in the embryonic optic nerve and at the chiasm of mice, *J. Comp. Neurol.* **223**:238–251.

Silver, J., and Hughes, A. F. W., 1973, The role of cell death during morphogenesis of the mammalian eye, *J. Morphol.* **140**:159–170.

Silver, J., and Robb, R. M., 1979, Studies on the development of the eye cup and optic nerve in normal mice and in mutants with congenital optic nerve aplasia, *Dev. Biol.* **68**:175–190.

Silver, J., and Sapiro, J., 1981, Axonal guidance during development of the optic nerve. The role of the pigment epithelia and other extrinsic factors, *J. Comp. Neurol.* **202**:521–538.

Silver, J., and Sidman, R. L., 1980, A mechanism for the guidance and topographic patterning of retinal ganglion cell axons, *J. Comp. Neurol.* **189**:101–111.

Stone, J., 1983, *Parallel Processing in the Visual System,* Plenum, New York.

Stone, J., and Rapaport, D. H., 1986, The role of cell death in shaping the ganglion cell layer population of the adult cat retina, in *Visual Neuroscience* (J. D. Pettigrew, K. J. Sanderson, and W. R. Levick, eds.), pp. 157–165, Cambridge University Press, Cambridge.

Stone, J., Campion, J. E., and Leicester, J., 1978, The nasotemporal division of retina in the Siamese cat, *J. Comp. Neurol.* **180**:783–798.

Strongin, A. C., and Guillery, R. W., 1981, The distribution of melanin in the developing optic cup and stalk and its relation to cellular degeneration, *J. Neurosci.* **1**(11):1193–1204.

Walsh, C., and Polley, E. H., 1985, The topography of ganglion cell production in the cat's retina, *J. Neurosci.* **5**:741–750.

Walsh, C., Polley, E. H., Hickey, T. L., and Guillery, R. W., 1983, Generation of cat retinal ganglion cells in relation to central pathways, *Nature (London)* **302**:611–614.

Webster, M. J., 1985, Cytogenesis, histogenesis and morphological differentiation of the retina, Ph.D., thesis, University of New South Wales, Sydney, Australia.

Webster, M. J., and Rowe, M. H., 1985, Development of retinal topography in albino and pigmented rats, *Neurosci. Lett. (Suppl.)* **19**:109.

Webster, M. J., and Rowe, M. H., 1988, Altered timing of the developmental processes in albino rat retina, *J. Comp. Neurol.,* in press.

Webster, M. J., Shatz, C. J., Kliot, M., and Silver, J., 1986, Abnormal differentiation of the optic stalk and its relation to axon misguidance in embryonic Siamese cat, *Soc. Neurosci. Abstr.* **16**:121.

Webster, M. J., Shatz, C. J., Kliot, M., and Silver, J., 1988, Abnormal pigmentation and unusual morphogenesis of the optic stalk may be correlated with retinal axon misguidance in embryonic Siamese cat, *J. Comp. Neurol.* **269**:592–611.

Williams, R. W., Bastiani, M. J., Lia, B., and Chalupa, L. M., 1986, Growth cones, dying axons and developmental fluctuations in the fiber population of the cat's optic nerve, *J. Comp. Neurol.* **246**:32–69.

Wise, R. P., and Lund, R. D., 1976, The retina and central projections of heterochromic rats, *Exp. Neurol.* **51**:68–77.

# Topographic Organization of the Visual Pathways

5

## Alan D. Springer

## 5.1. INTRODUCTION

### 5.1.1. Retinal Growth

The position of ganglion cells within a flat-mounted retina can be described with polar coordinates $(r, \theta)$. $r$ represents distance from the optic disc (origin), and $\theta$ represents the angle formed by a line connecting a retinal ganglion cell and the origin with respect to a fixed reference line (i.e., polar axis). Thus, $\theta$ denotes the position of a retinal ganglion cell along a circumferential line and $r$ denotes the position of a retinal ganglion cell along a radial line. For vertebrates such as goldfish, $r$ not only represents distance from the optic disc, but it also indicates the relative birthday of a retinal ganglion cell. The first retinal ganglion cells to have become postmitotic are closest to the optic disc and the most recently generated retinal ganglion cells are near the retinal margin. Such animals are never totally postembryonic, and their retinas continue to grow, much like a tree, by the addition of annuli of cells to the retinal margin (Johns, 1977; Meyer, 1978).

### 5.1.2. Retinotectal Topography

Axons arising from the ganglion cells form the optic fiber layer in the retina, and then, as they leave the eye, they form the optic nerve head. Between the optic nerve head and the optic chiasm, the axons form the optic nerve, and between the optic chiasm and the brain (e.g., optic tectum) they form the optic tract. In some species (e.g., goldfish), the main optic tract divides into two distinct optic tracts (brachia) just before the retinal ganglion cell

---

**Alan D. Springer** • Department of Anatomy, New York Medical College, Valhalla, New York 10595.

axons enter the optic tectum. The organization of the optic axons may be constant as they traverse several of these locations. However, on the assumption that the optic axons terminate topographically in a target, they must undergo some form of reorganization somewhere between the retina and their target. Thus, the organization that the axons manifest at the beginning of the visual pathway is inadequate for achieving a topographic distribution of their terminals within the target.

Retinal ganglion cell axons terminate, in varying numbers, in numerous brain nuclei, among which are the optic tectum, lateral geniculate nucleus, suprachiasmatic nucleus, and assorted pretectal nuclei. All vertebrates do not possess the identical (i.e., homologous) retinorecipient targets. This chapter emphasizes the behavior of the retinal ganglion cell axons with respect to the optic tectum, particularly in goldfish. Deployment of the retinal ganglion cell axon terminals within the contralateral optic tectum mirrors the relative position of the cells from which they originated. Ganglion cells that are located closest to the optic disc have axon terminals that terminate in the center of the tectum and cells that are situated near the retinal margin have axons that terminate along the edges of the tectum. Thus, the terminal map in the tectum is organized concentrically, just as the cellular map is in the retina. Such concentrically organized retinal terminal maps have also been observed in one pretectal nucleus of goldfish (Springer and Mednick, 1985b). Other thalamic and pretectal nuclei in fish receive projections from ganglion cells distributed throughout the retina, or from restricted parts of the retina (Springer and Mednick, 1984, 1985d; Presson *et al.*, 1985). However, the precise topography of the retinal axons within these nuclei remains to be determined.

The relative, but not absolute, circumferential position ($\theta$) of the ganglion cells is also mirrored in the tectum by their axon's terminals. A large number of electrophysiological and anatomical studies have shown that ganglion cells positioned rostrally in the retina (i.e., nasally) project to the caudal tectum, while ganglion cells situated caudally in the retina (i.e., temporally) project to the rostral tectum. Similarly, cells in dorsal retina project to the ventrolateral tectum, while cells in ventral retina project to the dorsomedial tectum. Apparently, the cellular map and the terminal map are out of register by approximately 180° with respect to both the rostrocaudal and dorsoventral tectal axes.

One hypothetical way to reconcile the differences in the orientation of the cellular map in the retina and the terminal map in the tectum is to make a seemingly reasonable assumption: optic axon organization within the visual pathways between the retina and tectum mirrors the topography of their somata within the retina. Namely, the retinal origin maps to the center of the pathway (i.e., axons from central cells are located in the center of the pathway) and $\theta$ in the visual pathway maps as it does in the retina. Rotating the visual pathway by 180° would result in the retinal and tectal maps coming into register with one another. This simple solution to reconciling the two maps is conceptually appealing. However, as will be seen below, implicit in this solution to reconciling the two maps are several erroneous assumptions about how optic axons are organized in the visual pathways and the route by which they enter the optic tectum.

Because the axons of ganglion cells terminate topographically in the tectum, as well as in other retinorecipient targets, many investigators have sought to understand the arrangement of the axons between the ganglion cells and their targets in a variety of different species. Organization of the axons in the visual pathways has been examined in invertebrates (Meinertzhagen, 1976; Chamberlain and Barlow, 1982) and in nonmammalian vertebrates such as fish (Scholes, 1979; Bodick and Levinthal, 1980; Easter *et al.,* 1981; Bunt, 1982; Rusoff, 1984; Springer and Mednick, 1985c; Springer and Mednick, 1986b,c) and frogs (Fawcett, 1981; Reh *et al.,* 1983; Scalia and Arango, 1983; Taylor, 1987). The organization of the optic axons within the optic nerve or tract has also been examined in mammals such as the cat (Horton *et al.,* 1979; Aebersold *et al.,* 1981; Torrealba *et al.,* 1982; Walsh and Guillery, 1984; Naito, 1986), in the optic nerve of embryonic monkeys (Williams and Rakic, 1985), and in the rat (Yamadori, 1981). Several studies have examined optic nerve organization in birds (Bunt and Horder, 1983; Ehrlich and Mark, 1984), urodeles (Bunt and Horder, 1983), and reptiles (Bunt and Horder, 1983).

Axonal organization, from the retina to the optic tectum, has been examined most comprehensively and is best understood in nonmammalian species such as goldfish (Easter *et al.,* 1984; Springer and Mednick, 1986b,c). The organization is less well understood in mammals for several reasons. The mammalian visual system is appreciably more complex than the fish visual system. In the goldfish (Springer and Gaffney, 1981) and oscar (*Astronotus ocellatus;* Springer and Mednick, 1985a), the optic nerves do not interdigitate at the optic chiasm, and the optic pathways are virtually crossed with respect to the optic tectum. Thus, both the nasal and temporal retina project to the contralateral tectum. In primates, there is a difference in the projections of nasal and temporal retinal ganglion cells, with the latter projecting ipsilaterally and the former projecting contralaterally. For many other mammals, the crossed retinal projection arises from ganglion cells situated throughout the entire retina, while the ipsilateral projection comes from ganglion cells restricted to the temporal and ventral retina (Drager, 1985; see Chapters 4 and 6 in this volume). Since mammals have several major retinorecipient targets and partially decussated visual pathways, it appears inevitable that the organization as well as the reorganizations that mammalian optic axons undergo will be significantly more complex than those that occur in fish. However, the rules that govern how optic axons behave in fish will hopefully provide the rudimentary guidelines that determine how mammalian optic axons behave.

Understanding the arrangement of the ganglion cell axons in the visual pathways, as well as the rearrangements that they must undergo prior to terminating, should provide some insights into the mechanisms that operate in the course of visual system development. An orderly arrangement of the axons within the pathways could account, either completely or in part, for the orderly map of their terminals within the tectum. Perhaps rather simple or minimal developmental mechanisms could influence where in the target the axon terminals ultimately deploy (Horder and Martin, 1978). A disordered, or random, organization of the axons within the pathways would have different implications as to the mechanisms that govern orderly terminal map formation in the course of development. A seemingly complex mechanism might be required to direct the axons to their terminal zones in the target if the axons were normally randomly organized in the pathways. Such mechanisms could involve unique axon–target cell affinities (Attardi and Sperry, 1963) or axons deploying along gradients that exist in the target (Gaze and Keating, 1972; Fraser, 1980).

## 5.2. DEVELOPMENT OF THE GOLDFISH RETINA

### 5.2.1. Retinal Ganglion Cell Histogenesis

In goldfish (Sharma and Ungar, 1980), as well as in other fish (Lyall, 1957; Blaxter and Jones, 1967; Wagner, 1974; Grun, 1975) and mammals (Donovan, 1966; Rapaport and Stone, 1983), the ganglion cells in the central, dorsal retina become postmitotic first. As the goldfish retina continues to grow, ganglion cells are added appositionally around its margin (Johns, 1977; Meyer, 1978). Such retinal growth probably continues for at least several years following hatching. Of particular significance is that the ganglion cells are organized *concentrically* with respect to date of birth. Older cells are located near the optic disc and recently born cells are located near the retinal margin. This is in contrast to the mammalian retina in which the developmental period is relatively brief and many ganglion cells are born simultaneously at all retinal eccentricities (e.g., Kliot and Shatz, 1982; Drager, 1985; Sengelaub *et al.*, 1986).

### 5.2.2. Optic Fiber Layer Organization

Axons of newly generated ganglion cells (in fish) course along the vitreal surface of the optic fiber layer toward the optic disk (Bodick and Levinthal, 1980; Bunt, 1982; Easter *et al.*, 1984). The optic fiber layer therefore becomes chronologically laminated as a consequence of new axons growing along the surface of prior generated axons. Chronological lamination of the optic fiber layer can be interpreted in different ways. New ganglion cell axons appear to grow generally along the surface of preexisting axons rather than burrowing between older axons. This suggests that newer optic fibers tend to follow a path of least resistance as they grow toward the optic disc. Alternatively, chronological lamination may reflect that the axons of newly generated cells have an affinity for axons from neighboring, previously generated ganglion cells. Such affinities, if they exist, may be expressed not only between the axons from neighboring cells that were born at slightly different times, but also between the axons of neighboring cells that were born at the same time. The latter cells are situated along a retinal circumferential line ($\theta$), while the former cells are situated along a retinal radius ($r$).

### 5.2.3. Optic Nerve Head Organization

#### 5.2.3.1. Mapping of Axonal Chronology

In contrast to the concentric organization of the ganglion cells, the optic axons are organized *eccentrically* in the optic nerve head (Bunt, 1982; Springer and Mednick 1986b). Axons of cells close to the origin of the retina (optic disc) are not located in the center of the optic nerve head. Instead, they are positioned along the dorsal surface of the nerve head (Fig. 5.1B). Axons from cells born later (i.e., farther from the origin) are added to the ventral surface of the optic nerve head (Fig. 5.1B). Repetition of this process leads to a chronological lamination of the optic nerve head.

The organization at the optic nerve head represents a continuation of the chronological organization of the optic fiber layer. This explanation of the organization of the optic nerve head is applicable to the axons of ventral retinal ganglion cells as well. The optic fiber layer contains a slit that extends from the optic disc to the ventral retinal pole. This slit is the embryonic fissure. Because of it, axons from cells that lie on opposite sides of the fissure come to occupy opposite sides of the optic nerve head (Fig. 5.1B).

It is worth considering what possible alternatives there are to an eccentrically organized optic nerve head. Eliminating the embryonic fissure from the retina would potentially allow the axons to form an optic nerve head that is concentrically organized. However, because of chronological lamination of the optic fiber layer, the oldest axons would be at the perimeter of the nerve head and the youngest axons would be in its center. This form of organization for the goldfish optic nerve, although incorrect, was proposed by Roth (1974). In addition, he speculated that the axons in the center of the nerve underwent a radial displacement. Younger axons were thought to move from the center toward the periphery of the nerve, while older axons moved from the periphery toward the center of the nerve. This transformation resulted in a nerve that was concentrically organized in the same way that the ganglion cells are organized.

It should be apparent that the embryonic fissue in goldfish forces a transition from the concentric to the eccentric organization (Fig. 5.1B). The optic fiber layer and the embryonic fissure are analogous to several fully opened (circular) and stacked Oriental fans having slits that extend from their origins to one of their edges. Fans with small spokes are at one end of the stack (i.e., sclerad) and those with large spokes are at the other end of the stack (i.e., vitread). Within a fan, each spoke represents a single axon. Thus, fans with small spokes represent the short axons from central ganglion cells (i.e., origin) and those with long spokes represent the long axons of pe-

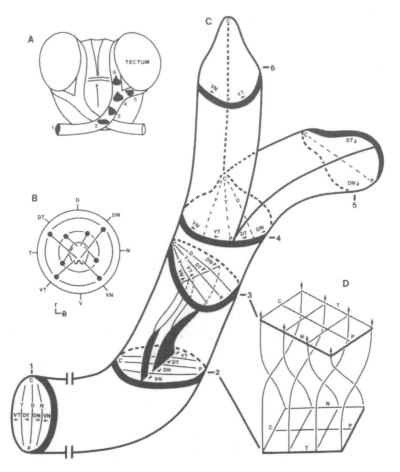

**Figure 5.1.** Summary diagrams of the organization and the topographical transformations of retinal ganglion cell (RGC) axons within the visual pathways of goldfish. (A) Frontal view of the goldfish brain with the telencephalon removed. The shape of sections (in black) that are perpendicular to the longitudinal axis of the pathway are presented at six different points: 1, optic nerve head; 2, optic chiasm; 3, main optic tract postchiasmatically; 4, main optic tract before it bifurcates; 5, ventral optic tract; 6, dorsal optic tract. (B) Diagram of the retina (outer circle) and optic nerve head (inner oval). Two annuli of RGCs (black circles) are shown. The inner annulus represents central, older RGCs, and their axons are dorsal in the optic nerve head. The outer annulus represents peripheral, younger RGCs, and their axons are ventral in the optic nerve head. Thus, the chronological feature of RGC axons maps as laminae along the dorsoventral axis of the optic nerve head. Retinal sector (θ) maps along the nasotemporal axis of the optic nerve head. The organization of the RGC axons within the optic nerve head is elaborated in C1. D, dorsal; V, ventral; N, nasal; T, temporal; DT, dorsotemporal; VT, ventrotemporal; DN, dorsonasal; VN, ventronasal. (C) Diagram showing the RGC axon organization and transformations between the eye and the tectum. The thick, black edges of sections 1–4 and 6 face rostrally and the thick black edge of section 5 faces dorsally. 1. The parameter $r$, central–peripheral (CP) retinal axis, maps along the dorsoventral axis of the optic nerve head. Ventrotemporal (VT) RGC axons are along the temporal (caudal) edge and ventronasal (VN) RGC axons are along the nasal (rostral) edge of the optic nerve head. Dorsal RGC axons are in the center of the optic nerve head, with dorsotemporal (DT) RGC axons on the temporal and dorsonasal (DN) RGC axons on the nasal side of the center of the optic nerve head. The arrowheads indicate the internal organiza-

ripheral cells. As the optic axons approach the optic nerve head, the axonal array begins to close up, much like a fan. After closing, the sides of the fan represent the axons that came from cells bordering the embryonic fissure. The axonal array (fan) now bends at a right angle as the axons pass into the optic disc. Adding the spokes from progressively larger fans (newer axons) builds the optic disc up in a dorsal–ventral direction.

### 5.2.3.2. Mapping of Retinal Sector

The circumferential dimension of the retina is also represented topographically in the optic nerve head (Fig. 5.1B). As described above, axons from cells situated on either side of the embryonic fissure are on opposite sides of the optic nerve head (Fig. 5.2C). Axons from ventronasal cells are on the rostral edge of the nerve head and axons from temporal cells are on the

---

tion of the axons in a column that represents a sector of the retina. At the tip of each arrowhead are the most temporal RGC axons within a column and at the base of each arrowhead are the most nasal RGC axons within a column. Thus, the arrowheads point toward the location of the most temporal (T) RGC axons and away from the most nasal (N) RGC axons. The location of axons from RGCs located at the temporal, dorsal, and nasal retinal poles are indicated by T, D, and N, respectively. 2. The optic nerve crosses the midline and bends dorsally. The organization of the RGC axons at this point is identical to that shown in C1. However, central RGC axons (C) are now medial instead of dorsal, peripheral (P) RGC axons are now lateral instead of ventral, nasal (N) RGC axons are still rostral, and temporal (TC) RGC axons are still caudal. The orientation of the arrowheads has not changed. 3. Reorganizations occur after point 2 and are complete at point 3. The CP axis of the optic tract rotates in a caudal direction between points 2 and 3. The VN RGC axons remain along the same edge of the tract, but the NT axis is inverted with respect to the edge of the tract. The VT, DT, and DN RGC axons rotate about the DT axons such that the VT RGC axons move two compartments rostrally (between sections 2 and 3) and come to lie near the VN RGC axons. The DN RGC axons move two compartments caudally (between sections 2 and 3) and are at the caudal edge of the tract. The DT RGC axons maintain their previous position in the tract (i.e., one compartment away from the caudal edge of the tract). In addition to the above transformations, the direction of all arrowheads has inverted. This inversion allows all arrowheads to still point toward the most temporal (T) RGC axons, which are now in the center of the tract. The inversion of the arrowheads reflects the spatial inversion of the NT axis of the RGC axons within each column. 4. The organization of the optic tract prior to its bifurcation around nucleus rotundus. The CP axis has rotated further in a caudal direction (compared to C3 and C2). No further transformations in the organization of RGC axons occurs between C3 and C4. 5. The ventral optic tract peels off the main optic tract such that the most temporal (DT) RGC axons are dorsal and the most nasal (DN) RGC axons are ventral. 6. The dorsal optic tract continues from the main optic tract (C4) without any major changes. The most temporal (VT) RGC axons are lateral and the most nasal (VN) RGC axons are medial in the dorsal optic tract. (D) Diagram illustrating how the NT axis becomes inverted, but the CP axis remains unaltered between sections C2 and C3. The VN compartment is used to illustrate the transformation. Four chronological laminae of axons are represented along the CP axis. The most temporal (T) RGC axons of each lamina are at the rostral edge and the most nasal (N) RGC axons of each lamina are at the caudal edge of the compartment at the level of section 2. As the axons of each lamina pass from C2 to C3 they invert their positions with respect to the rostral edge of the tract. The axons of all laminae along the CP axis do not invert simultaneously, because they are generated at different times during the course of development. Reprinted with permission from Springer and Mednick (1986c).

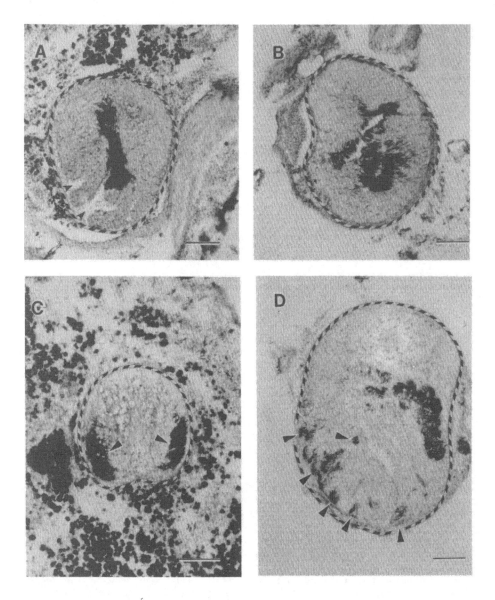

**Figure 5.2.** Sections are presented that are perpendicular to the longitudinal axis of the optic nerve head and optic nerve (enclosed by dashed lines). Dorsal is to the top. The scale for A–D is 100 μm. (A) Cobalt-filled optic axons in the right optic nerve head of a fish with cobalt applied to a slit in the peripheral dorsal retina. Nasal is to the right. The cobalt-filled axons are in the center and extend from the ventral edge of the optic nerve head (bottom) toward, but not all the way to, the dorsal edge of the nerve head (top). Arrowheads indicate clear areas that are glial septa. These septa are just beginning to invaginate the nerve. (B) Same fish as in (A) but the section is 300 μm more central. The straight arrangement of axons in (A) has become radically altered by an internal glial septum (clear area). The axons are now arranged in a crude Z shape. The structure adjacent to the indentation of the optic nerve on the left is the right retinal artery. (C) Cobalt-filled optic axons in the left optic nerve head of a fish with cobalt applied to a slit in the

caudal edge of the nerve. (Remember that goldfish eyes are positioned later-ally, and that the optic nerve passes medially toward the optic foramen as it leaves the eye.) Between the ventronasal and ventrotemporal axons are the axons from dorsonasal and dorsotemporal ganglion cells (Fig. 5.2A). Thus, if the retina is divided into four sectors, the axons arising from each sector occupy four columns that span the rostrocaudal axis of the optic nerve head. Within each column, axonal age maps along the dorsoventral axis of the optic nerve head. Therefore, the radial and circumferential aspects of the ganglion cell distribution are mapped orthogonally to one another in the optic nerve head.

### 5.2.4. Optic Nerve Organization

Topographical precision appears to deteriorate within a short distance of the optic nerve head of goldfish (Figs. 5.2B,D). The previously straight col-umns of axons begin to bend and break up into widely spaced clusters of axons. Topographical decomposition occurs for all four columns of axons. When ventral axons are examined in the optic nerve, midway between the optic disc and optic foramen, they no longer occupy exclusively the opposite sides of the nerve (Fig. 5.2D). In fact many axons travel in a plane that is perpendicular to the long axis of the nerve. Although a crude topography can be discerned, the organization looks chaotic. When only a few axons from peripheral ganglion cells are labeled, the topography is frequently impercep-tible and the axons appear to occupy random positions in the nerve. However, the arrangement for axons from central ganglion cells is more constant in that they occupy generally the dorsal part of the nerve throughout its intraorbital extent (Springer and Mednick, 1986b). Thus, it appears that the topography of the axons within the nerve becomes progressively more degraded for sub-sequent generations of axons.

Why does the goldfish optic nerve appear to be so chaotically organized? The answer to this question is related to the ophthalmic artery and glial septa. The ophthalmic artery joins the optic nerve along the nerve's temporal sur-face and then, near the eye, swings down to its ventral surface. An invagina-tion of the optic nerve is evident along the trajectory of the artery. The depth of this invagination is small near the optic nerve head but becomes deeper with increasing distance from the eye. Although the artery does not extend into the depths of the invagination midorbitally, glial septa and connective

---

peripheral ventral retina that was centered on the embryonic fissure. The filled axons are at the nasal (left arrowhead) and temporal (right arrowhead) edges of the optic nerve head. They extend from the ventral edge of the optic nerve head toward, but not all the way to, the dorsal edge of the optic nerve head. (D) Same fish as in (C) but the section is 465 μm more central. The fibers on the temporal side of the nerve (to the right) have shifted dorsally. The fibers on the nasal side (arrowheads) have broken up into numerous clusters. The structure adjacent to the indentation of the optic nerve on the right is the left retinal artery. Reprinted with permission from Springer and Mednick (1986b).

tissue do. They effectively deform the temporal side of the nerve, much like a finger pressed into dough (Fig. 5.3). One of these invaginations (primary septum or fold) extends all the way from the optic nerve head into, and slightly beyond, the optic foramen. In addition to the primary fold, numerous glial septa penetrate the rostral and caudal surfaces of the nerve, but to a lesser extent than the primary septum. The initially neatly organized columns of axons must traverse the septal intrusions and they do so by bending or by breaking up into subcolumns. In effect, the axons appear to be challenged or interrupted by a series of randomly oriented hurdles around and over which they must travel.

The functions of the invaginations and septa remain speculative. However, the invaginations appear to increase the surface area of the nerve much as sulci do in the mammalian cerebral cortex. The invaginations, as well as the greater nerve surface area that they produce, may render the optic axons more accessible to vascular elements. Septa are not evident at the optic nerve head or in the nerve once it enters the optic foramen (Fig. 5.3F). This is, in part, why the cross-sectional profile of the nerve is smaller at these locations than it is along its orbital extent. Thus, the septa are not present at the two points where the optic fibers are effectively anchored. Septa and invaginations may therefore serve to partly coil the optic axons. This additional elongation of the optic axons, as well as the cushioning produced by the septa, may serve to protect the optic axons when the eye rotates. In cichlid fish, the optic nerve distends when the eye rotates (Fernald, 1980). Cichlid fish have a regularly pleated optic nerve that compensates for optic nerve stretch. Goldfish do not have a pleated optic nerve and may therefore have adapted to the problem of optic nerve stretch in a much less elegant fashion.

Optic axons are organized similarly in both the goldfish and cichlid fish optic nerves (Scholes, 1979, 1981). Cichlid fish optic nerves are elongated enormously in the dorsoventral axis (the one that maps the age of the axons) (Anders and Hibbard, 1974; Scholes, 1979). The nerve, however, has many folds that run along its longitudinal axis and therefore folds back upon itself many times. Thus, the two sides of the cichlid optic nerve (the ones that have axons from ventronasal and ventrotemporal ganglion cells on opposite sides) are consistently parallel to one another. An absence of parallel rostral and caudal optic nerve sides characterizes the goldfish optic nerve. Therefore, the columnar organization of the goldfish nerve is distorted, but the axons are actually not organized randomly within it.

### 5.2.5. Organization in the Optic Foramen

The optic nerve's diameter diminishes as it passes into the optic foramen. In addition, the septa do not continue into the foramen. Thus, the appearances of the nerve in the optic foramen and at the optic nerve head are similar. Diminished optic axon organization within the intraorbital extent of the optic nerve, however, is more illusory than real. At the optic foramen the axons from ventral ganglion cells once again line the outer perimeter of the

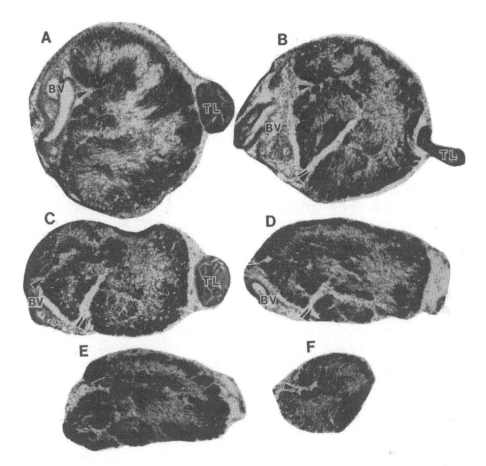

**Figure 5.3.** Sections (15 μm) through the right optic nerve of a fish that had cobalt applied to the severed left dorsal optic tract. Panel (A) is about 0.5 mm from the optic nerve head and panel (F) is within the optic foramen. The distances between sections are as follows: (A)–(B), 285 μm; (B)–(C), 345 μm; (C)–(D), 255 μm; (D)–(E), 255 μm; and (E)–(F), 375 μm. Dorsal is to the top, the central blood vessel (BV) is located temporally, and the tenacular ligament (TL) (Walls, 1942) is located nasally. The single arrowhead indicates the primary fold of the optic nerve and this fold is evident in all the sections. The double arrowhead indicates a secondary fold that is evident only in panels (B)–(D). The scale is 100 μm. (A) At this level the optic nerve has a primary fold (arrowhead) and the cobalt-filled axons are generally near the edges of the nerve. The outer contour of the nerve appears roughly as a backward C with respect to the primary fold. (B) At this level the nerve has both a primary (arrowhead) and a secondary fold (double arrowhead). The secondary fold causes the profile of the nerve to appear roughly as a backward E. (C) The location of cobalt-filled RGCs within the nerve is similar to that seen in (B), but the nerve is rotated slightly in a counterclockwise direction. (D) The primary and secondary folds are compressed, but the nerve is still similar in profile to that seen in (B) and (C). (E) The nerve is approaching the optic foramen. Its cross-sectional area is smaller than in (A) and the secondary fold is no longer evident. (F) The diameter of the nerve is now narrow because it is within the optic foramen. Cobalt-filled axons line the perimeter of the nerve as in (A). The nerve still has a primary fold and looks like a backward C. Reprinted with permission from Springer and Mednick (1986b).

nerve, and the axons from dorsal ganglion cells are sandwiched between the axons from ventral ganglion cells. The nerve, however, is distorted slightly by a small invagination on its caudal (temporal) surface (Fig. 5.3F). Reestablished columnar organization within the optic foramen indicates that the organization was not actually disrupted in the intraorbital part of the nerve.

### 5.2.6. Organization after Crossing the Midline

The optic nerve bends dorsally as it crosses the midline (Figs. 5.1C-2). This leads to a change in the position of the axons with respect to the animal's axes, but not with respect to one another. Thus, central ganglion cell axons are now positioned medially rather than dorsally, and peripheral ganglion cell axons are positioned laterally rather than ventrally. Peripheral ganglion cell axons are closest to the pial surface of the brain. Ventrotemporal and ventronasal ganglion cell axons are still located caudally and rostrally, respectively.

### 5.2.7. Organization in the Optic Chiasm

#### 5.2.7.1. Intercolumnar Rearrangements

Changes essential to aligning properly the axons with the optic tectum occur toward the caudal end of the optic chiasm, as the optic tract nears the wall of the diencephalon (Fig. 5.1A-2, 5.1A-3, 5.1C-2, 5.1C-3). Several different types of rearrangement occur. The first change unites the ventrotemporal with the ventronasal ganglion cell axons. Ventronasal ganglion cell axons remain at the rostral edge of the tract, but the ventrotemporal axons move from the caudal to the rostral side of the tract. This leaves the dorsotemporal and dorsonasal ganglion cell axons in the caudal half of the tract, with the former axons in the most caudal compartment. The second rearrangement involves an exchange in compartments that the dorsonasal and dorsotemporal ganglion cell axons occupy. Thus, the dorsonasal ganglion cell axons come to occupy the caudal compartment. Following this rearrangement, axons from temporal ganglion cells are in the center of the optic tract and are flanked on both sides by axons from nasal retinal ganglion cells (Figs. 5.1C-3). Although the optic tract rotates slightly to conform to the surface of the brain, axons from central ganglion cells remain close against the brain and axons from peripheral ganglion cells remain close to the pial surface.

#### 5.2.7.2. Intracolumnar Rearrangements

The changes described above actually occur simultaneously rather than sequentially. In fact, the changes can best be appreciated by dividing the optic tract into two main compartments, with one of these divided into three subcompartments (Figs. 5.1C-2). Rostrally is the compartment with the ventronasal ganglion cell axons and caudally is the compartment with the dorsonasal, dorsotemporal, and ventrotemporal ganglion cell axon subcompartments. At the interface between the two compartments, the ventronasal and dorsonasal ganglion cell axons appear to repel one another, much like identical poles on magnets. The ventronasal ganglion cell axons invert their organization within their compartment. Initially, the most temporal of the ventronasal ganglion cell axons (i.e., the axon of a cell closest to the embryonic fissure) was located at the rostral edge of the optic tract (Figs. 5.1C-2). In the course of the organizational inversion, the most nasal of the ventronasal ganglion cell axons (i.e., axon of a cell located at the nasal pole of the retina) comes to occupy the rostral edge of the tract (Figs. 5.1C-3).

The caudal compartment containing the dorsonasal, dorsotemporal, and ventrotemporal ganglion cell axons also inverts, but as a unit. This maneuver results in the transposition of the contents of the three caudal subcompartments. Thus, the most temporal of ventrotemporal and dorsotemporal ganglion cell axons face each other in the center of the tract (i.e., the temporal retinal pole maps to the center of the tract) (Figs. 5.1C-3). The most nasal ganglion cell axon in the dorsonasal subcompartment is initially in the rostral part of this subcompartment. However, when the contents of this compartment invert, this axon comes to be located at the caudal edge of the optic tract. Inversion of the axons within the two main compartments and within each of the subcompartments occurs prior to the division of the main optic tract into its dorsomedial and ventrolateral branches. Furthermore, the intracompartmental interaxonal rearrangements are essential for the proper alignment of the axons with the optic tectum.

### 5.2.8. Formation and Organization of the Optic Tracts

Further rotation of the main optic tract with respect to the brain results in the central ganglion cell axons being located caudally in the main optic tract (Figs. 5.1C-4). Peripheral ganglion cell axons are now located rostrally, close to the pia. At this point the optic axons have completed their reorganization and are poised to form the optic brachia. Splitting of the main optic tract occurs at the interface between the dorsotemporal and ventrotemporal compartments. Axons of ventral ganglion cells pass dorsally in the dorsomedial optic tract to enter the dorsal tectum (Fig. 5.1C-6). Axons of dorsal ganglion cells pass laterally to form the ventrolateral optic tract, and they enter the ventral tectum (Fig. 5.1C-5).

Temporal ganglion cell axons in both optic brachia are located close to the optic tectum and are interposed between the tectum and the nasal ganglion cell axons. Thus, temporal ganglion cell axons are closest to, and nasal ganglion cell axons are farthest from, the rostral tectum. This spatial relationship between temporal and nasal ganglion cell axons and the optic tectum probably accounts, at least in part, for temporal ganglion cell axons innervating rostral tectum and nasal ganglion cell axons innervating caudal tectum.

### 5.2.9. Multiple Maps in the Optic Tracts

Up to this point I have emphasized the axonal organization of the optic tracts with respect to their major target, the optic tectum. However, in the goldfish there are several other thalamic and pretectal retinorecipient targets that the optic tracts encounter prior to their reaching the tectum. One of these nuclei, the superficial pretectal parvicellular nucleus (SPP), is located at the site where the main optic tract bifurcates into its brachia. SPP receives optic axons from all retinal loci, and the axons terminate topographically in SPP (Springer and Mednick, 1985b). Temporal retinal ganglion cell axons pass around and over SPP and give off collateral branches to SPP prior to entering and terminating in the superficial retinorecipient lamina of the optic tectum.

In order for nasal ganglion cell axons to reach SPP they must travel through the middle optic tract compartments that contain the temporal ganglion cell axons. Thus, the compartments that normally contain only dorsotemporal and ventrotemporal ganglion cell axons are, after a fashion, "contaminated" with nasal ganglion cell axons. Other nontectal retinorecipient nuclei, to varying degrees, also attract axons from ganglion cells situated throughout the retina. As for the SPP, in order to reach their target, some optic axons must travel through optic tract compartments that are usually occupied by axons from ganglion cells in a different retinal sector. This serves to illustrate that there may be several, independent topographical overlays within the optic tracts. Each overlay may be very well organized. However, superimposing the overlays and assuming that there is only one map in the optic tracts can lead to the erroneous conclusion that the organization is not very precise.

### 5.3. OBSERVATIONS IN OTHER SPECIES

### 5.3.1. Lamination of the Optic Fiber Layer

In addition to several species of fish, chronological lamination of the retinal optic fiber layer has also been observed in the mouse (Hinds and Hinds, 1974) and the chicken (Rager, 1980; Krayanek and Goldberg, 1981). The macaque monkey optic fiber layer is also chronologically laminated in that the longest axons are closest to the vitreal surface of the optic fiber layer (Ogden, 1983a). However, such lamination is better developed in the temporal rather than the nasal retina of the owl monkey (Ogden, 1983b). *Xenopus* retina demonstrates poor chronological lamination (Taylor, 1987).

Whether the poor chronological lamination of the optic fiber layer in species such as *Xenopus* also characterizes the retina of other frogs such as *Rana pipiens* remains to be determined. Nevertheless the degree of chronological lamination within the optic fiber layer probably has a major impact on the nature of the axonal organization that is seen as the optic axons leave the eye. Species with a high degree of chronological lamination might be expected to have optic nerve organizations that have features in common. Those with poor chronological lamination may have a poorly organized optic nerve.

## 5.3.2. Organization of the Optic Nerve and Tract

### 5.3.2.1. Fish

Interspecies variability in axonal organization is most evident in the optic nerve. Both goldfish and cichlid fish have chronologically laminated optic nerves. In cichlid fish the optic nerve is ribbon-shaped, and the lamination is evident along the entire extent of the nerve. In goldfish, the sides of the optic nerve are distorted by connective tissue invaginations, but the chronological lamination still persists throughout the nerve.

### 5.3.2.2. Frogs

Frogs (*Rana pipiens*) have a very complex optic nerve organization (Reh *et al.*, 1983; Scalia and Arango, 1983). There is extensive crossing of axons as they leave the eye and enter the optic nerve. This results in the axons of central ganglion cells being in the center of the nerve and axons of peripheral ganglion cell axons being on the perimeter of the nerve. Thus, the optic axons are organized initially in a concentric fashion. Furthermore, axons of dorsal ganglion cells are located in the dorsal part of the nerve and axons of ventral ganglion cells are located in the ventral part of the nerve. In addition, the optic nerve contains a dual representation of the retina. A portion of the ganglion cell axons from the nasal and temporal retina cross over into the opposite side of the optic nerve. Therefore, axons of nasal and temporal ganglion cells are both located along the rostral and caudal edges of the nerve.

The concentric organization within the nerve changes progressively into an eccentric organization (Reh *et al.*, 1983). Axons from central ganglion cells move from the center of the nerve to its dorsal surface and axons from peripheral ganglion cells move to its ventral surface. Prior to reaching the optic tracts, the dual representations of the retina coalesce. Within the optic tracts, the organization is similar to that seen for the goldfish. Axonal chronology ($r$) is mapped orthogonally to retinal sector ($\theta$).

Optic axons in *Xenopus laevis* also cross extensively as they leave the eye. They form a concentrically organized nerve, but the nerve does not contain a dual representation of the temporal and nasal retina (Taylor, 1987). With increasing distance from the eye, the axons from cells in a particular part of the retina tend to spread out around the circumference of the nerve. Thus, there is a progressive degradation in the map of the retinal circumference in the optic nerve. As for *Rana pipiens*, the organization in *Xenopus laevis* transforms progressively from a concentric to an eccentric one. Within the optic tracts, the axons are organized in the same manner as they are for *Rana* and goldfish.

Differences in the organization of the optic nerves of *Rana pipiens* and *Xenopus laevis* have been attributed to differences in the shape of their optic discs (Taylor, 1987). Although the optic nerves of goldfish and frogs are organized in seemingly different manners, the organizations of their optic tracts are essentially identical. In goldfish, an eccentric organization exists in the optic nerve, but in frogs an eccentric organization emerges some distance from the eye. Intercolumnar and intracolumnar reorganizations first occur at the caudal end of the optic chiasm in goldfish and are completed before the main optic tract divides. Intracolumnar rearrangements have not been documented in frogs, but that they occur is inevitable. Intercolumnar rearrangements also seem to occur in the vicinity of the optic chiasm of frogs (Reh *et al.,* 1983; Fawcett *et al.,* 1984).

### 5.3.2.3. Rodents

Appreciably less is known about the axonal organization of the optic nerve in mammals. Within the optic stalk of embryonic mice, the initial optic axons are arranged as they are in the goldfish (Silver, 1984). Ventronasal and ventrotemporal ganglion cell axons are on opposite sides of the optic stalk. Dorsotemporal and dorsonasal ganglion cell axons are sandwiched between the two columns of ventral ganglion cell axons. Later born ganglion cell axons may be added to the ventral surface of previously born axons. If so, then the nerve would take on an eccentric organization, with older ganglion cell axons along the dorsal surface of the nerve and later born ganglion cell axons along the ventral surface of the nerve.

The rat optic nerve is claimed to be organized in a concentric fashion that mirrors the position of the ganglion cells within the retina (Yamadori, 1981). Furthermore, this organization is believed to extend into the optic tract. Rotation of the contents of the optic tract by approximately 180° allows the axons in the tract to align themselves properly with the tectum.

The claimed organization of the visual pathway in the rat seems to be based on an erroneous assumption. It is conceptually convenient to assume that rotation of the retinal map by 180° is all that is necessary to align the retinal and tectal maps (Scholes, 1979). However, this scheme only works in two circumstances: first, when the actual anatomical relationships between the optic tract and the tectum are ignored; and second, when the visual system is composed of no more than four axons.

Figure 5.4 illustrates a hypothetical visual system that contains four ganglion cell axons (labeled N, D, T, and V) that are organized concentrically in the optic pathways. In addition, in this model the optic pathway enters the center of the tectum. In this way the axons of central and peripheral ganglion cells are aligned with the center and the periphery of the tectum, respectively. Rotating the pathway clockwise (Fig. 5.4-1) leads to the proper positioning of the dorsal and ventral ganglion cell axons, but not the nasal and temporal ganglion cell axons. However, this rotation would effectively align all four axons with the tectum if the tract entered the tectum from its ventral surface (which of course it does not). Rotating the pathway counterclockwise (Fig. 5.4-2) leads to the correct positioning of the nasal and temporal ganglion cell axons, but not the dorsal and ventral ganglion cell axons. Rotating the dorsal and ventral ganglion cell axons clockwise and the nasal and temporal ganglion

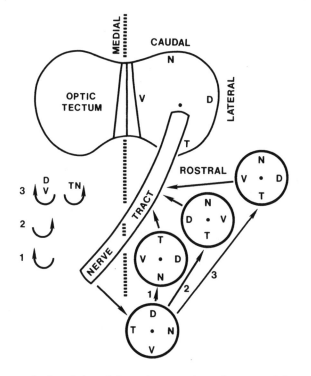

**Figure 5.4.** Diagram of a dorsal view of the optic nerve, the optic tract, and the optic tectum that illustrates a hypothetical organization and three different types of reorganizations of retinal ganglion cell (RGC) axons in the visual pathways. The circles represent sections through the optic nerve or tract. The dorsal surface of each section is to the top and the rostral edge of each section is to the right. The singular optic tract enters the dorsal surface of the tectum at its center. The arrangement of RGC axons in the optic nerve, tract, and tectum is assumed to be concentric with older RGC axons in the center (dot) and younger RGC axons toward the edges of each structure. A 90° clockwise rotation of all the axons (1) aligns the ventral (V) and dorsal (D), but not the nasal (N) and temporal (T) RGC axons, with their respective termini in the tectum. Transformation 1 would lead to the proper alignment of optic axons with the optic tectum if the optic tract enters the ventral, rather than the dorsal, surface of the tectum. A 90° counterclockwise rotation of all the axons (2) aligns the N and T, but not the D and V, RGC axons with their respective termini in the tectum. In order to align the D, V, N, and T axons with their termini in the tectum, it is necessary to rotate the D and V axons 90° clockwise and the N and T axons 90° counterclockwise (3). Transformation 3 leads to the proper alignment without disturbing the original nearest-neighbor relationships that exist in the nerve. However, this transformation is effective only when four axons are used in the model. This transformation severely disrupts nearest-neighbor relationships when eight axons are used in the model. Reprinted with permission from Springer and Mednick (1986b).

cell axons counterclockwise leads to the correct positioning of all four axons (Fig. 5.4-3). In addition, this latter maneuver leaves unperturbed the nearest-neighbor relationship that existed prior to the rotations.

Extending the double rotation maneuver to a set of eight axons shows the limitations of this maneuver (Fig. 5.5A). Again, the ventral and dorsal gan-

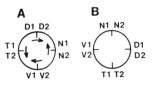

**Figure 5.5.** (A) A hypothetical concentrically organized optic nerve with eight axons. They undergo the third type of transformation that is described in Fig. 5.4. Dorsal and ventral axons (D1, D2, V1, and V2) rotate in a clockwise direction. Temporal and nasal axons (T1, T2, N1, and N2) rotate in a counterclockwise direction. The consequences of these movements are shown in (B). The nearest-neighbor relationships that existed in (A) are disrupted in (B).

glion cell axons rotate clockwise and the nasal and temporal ganglion cell axons rotate counterclockwise. Now, however, the nearest-neighbor relationships that existed prior to the rotations are severely disrupted (Fig. 5.5B). For example, axon D1 was flanked initially by axons T1 and D2 (Fig. 5.5A). Following the rotations axon D1 is flanked by axons D2 and N2 (Fig. 5.5B). Now there are two axons interposed between axon D1 and its former neighbor, axon T1. As more axons are added to the pathway, the distance between axons that were neighbors before the rotations becomes progressively greater following the rotations. If the pathway comprises 800 axons, following the rotations 400 axons would be interposed between an axon and one of its former neighbors. Such potentially large distances between the axons of neighboring ganglion cells are not seen in the visual pathways of goldfish (Springer and Mednick, 1986b,c). Thus, rotating the ganglion cell axon array or dividing the array into two sets and rotating each set in different directions are not satisfactory means by which to align properly a concentrically organized axonal array with the optic tectum.

### 5.3.2.4. Cat

The cat optic nerve did not appear to have any great degree of organization when a small number of axons were retrogradely labeled (Horton *et al.*, 1979). However, a recent study labeled a larger number of axons that arose from retinal ganglion cells situated in discrete regions of the retina (Naito, 1986). Axons from ganglion cells in the area centralis and pericentral retina were found along the lateral edge of the optic nerve several millimeters behind the eye. Retinal maturation in the cat begins at the area centralis and spreads out toward the periphery of the retina (Robinson, 1987). Therefore, it is likely that the first axons to enter the optic stalk originate from ganglion cells in the area centralis. Axons from the middle part of the upper and lower retina are found in the middle of the optic nerve and axons from ganglion cells in the periphery of the upper and lower retina are situated in the medial part of the optic nerve (Naito, 1986). In addition, axons from ganglion cells in the upper (dorsal) retina are positioned in the dorsolateral part of the optic nerve, and axons from ganglion cells in the lower (ventral) retina are positioned in the ventromedial part of the optic nerve. Axons of temporal and nasal ganglion cells are situated in the lateral and medial parts of the optic nerve, respectively.

Several aspects of the organization of the cat's optic nerve are worth noting. First, it appears to be organized eccentrically. The area centralis is represented toward the lateral edge of the optic nerve and the peripheral retina is represented toward the medial edge of the optic nerve. Although the cat retina develops generally in a central–peripheral direction, ganglion cells are born simultaneously throughout the retina. Thus, the cat retina does not develop in a strict central–peripheral direction as does the fish retina. Nevertheless, it seems that the cat optic nerve is roughly organized in a chronological fashion. A second feature worth noting is that ganglion cell axons representing the dorsal–ventral axis of the retina are oriented orthogonally (dorsoventrally) to the mediolateral axis of the optic nerve (i.e., the axis that maps the chronology of the axons).

Axons in the cat optic nerve are more tightly clustered close to the eye but then begin to scatter in the nerve (Naito, 1986). This scattering occurs before the axons reach the optic chiasm. Concomitant with the scattering of the axons is a reorganization of the axons from dorsal and ventral retina. Dorsal ganglion cell axons, originally in the dorsal part of the nerve, move to the ventral part of the nerve, and ventral ganglion cell axons, originally in the ventral part of the nerve, pass to the dorsal part of the nerve. It would therefore appear that a major reorganization in the ganglion cell axon array takes place *before*, rather than at or beyond, the optic chiasm.

Chronological organization is also a feature of the cat optic tract. Medium-sized ganglion cells are produced early in the course of retinal histogenesis and their axons are farthest from the pial surface of the optic tract (Walsh and Guillery, 1984). The last wave of ganglion cell birth includes predominantly small cells, and their axons are closest to the pial surface. The optic tract is divided into two compartments that contain the axons of either dorsal or ventral ganglion cells. Unlike the optic tract of fish and frogs, the axons of nasal and temporal ganglion cells in the cat overlap one another (Torrealba *et al.*, 1981, 1982). Thus, the organizational features of the optic tract that several species have in common are (1) chronological lamination of the optic tract and (2) the optic tract contains two compartments that contain the axons of either dorsal or ventral ganglion cells.

### 5.3.2.5. Monkey

Only one study has examined the organization of the optic nerve in fetal monkeys (Williams and Rakic, 1985). Growing axons changed their immediate neighbors, suggesting that the monkey optic nerve is not very well organized. This result indicates that topography may not exist on an axon–axon level (i.e., microtopography). However, this result does not preclude the possibility that the monkey nerve is organized in a generally chronological manner and that axons from ganglion cells in distinct retinal regions occupy different regions of the optic nerve (i.e., macrotopography). Microtopographical organization, if it exists, is probably established by the direct interaction of the ganglion cell axon terminals with their target.

## 5.4. BOUNDARIES BETWEEN DORSAL AND VENTRAL RETINA

Segregation of dorsal and ventral ganglion cell axons into two distinct visual pathway compartments appears to occur in many, if not all, vertebrates. This dichotomy may be achieved by active, rather than passive, processes. For example, in goldfish the ventronasal and ventrotemporal ganglion cell axons are on opposite sides of the visual pathway. However, these two columns of axons come together postchiasmatically. For this to occur, the ventrotemporal ganglion cell axons move from one side of the tract to the other. In the course of moving toward the ventronasal ganglion cell axons the ventrotemporal ganglion cell axons interdigitate with dorsal ganglion cell axons that are moving away from the ventronasal ganglion cell axons. Although the ventrotemporal ganglion cell axons come into close contact with dorsal ganglion cell axons, such a close encounter does not lead to the misrouting of either the dorsal or the ventrotemporal ganglion cell axons into the incorrect optic tract compartment. In the cat, dorsal and ventral ganglion cell axons may also interdigitate with one another as they pass from one side of the optic nerve to the other (Naito, 1986).

Separation of dorsal and ventral ganglion cell axons could suggest that dorsal and ventral ganglion cell axons have mutual disaffinities for one another. Alternatively, there may be unique affinities between the axons of ventral ganglion cells and similar affinities between the axons of dorsal ganglion cell. There is some evidence in goldfish that ventral ganglion cell axons avoid dorsal ganglion cell axons within the retina (Morel *et al.*, 1986). The goldfish iris contains two pigmentation marks that extend into the tissue beneath the sclera (Springer and Mednick, 1986a) (Fig. 5.6). These marks are conveniently correlated with the boundary between dorsal and ventral retina. Ganglion cells situated superior to the pigmentation marks have axons that project to ventral tectum and ganglion cells situated inferior to the marks have axons that project to dorsal tectum. Although most ganglion cell axons travel in radially toward the optic disc, some ganglion cell axons first travel parallel to the retinal margin and then turn radially to course toward the optic disc (Cook, 1982). Such circumferential ganglion cell axons bend radially toward the optic disc at two very specific locations in the retina. They bend radially upon reaching either the nasal or temporal boundary between dorsal and ventral retina (Morel *et al.*, 1986; Fig. 5.7). Circumferential axons appear to originate only from a small number of retinal ganglion cells situated in ventral retina. These axons travel between the optic fiber layer and the retinal ganglion cells and do not enter into dorsal retina. This finding suggests that ventral ganglion cell axons actively avoid aggregating with dorsal ganglion cell axons. It would therefore appear that axon–axon interactions based on

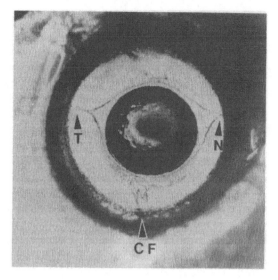

**Figure 5.6.** Photograph of the right eye of a goldfish illustrating the pigmentation marks ("darts") on the iris (arrowheads). Darts are evident temporally (T) and nasally (N). They extend under the sclera for about 1 mm. The remnant of the choroid fissure (CF) is visible on the iris and under the sclera. Reprinted with permission from Springer and Mednick (1986a).

putative molecular factors may be the basis for some of the topography that is observed in the optic pathways.

## 5.5. CONCLUSIONS

The organization of the optic axons in the goldfish retina, optic nerve head, optic nerve, and optic tract is inexorably related to the temporal sequence of axonal generation. More recently generated axons course over their predecessors. Repetition of this process leads to a chronological lamination of the optic axons. The mechanisms that mediate chronological lamination are not clear. However, the process may be totally passive such that new

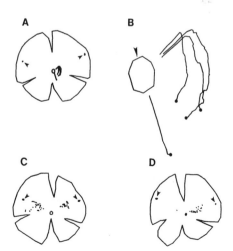

**Figure 5.7.** (A) Drawing of a flat-mounted retina showing the axons of three cells with circumferential axons and one cell with a radial axon. The arrowheads point to holes in the retina that indicate the location of the temporal (right) and nasal (left) retinal poles. (B) Enlargement of the axons shown in (A). The arrowhead points to the optic disc. (C) and (D) Two flat-mounted retinas in which the locations of the bending points of circumferential axons are plotted. Most of the axons change their course and travel radially very close to either the nasal or temporal boundary between the dorsal and ventral retina.

axons simply follow the path of least resistance and course over their older counterparts rather than in-between previously generated axons. Chronological lamination could also occur if there were affinities between the axons of neighboring retinal ganglion cells. Thus, newly generated axons may actively aggregate with the axons of neighboring cells rather than with the axons of distant cells.

Different mechanisms are necessary to explain the behavior of the axons as their arrangement is transformed at the optic chiasm (in the case of fish). There may be some sort of macromolecular template on the surface of the optic pathway that guides successive generations of axons through the various transformations. Such a template may only exist for the first set of axons that are generated and disappear thereafter. Successive generations of axons may rearrange themselves appropriately by following their nearest neighbors in the previously generated set of axons. Thus, this latter hypothesis shifts the template from an external source to the retinal ganglion cell axons themselves. The original template is probably more complex than a single macromolecular gradient. This conclusion is based on the complexity of the rearrangements that occur. It is difficult to understand how a single gradient could account for some axons remaining in the same compartment throughout the pathway (i.e., ventronasal axons), while other axons change their compartments (i.e., dorsonasal and ventrotemporal axons).

Differential interfiber affinities could, in part, explain the rearrangements that occur at the goldfish optic chiasm. Thus, a strong affinity between the ventrotemporal and ventronasal axons could initiate the rearrangements once the axons have entered the optic foramen. Further affinities must also be assumed: a weak affinity or a disaffinity between dorsonasal and ventronasal axons and a strong dorsonasal–dorsotemporal–ventrotemporal interaxonal affinity. Future research will hopefully determine whether such interaxonal affinities are the basis for the retinal ganglion cell axonal rearrangements.

ACKNOWLEDGMENT. This research was supported by grant EY-03552 from the National Eye Institute.

## 5.6. REFERENCES

Aebersold, H., Creutzfeld, O. D., Kuhnt, U., and Sanides, D., 1981, Representation of the visual field in the optic tract and optic chiasma of the cat, *Exp. Brain Res.* **42**:127–145.

Anders, J., and Hibbard, E., 1974, The optic system of the teleost *Cichlasoma biocellatum, J. Comp. Neurol.* **158**:145–154.

Attardi, D. G., and Sperry, R. W., 1963, Preferential selection of central pathways by regenerating optic fibers, *Exp. Neurol.* **7**:46–64.

Blaxter, J. H. S., and Jones, M. P., 1967, The development of the retina and retinomotor responses in the herring, *J. Mar Biol. Assoc. U.K.* **47**:677–697.

Bodick, N., and Levinthal, C., 1980, Growing optic nerve fibers follow neighbors during embryogenesis, *Proc. Natl. Acad. Sci. U.S.A.* **77**:4374–4378.

Bunt, S. M., 1982, Retinotopic and temporal organization of the optic nerve and tracts in the adult goldfish, *J. Comp. Neurol.* **206:**209–226.

Bunt, S. M., and Horder, T. J., 1983, Evidence for an orderly arrangement of optic axons within the optic nerves of the major nonmammalian vertebrate classes, *J. Comp. Neurol.* **213:**94–114.

Chamberlain, S. C., and Barlow, R. B. Jr., 1982, Retinotopic organization of lateral eye input to *Limulus* brain, *J. Neurophysiol.* **48:**505–520.

Cook, J. E., 1982, Errant axons in the normal goldfish retina reach retinotopic tectal sites, *Brain Res.* **250:**154–158.

Donovan, A., 1966, The postnatal development of the cat's retina, *Exp. Eye Res.* **5:**439–453.

Drager, U. C., 1985, Birth dates of retinal ganglion cells giving rise to the crossed and uncrossed optic projections in the mouse, *Proc. R. Soc. London Ser. B* **224:**57–77.

Easter, S. S. Jr., Rusoff, A. C., and Kish, P. E., 1981, The growth and organization of the optic nerve and tract in juvenile and adult goldfish, *J. Neurosci.* **1:**793–811.

Easter, S. S. Jr., Bratton, B., and Scherer, S. S., 1984, Growth-related order of the retinal fiber layer in goldfish, *J. Neurosci.* **4:**2173–2190.

Ehrlich, D., and Mark, R., 1984, The course of axons of retinal ganglion cells within the optic nerve and tract of the chick (*Gallus gallus*), *J. Comp. Neurol.* **223:**583–591.

Fawcett, J. W., 1981, How axons grow down the *Xenopus* optic nerve, *J. Embryol. Exp. Morphol.* **65:**219–233.

Fawcett, J. W., Taylor, J. S. H., Gaze, R. M., Grant, P., and Hirst, E., 1984, Fibre order in the normal *Xenopus* optic tract, near the chiasma, *J. Embryol. Exp. Morphol.* **83:**1–14.

Fernald, R. D., 1980, Optic nerve distention in a cichlid fish, *Vision Res.* **20:**1015–1019.

Fraser, S. E., 1980, A differential adhesion approach to the patterning of nerve connections, *Dev. Biol.* **79:**453–464.

Gaze, R. M., and Keating, M. J., 1972, The visual system and "neuronal specificity," *Nature (London)* **237:**375–378.

Grun, G., 1975, Structural basis of the functional development of the retina in the cichlid *Tilapia leucosticta* (Teleostei), *J. Embryol. Exp. Morphol.* **33:**243–257.

Hinds, J. E., and Hinds, P. L., 1974, Early ganglionic cell differentiation in the mouse retina: An electron microscopic analysis utilizing serial sections, *Dev. Biol.* **37:**381–416.

Horder, T. J., and Martin, K. A. C., 1978, Morphogenetics as an alternative to chemospecificity in the formation of nerve connections, In *Cell–Cell Recognition* (A. S. G. Curtis, ed), pp. 275–358, Cambridge University Press, Cambridge.

Horton, J. C., Greenwood, M. M., and Hubel, D. H., 1979, Non-retinotopic arrangement of fibres in cat optic nerve, *Nature (London)* **282:**720–722.

Johns, P. R., 1977, Growth of the adult goldfish eye. III. Source of the new retinal cells, *J. Comp. Neurol.* **176:**343–358.

Kliot, M., and Shatz, C. J., 1982, Genesis of different retinal ganglion cell types in the cat, *Neurosci. Abstr.* **8:**815.

Krayanek, S., and Goldberg, S., 1981, Oriented extracellular channels and axonal guidance in the embryonic chick retina, *Dev. Biol.* **84:**41–50.

Lyall, A. H., 1957, The growth of the trout retina, *Q. J. Microsc. Sci.* **98:**101–110.

Meinertzhagen, I. A., 1976, The organization of perpendicular fibre pathways in the insect optic lobe, *Philos. Trans. R. Soc. London Ser. B* **274:**555–596.

Meyer, R. L., 1978, Evidence from thymidine labeling for continuing growth of retina and tectum in juvenile goldfish, *Exp. Neurol.* **59:**99–111.

Morel, K. D., Mednick, A. S., and Springer, A. D., 1986, A retinal ganglion cell found only in the ventral retina has a unique intraretinal axonal trajectory, *Neurosci. Abstr.* **12:**636.

Naito, J., 1986, Course of retinogeniculate projection fibers in the cat optic nerve, *J. Comp. Neurol.* **251:**376–387.

Ogden, T. E., 1983a, Nerve fiber layer of the macaque retina: Retinotopic organization, *Invest. Opthalmol. Vis. Sci.* **24:**85–98.

Ogden, T. E., 1983b, Nerve fiber layer of the owl monkey retina: Retinotopic organization, *Invest. Ophthalmol. Vis. Sci.* **24:**265–269.

Presson, J., Fernald, R. D., and Max, M., 1985, The organization of retinal projections to the diencephalon and pretectum in the cichlid fish, *Haplochromis burtoni, J. Comp. Neurol.* **235:**360–374.

Rager, G., 1980, Development of the retinotectal projection in the chicken, *Adv. Anat. Embryol. Cell Biol.* **63:**1–92.

Rapaport, D. H., and Stone, J., 1983, The topography of cytogenesis in the developing retina of the cat, *J. Neurosci.* **3:**1824–1834.

Reh, T. A., Pitts, E., and Constantine-Paton, M., 1983, The organization of the fibers in the optic nerve of the normal and tectum-less *Rana pipiens, J. Comp. Neurol.* **218:**282–296.

Robinson, S. R., 1987, Ontogeny of the area centralis in the cat, *J. Comp. Neurol.* **255:**50–67.

Roth, R. L., 1974, Retinotopic organization of goldfish optic nerve and tract, *Anat. Rec.* **178:**453.

Rusoff, A. C., 1984, Paths of axons in the visual system of perciform fish and implications of these paths for rules governing axonal growth, *J. Neurosci.* **4:**1414–1428.

Scalia, F., and Arango, V., 1983, The anti-retinotopic organization of the frog's optic nerve, *Brain Res.* **266:**121–126.

Scholes, J. H., 1979, Nerve fibre topography in the retinal projection to the tectum, *Nature (London)* **278:**620–624.

Scholes, J. H., 1981, Ribbon optic nerves and axonal growth patterns in the retinal projection to the tectum, *Br. Soc. Dev. Biol. Symp.* **5:**181–214.

Sengelaub, D. R., Dolan, R. P., and Finlay, B. L., 1986, Cell generation, death, and retinal growth in the development of the hamster retinal ganglion cell layer, *J. Comp. Neurol.* **246:**527–543.

Sharma, S. C., and Ungar, F., 1980, Histogenesis of the goldfish retina, *J. Comp. Neurol.* **191:**373–382.

Silver, J., 1984, Studies on the factors that govern directionality of axonal growth in the embryonic optic nerve and at the chiasm in mice, *J. Comp. Neurol.* **223:**238–251.

Springer, A. D., and Gaffney, J. S., 1981, Retinal projections in the goldfish: A cobaltous-lysine study, *J. Comp. Neurol.* **203:**401–424.

Springer, A. D., and Mednick, A. S., 1984, Selective innervation of the goldfish suprachiasmatic nucleus by ventral retinal ganglion cell axons, *Brain Res.* **323:**293–296.

Springer, A. D., and Mednick, A. S., 1985a, Retinofugal and retinopetal projections in the cichlid fish *Astronotus ocellatus, J. Comp. Neurol.* **236:**179–196.

Springer, A. D., and Mednick, A. S., 1985b, Topography of the retinal projection to the superficial pretectal parvicellular nucleus of goldfish: A cobaltous-lysine study, *J. Comp. Neurol.* **237:**239–250.

Springer, A. D., and Mednick, A. S., 1985c, Topography of the goldfish optic tracts: Implications for the chronological clustering model, *J. Comp. Neurol.* **239:**108–116.

Springer, A. D., and Mednick, A. S., 1985d, A quantitative study of the relative contribution of different retinal sectors to the innervation of various thalamic and pretectal nuclei in goldfish, *J. Comp. Neurol.* **242:**369–380.

Springer, A. D., and Mednick, A. S., 1986a, Relationship of ocular pigmentation to the boundaries of dorsal and ventral retina in a nonmammalian vertebrate, *J. Comp. Neurol.* **245:**74–82.

Springer, A. D., and Mednick, A. S., 1986b, Retinotopic and chronotopic organization of the goldfish retinal ganglion cell axons throughout the optic nerve, *J. Comp. Neurol.* **247:**221–232.

Springer, A. D., and Mednick, A. S., 1986c, Simple and complex retinal ganglion cell axonal rearrangements at the optic chiasm, *J. Comp. Neurol.* **247:**233–245.

Taylor, J. S. H., 1987, Fibre organization and reorganization in the retinotectal projection of *Xenopus, Development* **99:**393–410.

Torrealba, F., Guillery, R. W., Polley, E. H., and Mason, C. A., 1981, A demonstration of several independent, partially overlapping retinotopic maps in the optic tract of the cat, *Brain Res.* **219:**428–432.

Torrealba, F., Guillery, R. W., Eysel, U., Polley, E. H., and Mason, C. A., 1982, Studies of retinal representations within the cat's optic tract, *J. Comp. Neurol.* **211:**377–396.

Wagner, H.-J., 1974, Development of the retina of Nannacara anomala (Regan) (Cichlidae, Teleostei) with special reference to regional variations of differentiation, *Z. Morphol. Tiere* **79:**113–131.

Walls, G. L., 1942, The Vertebrate Eye and Its Adaptive Radiation, Bloomfield, New Jersey: Cranbrook Institute of Science.

Walsh, C., and Guillery, R. W., 1984, Fibre order in the pathways from the eye to the brain, *Trends Neurosci.* **7:**208–211.

Williams, R. W., and Rakic, P., 1985, Dispersion of growing axons within the optic nerve of the embryonic monkey, *Proc. Natl. Acad. Sci. U.S.A.* **82:**3906–3910.

Yamadori, T., 1981, An experimental anatomical study on the topographic termination of the optic nerve fibers in the rat, *J. Hirnforsch.* **22:**313–326.

# Routing of Axons at the Optic Chiasm

**6**

## Ipsilateral Projections and Their Development

### Sally G. Hoskins

### 6.1. INTRODUCTION

The central mystery of brain development is how the precise point-to-point connectivity of billions of neurons is achieved. An important aspect of this problem concerns the pathfinding behavior of axons: What determines the branch taken at each fork of a pathway? In the developing visual system, retinal ganglion cell axons from the two eyes intersect below the diencephalon, forming the optic chiasm. Here some axons continue in the same direction to innervate the opposite side of the brain, while others turn to innervate the same side, making the chiasm a major trajectory choice point. Throughout the developing nervous system, growing axons are confronted with a series of such choice points as they move toward and synapse with specific targets. The rules governing such choices are poorly understood. Understanding how axons choose which way to go in the optic chiasm may help to elucidate how fibers in general find their way to their targets. In the developing visual system of vertebrates, large numbers of fibers are confronted with the same ipsilateral/contralateral choice; the choice is made in a structure physically separate from the brain and free of other fiber tracts, and the eyes in many species are accessible for experimental manipulation during embryogenesis. Thus, this system is a powerful one for exploration of the rules that regulate laterality choice.

The importance of the crossing pattern at the optic chiasm being made precisely and reproducibly is underscored by the fact that properly organized

**Sally G. Hoskins** • Department of Biological Sciences, Columbia University, New York, New York 10027; *present address:* Department of Biology, City College of New York, New York, New York 10031.

**113**

binocular vision in vertebrates depends on this partial decussation* of the optic nerve (reviewed in Polyak, 1957). In primates, for example, fibers of the nasal hemiretina cross, and fibers of temporal hemiretina project ipsilaterally (reviewed in Hughes, 1977; Rodiek, 1979). Such a crossing pattern brings together in the optic tract fibers from each eye which view the same regions of binocular visual space. Axons from the two eyes subsequently form orderly maps in target structures in the thalamus and midbrain.

The maps are aligned so that adjacent regions of the targets receive input from axons of each eye, representing a particular coordinate in binocular visual space. The laterality decision is the first in a hierarchy of choices. Development of an accurately wired binocular visual system requires that in addition axons make accurate choices of target nucleus and of subregion or lamina within the target nucleus in which to terminate. It is not understood whether the laterality decision is made actively, perhaps influenced by local environmental conditions, or whether fibers are in some way "programmed" for a particular route at the chiasm. The development of the ganglion cell layer and the subsequent formation of retinotopic maps in CNS targets have been scrutinized at length in both lower vertebrates and mammals, in an effort to reveal developmental mechanisms that underlie laterality choices and subsequent target selection, and the constraints within which such mechanisms operate.

Partial decussation of retinofugal fibers in the chiasm is characteristic of vertebrate visual systems, and there is a rough correlation between amount of binocular visual field and size of the retinal region containing ipsilaterally projecting fibers (see Hughes, 1977, for review). So, for example, monkeys have frontally directed eyes and 50% of their retinal fibers project ipsilaterally, while mice, with lateral eyes and less binocular overlap, have only about 10% of the retina involved in ipsilateral projection (Dräger and Olson, 1980). While visual system development has been analyzed extensively in a variety of species (reviewed in Stone *et al.*, 1984; Shatz and Sretavan, 1986), the question of laterality determination remains unresolved. In particular, it is not clear why particular fibers cross, while others project ipsilaterally. Such choices could be made in response to side-specific cues at choice points. Alternatively, axons themselves may somehow be labeled as to their laterality. In nonprimates, neighboring cells in the same retinal region often project to opposite sides of the brain, suggesting that they have such operationally defined distinct "laterality labels."

The idea that axons might be intrinsically programmed for projection to specific targets derives in large part from the work of Roger Sperry in the developing visual systems of lower vertebrates. Sperry demonstrated that amphibians with one optic nerve cut and the eye rotated 180° behaved, upon regeneration, as if the nerve fibers had reconnected to their previous tectal

---

*While strictly speaking, fibers "decussate" only if they form an X pattern, the term has commonly been used in neurobiology to describe fibers, either individually or in groups, crossing from one side of the brain to the other. The latter sense is used throughout this chapter.

target sites despite the fact that such a "wiring pattern" now produced maladaptive, reversed behavior (Sperry, 1944, 1945). The prevailing idea of the time was that regeneration involved an initial phase of random reconnection followed by a "tuning up" of the projections brought about by experience (reviewed in Weiss, 1960). Sperry's result seemed more consistent with the concept that connections between growing axons and their targets were established based on "specific cytochemical affinities" which arose in neurons "via self-differentiation, induction through terminal contacts, and embryonic gradient effects" (Sperry, 1963, p. 703). Additionally, a subsequent series of experiments showed that regenerating axons from a half-eye "bypassed" available tectal synaptic space while regenerating to their usual topographically appropriate positions in the optic tectum (Attardi and Sperry, 1963; Arora and Sperry, 1962). Axons behaved as if programmed for connection at specific sites, apparently refusing to innervate inappropriate regions even if such available space was encountered early in the regeneration process and before the axons reached their normal targets. Such results argued against connection specificity being achieved via mechanisms based on outgrowth timing or fiber interactions, and in favor of a specific matching function which ensured topographically appropriate reconnection (reviewed in Gaze, 1978; Fraser and Hunt, 1980; Hollyday and Grobstein, 1981). The idea that neurons are individually distinguishable at the cytochemical level has strongly influenced subsequent work on the developing visual system. For the developing ipsilateral projection, it is clear that the neurons of interest are differentially distributed in the retina and their axons terminate in specific patterns in their targets. The nature of the "individual identification tags" (Sperry, 1963, p. 70), and the rules that govern the way in which they function to produce the laterality choice which is fundamental to binocular vision, are not yet understood.

Perhaps the strongest example of the importance of accurate "laterality choice" is provided by the Siamese cat, a temperature-sensitive albino mutant in which reproducible mapping errors between the retina and dorsal lateral geniculate nucleus (dLGN) can be attributed to a local failure of laterality determination and concomitant misrouting of a subset of retinal ganglion cell fibers (Guillery, 1969). In the Siamese cat, as in all other albino mammals studied, too many fibers of temporal retina cross in the optic chiasm. Intriguingly, the misrouted fibers terminate in an orderly way in appropriate (temporal retina-recipient) regions of the wrong (contralateral) side of the brain (reviewed in Guillery *et al.*, 1974). Thus, the albino mutation alters the laterality choice made by a subset of fibers in temporal retina but does not prevent the fibers from finding the appropriate layer of the dLGN nor from making patterned and functional connections. Mechanisms or labels that influence laterality decision, target choice, and synapse site can apparently be affected independently. (See Chapter 4 in this volume.)

Thus, the decussation pattern of the developing optic nerve has important implications with regard to function of a binocular visual system. The goal of this chapter is to provide an overview of the development of the ipsilaterally projecting axon population, viewed in the context of its role in binocular vision. We begin with a consideration of the organization of the retina with respect to projection laterality and birthdates of ipsilaterally projecting cells. While the details of retinal organization vary in different species, several general patterns emerge (see Fig. 6.1). In particular, it is clear that in most species, with the exception of primates and megachiropteran bats, discussed below, ipsilaterally and contralaterally projecting ganglion cells are intermingled in bilaterally projecting retinal regions. Such a situation puts some constraints on the types of mechanism that can plausibly be proposed to explain the routing. For example, a pioneer fiber mechanism whereby later developing fibers fasciculate with and follow neighboring axons which have already established projections (Bate, 1976) will not explain decussation patterns in nonprimates. Instead, it seems likely that individual ganglion cells generate axons which behave as if they are uniquely distinguished from one another, or labeled, in some way that correlates with projection laterality. Much of the research on retinal differentiation and laterality determination is, in effect, a search for a more precise definition of the distinguishing characteristics of retinal ganglion cells, characteristics inferred chiefly from their patterns of projection. Many of the experiments aimed at ascertaining the "rules" governing the development of ipsilaterally projecting axons, their laterality choice, and their ability to make synapses in appropriate target regions involve perturbing particular conditions and assessing alterations in the ipsilateral projections subsequently formed. Common approaches involve producing environmental perturbations, for example, disrupting the usual situation at the optic chiasm by removing one eye or studying mutants with

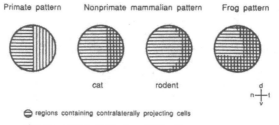

**Figure 6.1.** Schematic representation of main classes of retinal organization for ipsilateral and contralateral projection (see text for discussion). Shading denotes retinal regions containing contralaterally projecting cells and/or ipsilaterally projecting cells. In retinas organized along the "primate pattern," cells with differing laterality of projection originate from virtually nonoverlapping regions. In all other species examined, ipsilaterally projecting ganglion cells lie in retinal regions which also contain contralaterally projecting cells. The relative distributions of ipsilaterally and contralaterally projecting cells in bilaterally projecting retinal regions is not represented in the figure. In cat, most cells of temporal retina project ipsilaterally, while in rodents and amphibians, most cells of temporal retina project contralaterally. In addition, it should be noted that in rodent and frog retinas, small numbers of ipsilaterally projecting cells (not shown in the figure) are found throughout the retina. Such cells are extremely rare in cat and are not found in primates.

ocular size disparities; temporal perturbations, for example, challenging the projection to develop earlier or later than usual; and spatial perturbations, for example, embryonic eye rotations or grafts. Underlying all such approaches is the general question of intrinsic, "programmed" versus extrinsic, environmentally controlled axon development.

The developing visual system lends itself to a variety of types of manipulation. Some of the early experiments focused on the laterality decision in mammals investigated the role of interactions between fibers from the two eyes by removing one eye at various developmental times. Subsequent work has revealed that such studies may alter other relevant aspects of normal development by interfering with competition between fibers at the targets and with cell death. In lower vertebrates, the role of outgrowth timing in determining projection patterns can be investigated by forcing the projection to regenerate or by inducing it to develop prematurely. Species in which mutation alters the normal balance of fibers in the chiasm can be used to test the effects of more subtle alterations of chiasm structure. The existence of labels and the acquisition of retinal polarity with regard to ipsilaterally and contralaterally projecting regions can be examined through grafting and eye rotation experiments in embryos. With hindsight it is clear that studies on the mechanisms regulating development of ipsilateral retinofugal projections are most interpretable when done in species in which the parameters that apply to the normal case are well defined. That is, one would like to know the numbers of ipsilaterally projecting cells, their distribution, birthdates, and stages of ingrowth before beginning to perturb the system. Historically, the experimental perturbations have often preceded complete analysis of normal parameters, leading to erroneous interpretations in some cases. It is also clear that technical advances have been influential, with more sensitive tracing methods allowing more precise descriptive analyses. Clearly, as techniques continue to evolve, our view of the developing visual system will change further.

## 6.2. ORGANIZATION OF THE RETINA WITH RESPECT TO PROJECTION LATERALITY

### 6.2.1. The Primate Pattern

Early studies of projection laterality involved lesioning the optic tract and examining the resulting patterns of degeneration in the ipsilateral eye. More recently, this approach has been superseded by the use of retrogradely transported fluorescent or histochemically detectable compounds. In primates, either optic tract section or injection of a retrogradely transported marker into ipsilateral recipient zones shows a sharp boundary or "line of decussation" in the retina, separating ipsilaterally projecting from contralaterally projecting regions. Temporal retinal ganglion cells project ipsilaterally, cells in nasal retina project contralaterally, and there is only a narrow vertical strip, subtending approximately 0.5° of visual space, which contains intermingled ipsilaterally and contralaterally projecting ganglion cells (Bunt *et al.*, 1977; Stone *et al.*, 1973). Thus, the line of decussation in the monkey retina separates a region within which virtually every retinal ganglion cell has an "ipsilateral label" from a region exclusively composed of contralaterally projecting cells. Such complete decussation of retinofugal projections has as a consequence that the superior colliculus on one side of the brain receives projections from both eyes but receives visual input only from the contralateral hemifield (see Guillery, 1982, for review). Until recently, this organization of retinofugal projections had been seen only in primates. In nonprimate mammals, typically there is a sharp line of decussation for retinogeniculate fibers but not for retinocollicular ones. Thus, in nonprimate mammals and lower vertebrates, each superior colliculus (optic tectum) receives input from all regions of the visual field of the contralateral eye. Recent studies of megachiropteran bats show them to have a "primate pattern" of collicular projections (Pettigrew, 1986). Collicular injections of HRP label cells in both retinas of *Pteropus* spp., but only on one side of a vertical decussation line. Similar experiments in microchiropteran bats reveal a "lower vertebrate" pattern, with collicular injections labeling cells in all regions of contralateral retinas but no cells of ipsilateral retinas. These results raise the possibility that primates and bats are evolutionarily linked (Pettigrew, 1986).

### 6.2.2. The Nonprimate Mammalian Pattern

In the cat, retrograde transport of HRP from retina-recipient regions of the lateral geniculate nucleus (LGN) on one side of the brain similarly reveals a sharp line of decussation in the retina. In retinas ipsilateral to the injection site, stained ganglion cells are found in temporal retina, separated from unstained contralaterally projecting cells of nasal retina by a vertical decussation line (Stone *et al.*, 1978; Cooper and Pettigrew 1979a). This line is "sharper" in ipsilateral than contralateral retinas; very few stained cells are found nasal to the line in ipsilateral retinas; however, a small but significant number of large cells in contralateral retinas lie temporal to the line. This variation reflects the fact that in cat retina different classes of cells obey slightly different decussation rules. X cells as a class come closest to following the primate pattern, in that essentially all X cells of temporal retina project ipsilaterally and all X cells of nasal retina cross. Only within a 1° vertical strip at the midline of the retina is there intermixing of ipsilaterally and contralaterally projecting X cells. Y cells, in contrast, have a decussation line 1–2° temporal to the line respected by X cells, and 5% of all Y cells of temporal retina project contralaterally. For W cells, the laterality choice correlates with physiological properties of the cells; tonic W cells of temporal retina project ipsilaterally, while the rest of the W cells are mainly phasic, and project contralaterally (Stone and Fukuda, 1974). With regard to the LGN target then, large regions of cat retina project in an all-or-none fashion, with temporal cells projecting ipsilaterally and nasal cell contralaterally, and a small narrow strip of retina contains intermingled ipsilaterally and contralaterally projecting cells. With regard to other retinofugal projections, however, more heterogeneity is seen. In contrast to the situation in the monkey, temporal retina in the cat contains a substantial number of cells that persist after section of the ipsilateral optic tract, which project to the contralateral rostral superior colliculus and the medial interlaminar nucleus (Stone, 1966; Feldon *et al.*, 1970; Sanderson, 1971; Harting and Guillery 1976).

Retinal organization in other mammals in general resembles the "cat pattern" rather than the "primate pattern," with the exception of megachiropteran bats, discussed above. In the mouse, there is no line of decussation dividing the retina into essentially nonoverlapping "ipsilaterally projecting" or "contralaterally projecting" regions. Instead, HRP backfills of the optic tract reveal that the vast majority of the ipsilaterally projecting cells lie in a crescent shaped area of temporal retina which also contains cells which project contralaterally (Dräger and Olson, 1980). In addition, a few ipsilaterally projecting cells (approximately 4% of the total number) lie outside the temporal crescent, interspersed with contralaterally projecting cells of nontemporoventral retina. The distribution of ipsilaterally projecting cells in the mouse is such as to subdivide the retina into a bilaterally projecting temporal crescent and a largely contralaterally projecting nontemporal region. As mice have laterally positioned eyes, the line defining the edge of the temporal crescent also appears likely to define the edge of binocular field.

As in the mouse, the vast majority of ipsilaterally projecting cells in the rat retina lie in the lower temporal crescent (Jeffery, 1984; Dreher *et al.*, 1985). For example, 96% of the cells labeled in ipsilateral retinas by injections of HRP to the optic tract lie in this region (Dreher *et al.*, 1985). Nevertheless, 75% of the cells in the temporal crescent project contralaterally (Cowey and Perry, 1979; Reese and Jeffery, 1983), and in contrast to the situation in cat, at least some of the contralaterally projecting cells innervate the dLGN (Lund *et al.*, 1974). Physiological mapping of the projections of temporal retina suggests that many of the contralaterally projecting cells lie along the central border of the temporal crescent. Similar studies in ferret reveal approximately 6000 ipsilaterally projecting retinal ganglion cells, almost all of which lie in temporal retina (Morgan *et al.*, 1987). Like the cat retina, the ferret retina has a sharp decussation line with most cells of nasal retina crossing and most cells of temporal retina projecting ipsilaterally. Only 1% of HRP-labeled neurons in ipsilateral retinas are found outside the crescent. As in the mouse and rat, temporal retina in ferrets also contains contralaterally projecting cells, though these are only a small percentage of the total (less than 10%). In the rabbit, HRP injection into the optic tract, or lesioning of one optic tract, reveals ipsilaterally projecting ganglion cells in a 3–3.5 mm wide strip at the far temporal retinal periphery (Provis and Watson, 1981). As in other nonprimates, ipsilateral and contralaterally projecting cells are interspersed within the temporal region. HRP backfilling of chipmunk optic tract reveals ipsilaterally projecting cells in a crescent at the temporal periphery, occupying approximately 25% of the total retinal area, and interspersed with contralaterally projecting cells (Wakakuwa *et al.*, 1985b). Thus, while in cat and ferret a minority of cells of temporal retina project contralaterally, in rabbit and chipmunk a minority project ipsilaterally.

### 6.2.3. The Lower Vertebrate Pattern

In frog, the organization of ipsilaterally and contralaterally projecting regions of retina parallels that seen in nonprimate mammals, with a portion of retina which projects almost exclusively contralaterally and another region which contains intermingled ipsilaterally and contralaterally projecting cells. In contrast to the mammalian cases, however, the relative proportion of retina devoted to bilateral projection has increased significantly. In *Xenopus laevis*, lesion studies and retrograde labeling with HRP indicate that much of the periphery as well as a large proportion of central retina projects bilaterally (Hoskins and Grobstein, 1980; Kennard, 1981; Hoskins and Grobstein 1985a). Thus, rather than being confined to a temporoventral crescent, ipsilaterally projecting cells are found in most retinal regions, with only naso-dorsal retina and a small portion of retina around the optic nerve head virtually devoid of ipsilaterally projecting cells. This distribution may reflect the extensive binocular field of *X. laevis*. Since at the completion of metamorphosis the eyes are situated on top of the head, there is a larger region of binocular field in *Xenopus* than in animals with frontal (cat, monkey) or lateral (rabbit, chipmunk, rat, mouse) eyes (Grobstein and Comer, 1977).

Overall, there is a fairly reliable correlation in mammals between degree of binocular overlap of the visual fields of the eyes and relative proportion of the retina containing ipsilaterally projecting fibers (see Hughes, 1977, for review). Animals with lateral eyes tend to have ipsilaterally projecting cells confined to the far temporal periphery, while in animals with frontal eyes or substantial binocular overlap, 50% of retinal area may contain ipsilaterally projecting cells. In amphibians other than frogs as well there is a tendency for more frontal eye position to correlate with more extensive ipsilateral projections although the retinal locations of ipsilaterally projecting cells have not been examined. So, for example, the alpine newt *Triturus alpestris* has a small ipsilateral projection early in development, and the projection becomes much denser and more extensive during metamorphosis, as the eyes move to more frontal positions (Rettig *et al.*, 1981). In a comparison of three species of lungless salamanders, a qualitative correlation was seen between more frontal eye position and density and extent of ipsilateral projections (Rettig and Roth, 1982). However, in the caecilian *Ichthyophis kohtaoensis*, a substantial ipsilateral projection to the thalamus is seen in an animal with quite lateral eyes, in which there is little if any binocular overlap (Himstedt and Manteuffel, 1985), and in barn owls, in which visual fields of the two eyes overlap significantly, the retinofugal projections to both thalamus and tectum are completely crossed (Bravo and Pettigrew, 1981). It should be noted that technical improvements in anterograde tracers have in some systems allowed detection of ipsilateral projections in species previously believed to lack them. In the chick, for example, an ipsilateral projection that is mostly later eliminated by cell death is seen transiently early in embryonic development (O'Leary *et al.*, 1983).

## 6.3. ONTOGENY OF IPSILATERALLY PROJECTING CELLS

The developmental time periods during which ganglion cells are born are important parameters for experiments designed to interfere with particular aspects of axonal development. The issue of retinal cytogenesis is discussed in detail elsewhere in this volume (see Chapters 1–3); in this section the issue of neuronal birthdating with regard to the production of ganglion cells is discussed briefly. While the overall distribution of ipsilaterally and contralaterally projecting ganglion cells is quite similar among species, with ipsilaterally projecting cells consistently concentrated in binocular regions of retina, the developmental time periods in which these cells are generated are variable between species (see Fig. 6.2). Two main patterns of cell generation are seen. Ipsilaterally and contralaterally projecting ganglion cells may be generated simultaneously, or there may be a substantial temporal delay between the birth of the first contralaterally projecting cells and birth of the first ipsilaterally projecting cells. When they are generated during the same stages, it is reasonable to expect that ipsilaterally and contralaterally projecting cells encounter similar environmental conditions during their early development. When they are born at quite different developmental time points, this is less likely to be the case.

In the monkey, all ganglion cells are born early in development, over a 2 month period that ends at the midpoint of gestation, and no cells are produced postnatally (Rakic, 1977). In the cat, most retinal ganglion cells are produced during a 2 week period at the end of the first half of gestation, and

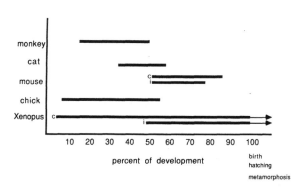

**Figure 6.2.** Birthdates of ganglion cells expressed as percentage of embryonic development. Bars represent the developmental time period over which retinal ganglion cells are born. For the mouse and *Xenopus,* birthdates for the ipsilateral (i) and contralaterally (c) projecting populations are noted separately. Arrows on the *Xenopus* bars indicate that cell addition continues postmetamorphically. Data from Rakic, 1977 (monkey); Walsh *et al.,* 1983 (cat); Dräger, 1985b (mouse); Kahn, 1973 (chick); Hollyfield, 1971; Hoskins and Grobstein, 1985b (*Xenopus*).

there is apparently no temporal delineation between the generation of ipsilaterally projecting and contralaterally projecting cells (Walsh *et al.*, 1983; Walsh and Polley, 1985; see also Chapter 1 in this volume). Within the ipsilaterally projecting population, the earliest ganglion cells to be produced are medium sized cells, the class which provides the most input to the dLGN and has axons with the sharpest line of decussation. Cells are generated in a rough central-to-peripheral gradient, with older cells found in central retina. In the mouse, as in the cat, central retina is generated before peripheral retina, and contralaterally and ipsilaterally projecting cells are generated simultaneously, beginning midway through embryonic development. During the first days of their production, the ipsilaterally and contralaterally projecting cells are born in distinct retinal regions, separated by a region of retina which will become the future midline of the visual field (Dräger, 1985b). A few days later, distinct populations of contralaterally and ipsilaterally projecting cells are generated simultaneously in temporoventral retina. In the mouse, as in the cat, it appears that the first ganglion cells born are ones that will project to the LGN (Walsh *et al.*, 1983; Dräger, 1985b).

Thus, the generation of ipsilaterally projecting retinal ganglion cells is similar in cat and mouse, with ipsilaterally and contralaterally projecting ganglion cells produced during overlapping time periods. In *Xenopus*, in contrast, a different pattern of cell generation is seen. Retina produced during the first half of premetamorphic development contains only contralaterally projecting cells. Ipsilaterally projecting cells are first seen as metamorphosis begins, and both contralaterally and ipsilaterally projecting cells are produced throughout the metamorphic period as well as postmetamorphically (Beach and Jacobson, 1979b; Hoskins and Grobstein, 1985b).

### 6.3.1. Perturbations of Developing Ipsilateral Projections

Observations about the organization of ipsilaterally and contralaterally projecting cell populations in the retina led to questions about the development of the observed organization. A popular approach in early work involved removing one eye from embryos or neonates and assessing changes induced in the projection from the remaining eye using anterograde tracing techniques, such as injection of $^3$H-proline and subsequent autoradiography. Typically, what was seen was a larger-than-normal ipsilateral projection from the remaining eye (reviewed in Lund, 1978). Initially, such results were taken as indicating that fibers from "inappropriate" retinal regions (e.g., nasal retina of the rat) had made erroneous crossing choices at the optic chiasm. Such interpretations partially reflect the fact that the true distribution of ipsilaterally projecting retinal ganglion cells, and in particular the normal existence, in many species, of a small number of such cells in regions outside the temporal crescent, was not known when the original eye removal studies were performed. As a consequence, any ipsilateral projection formed by fibers from nasal retina was considered aberrant. The putative "mistakes" in the one-eyed animals were in turn used as evidence for the idea that interaction between fibers from the two eyes in the optic chiasm were, in normal animals, important for proper routing (Lund *et al.*, 1973; Frost and Schneider, 1979). In the one-eyed cases, it was argued, the absence of fibers with which to interact at the chiasm caused remaining fibers to err, first by projecting to the wrong side of the brain, and once there by sprouting into regions larger than those they would normally innervate. While this interpretation appeared for some time to be consistent with experimental results in a wide variety of systems, further analysis of visual system organization revealed that in addition to the question of the actual retinal origins of ipsilaterally projecting fibers, the roles played by additional developmental processes had been overlooked. Studies of hamster retina revealed that removal of one eye at birth decreased the amount of cell death seen in the remaining eye (Sengelaub and Finlay, 1981). Such a result suggests that the amount of normally occurring cell death seen during development may be related to a competition among ingrowing retinal axons for some sort of trophic factor present at the targets, or for target space itself. Removal of 50% of the input to a target allows a proportion of the remaining fibers, which would normally have died, to instead be stabilized and survive. The implication for eye removal studies is that the increased ipsilateral projection seen in enucleates may reflect an increased number of retinal ganglion cell axons in the target, rather than misrouting of the usual number of ipsilaterally projecting axons. In support of such an interpretation is the finding that cells are "saved" preferentially in temporal retina of the hamster, the usual ipsilaterally projecting region (Sengelaub and Finlay, 1981). In tandem with the new information on retinal ganglion cell death have been recent studies on the initial development of ipsilateral projections, showing that contralateral projections typically arrive at their targets before the ipsilateral projections, and that in many species the ipsilateral and contralateral projections from the two eyes form overlapping projections during development. The overlap gradually decreases either pre- or postnatally both through the "weeding out" process of axonal death and by the withdrawal or shrinkage of axonal arborizations back into restricted territories (reviewed in Shatz and Sretavan, 1986). Given these new aspects of the development of ipsilateral projections, eye removal experiments can be reinterpreted. It appears that eye removal at a particular developmental day "arrests" the developing projection at the embryonic state it has achieved by this day. For this reason, it makes sense to consider the results of recent eye removal experiments in the context of what is known about the timing of development of ipsilateral projections.

### 6.3.1.1. Environmental Perturbations: Eye Removal Studies

The availability of anterograde tracers allowed analysis of the development of ipsilateral and contralateral projections. Several general patterns emerge. In the visual systems of many mammals, developing ipsilateral and contralateral retinofugal projections initially overlap in their target nuclei and subsequently separate completely or partially to achieve the distribution characteristic of adults. Concurrent with the segregation, in many species, is death of retinal ganglion cells and consequent elimination of a substantial number of optic axons. Eye removal in developing animals often results in "expanded" ipsilateral projections to target nuclei, from larger than usual retinal regions. In the monkey, for example, the process of axon elimination and that of segregation of initially overlapping ipsilateral and contralateral projections overlap temporally. This observation led to the proposal of a "selective elimination hypothesis," suggesting that axon loss is normally brought about by competition among terminals for topographically appropriate target space in particular layers of the LGN (Rakic and Riley, 1983a). Such a mechanism would explain the expanded ipsilateral projections seen in one-eyed animals as due to a failure of the normal winnowing process, which cannot occur in the absence of the usual large number of "competing" axons from the missing eye. The hypothesis predicted further that in addition to preserving the "embryonic" expanded pattern of projection of retinal afferents from one eye throughout the entire LGN, eye removal should also prevent the loss of axons by preventing ganglion cell death. While removing one eye does decrease the amount of cell death in the remaining eye of the monkey, the effects are not sufficient to explain the observed alterations in ipsilateral projection. Thus, unilateral enucleation early in monkey development and well before segregation of projections normally begins provides the remaining eye with twice as much available target space, but only about a 35% increase in axon number is seen (Rakic and Riley, 1983b). Such a result suggests that where it exists, competition for synaptic space is only one of a number of factors which regulate the ultimate number of surviving axons. (See Chapters 7 and 8 in this volume.)

In the cat visual system, axon number appears less strongly linked to binocular interactions, but such interactions do influence axonal morphology. Eye removal early in development, at a stage before axons reach the optic chiasm and thus before any putative interactions between the two eyes could occur, does not increase the number of surviving axons in the remaining nerve (Sretavan and Shatz, 1986). Enucleation several weeks later, when axons have established bilateral projections, increases the number of surviving axons of the other nerve only slightly (approximately 12%; Williams *et al.*, 1983). While binocular interactions are apparently not a major determinant of ultimate axon number in the cat, they do appear to influence other aspects of axon development. In the early enucleation studies, the remaining eye, although able to make normal bilateral projections and to undergo the normal decrease in axon number, did not have a normal pattern of projection in the LGN. Rather than terminating in eye-specific territories, the axons of enucleates made diffuse projections throughout the LGN (Sretavan and Shatz, 1986). Such a result suggests that different aspects of the processes of axon initiation, outgrowth, target choice, and selection of ultimate synapse site differ with regard to the degree to which they depend on binocular interaction or axonal competition. Interactions between the two optic nerves are apparently not required for routing at the chiasm or for axon elimination, but such interactions are necessary for determining the subregions of a target in which particular axons should terminate. In particular, it is clear that the development of restricted terminal arbors does not necessarily require interactions with afferents from the other eye. HRP filling of individual axons in animals unilaterally enucleated before ganglion cells are born showed aborization patterns like those of axons in normal animals, and overall axon length unchanged. Since arbor architecture after later enucleations differs from this pattern, with axons showing extra branches and covering more area than usual, it is evident that more than one factor is relevant in influencing final arbor size and shape, and that the stage at which an eye is removed may be a significant variable (Sretevan and Shatz, 1986; see also Chapters 7 and 8 in this volume).

Whether or not cell death is brought about by interactions between fibers from the two eyes, this process clearly can play a role in sculpting patterns of ipsilateral projection. For example, in the hamster, optic projections develop postnatally (So *et al.*, 1978). In normal animals, studies using long-lived fluorescent retrograde tracers indicate that the number of labeled cells seen outside the ipsilateral temporal crescent shows a steady decrease as postinjection survival times increase (Insausti *et al.*, 1984). Many ipsilaterally projecting cells are found in nontemporal crescent on postnatal day 1, but only a few are seen 2.5 weeks later, apparently reflecting the usual thinning of the projection brought about by normally occurring cell death. Double label studies have shown that at least some of the ipsilaterally projecting cells in the nontemporal retina, which normally die, can be "rescued" by enucleation. As in other species, hamster enucleates have larger numbers of ipsilaterally projecting cells, both in the temporal crescent and outside it. The cells outside the temporal crescent represent a population which in the enucleate have survived since the day of birth but which in the two-eyed animal would die sometime during the first few postnatal weeks, presumably, in this case, due to competitive interactions among fibers from the two eyes. The persistent cells of nasal retina are ones that in early studies would have been termed "errors." Here it is clear that they do not result from an experimentally induced alteration in the mechanism of laterality determination, but rather represent a preservation of the embryonic state of the projection, possibly, in this case, due to interference with the process of cell death. The fact that the temporal crescent, which normally projects ipsilaterally, also has more labeled cells in enucleates than in normal animals indicates that naturally occurring retinal ganglion cell death is not a mechanism for the elimination of nontemporal crescent ipsilateral projection "errors" but rather a process that affects cells in all retinal regions.

With regard to the issue of "errors" in ipsilateral collicular projections in hamster enucleates, Thompson (1979) has shown that if one eye is removed from the neonate hamster, an increased projection forms from the remaining eye to the ipsilateral superior colliculus. Physiological analysis of the topography of the enlarged ipsilateral projection indicates that it is formed by axons from all retinal regions, which form two maps in the remaining colliculus, a normally oriented map in rostral colliculus (typical of the normal ipsilateral projection), and a nasotemporal reversed map in caudal colliculus, formed by axons from all retinal regions. Thus, as pointed out by Dräger (1985b), eye removal in hamster does produce a larger than usual projection to the ipsilateral superior colliculus, but not because of errors made by the usual ipsilaterally projecting population.

One troublesome aspect of eye removal experiments is that they may produce a population of degenerating fibers at the optic chiasm. Depending on when the eye is removed, the degenerating fibers may be those of the crossing projection made by the removed eye (and hence ipsilateral to the remaining eye) or may contain crossing and ipsilaterally projecting axons. Despite reports that degenerating fibers may "mark" their targets in a semi-permanent way (Schmidt, 1978), that terminal segments of axons may remain motile for some time after axotomy (Shaw and Bray, 1977), that axons may alter their trajectories when they encounter glial elements in their path (Silver, 1984), and that eye removal may substantially alter the shape of the suprachiasmatic ventricular recesses (Guillery, 1982), little attention has been focused on possible effects of degenerating debris on the routing of fibers in eye removal experiments. Eye removal is often viewed instead as simply a way to eliminate potential competitive interactions between fibers from the two eyes, either in the chiasm or at the target.

The effect of stage of enucleation and concomitant degeneration of fibers on the development of ipsilateral projections has been addressed in a series of eye removal experiments in the mouse, the results of which suggest that axonal debris may be a significant complicating factor. In mouse, the time of origin of ipsilaterally projecting cells (Dräger, 1985b) as well as the timing of axonal development (Silver and Sidman, 1980) and details of the segregation of ipsilateral and contralateral projections (Godement *et al.*, 1984) are known. Eye removal in neonates or late embryos, both with bilateral retinofugal projections, results in increased numbers of ipsilaterally projecting cells in the remaining eye, both from the temporal crescent and the rest of the retina. Eye removal at earlier stages, before a chiasm has formed, results in fewer ipsilaterally projecting cells in the remaining eye, both in the temporal crescent and outside it (Godement *et al.*, 1980, 1987; Dräger, 1985b). If eye removal interfered with cell death in this system, by decreasing competition among axons from the two eyes, one would expect to see the number of ipsilaterally projecting cells correlated inversely with the age of the animal at enucleation. The earlier an eye is removed, the sooner competition is decreased, and the more cells would be expected to survive and contribute to the ipsilateral projection. The difference in results after early and late enucleation may be more related to the presence of degenerating crossing and ipsilaterally projecting axons in the optic chiasm at late embryonic stages, and their absence in mice enucleated before fibers have reached the chiasm. Dräger suggests in addition that enucleation may cause release or neuronal survival factors (Varon *et al.*, 1984) from the cut axons, and that these may support survival of extra retinal ganglion cells in the older embryos and neonates.

The complexity of the relationships between cell death, axonal interactions in thalamic and collicular terminal fields, and the sculpting of adult patterns of projection is exemplified by the situation in rat, where it appears that cell death is regulated differently in pigmented and albino animals. At birth in the rat, projections from both eyes overlap in the dLGN and SC, and many ipsilaterally projecting cells are found in nontemporal retina (Jeffery, 1984). During the first few days after birth, about 50% of the ganglion cells die (Lam *et al.*, 1982; Perry *et al.*, 1983), and terminals retract to form the adult pattern, with ipsilateral projections eventually occupying the centromedial dLGN, the anterior border of the SC, and a small amount of pretectum (Jeffery, 1984). (See Table 6.1.) Contralaterally projecting cells are found in all retinal regions, while ipsilaterally projecting cells originate from lower temporal retina (approximately 25% of retinal area; Cowey and Perry, 1979). In a study of axon elimination in albino rats, it was found that interactions between fibers from the two eyes appear to affect the rate of axon elimination but not the ultimate number of surviving axons, which is the same in enucleates and in normal animals (Lam *et al.*, 1982). In the first few days after eye removal in albino neonates, fewer axons are seen in the remaining optic nerve than in controls; however, in adults axon numbers are similar in the two groups. In pigmented rats, in contrast, enucleation on the day of birth saves some but not all of the cells which would normally die (Jeffery, 1984). In pigmented rats unilaterally enucleated at birth there is an increased number of ipsilaterally projecting cells in both the usual temporal regions and in nontemporal retina. As in other systems, early enucleation results in a larger ipsilateral projection from the remaining eye than does later enucleation, supportive of the hypothesis that eye removal stabilizes the embryonic state of projections by interfering with the normal process of cell death, which would normally remove many of the supernumerary projections during the first few days of postnatal development (Jeffery and Perry, 1982). The proportion of dLGN target occupied by ipsilaterally projecting fibers decreases over the same time period that cell death occurs, supportive of the idea that it is death, rather than fiber retraction or loss of branches, which sculpts the mature pattern of projection (Jeffery, 1984). Branching of fibers appears to contribute minimally to normal ipsilateral projections and to show little or no increase after enucleation (Jeffery and Perry, 1982).

An interesting aspect of laterality determination and the specification of axons for particular targets is that regions of retina do not in general seem to be specified in an "all-or-none" fashion, except perhaps in primate retina where retinal position correlates directly with laterality of projection. In non-primates, laterality labels may differ for ganglion cells of different class (e.g., X, Y, and W cells of the cat), or among cells labeled for different targets (LGN or colliculus). In the rabbit, it appears that eye removal affects the behavior of different subgroups of axons in different ways. About 10% of rabbit retinal ganglion cell axons project ipsilaterally (Giolli and Guthrie, 1969), terminating in the dorsal and ventral lateral geniculate nuclei and the intergeniculate leaflet (IGL). In the dLGN, ipsilateral and contralateral projections overlap

## Table 6.1. Interspecies Comparison of Ipsilateral Projections in the Context of Visual System Ontogeny

| Species | Gestation time | Approximate percentages of retina containing ipsilaterally projecting cells | Birthdates of retinal ganglion cells (as % gestation) | Period of cell death | Stages of overlap and segregation of developing ipsilateral and contralateral projections | Approximate percentages of optic axons that project ipsilaterally | In the ipsilaterally projecting temporal crescent (or equivalent), approximately what percentages of ganglion cells project contralaterally? |
|---|---|---|---|---|---|---|---|
| Monkey | 162 days | 50% (Ref.1) | E38–E80 (Ref. 2) (18–48%) | E95–E120 (Ref. 3) | Projections overlap before E90; segregate during E90–E120 (Ref. 4) | 50% (Ref. 5) | Virtually none (Ref. 6) |
| Cat | 65 days | 40% (Ref. 7) | E20–E35 (Ref. 8) (31–54%) | E45–postnatal (Ref. 9) | Projections overlap at E35; segregation begins at E47 (Ref. 10) | 40% (Ref. 11) | 25% (Ref. 12) |
| Ferret | 41 days | 15% (Ref. 13) | | | Overlap on day of birth; segregate D0–D9 (Ref. 14) | 10–12% (Ref. 15) | Less than 10% (Ref. 16) |
| Rat | 21 days | 25% (Ref. 17) | | First 5–10 postnatal days (Ref. 18) | Projections overlap E17–E20; segregation begins by E21 (Ref. 19) | 3–5% (Ref. 20) | 75% (Ref. 21) |

| Mouse | 21 days | 20% (Ref. 22) | E11–E16 (Ref. 23) (52–76%) | First 2 postnatal weeks (Ref. 24) | Overlap from E16–P3, then segregate (Ref. 25) | 2–3% (Ref. 26) | 85% (Ref. 27) |
|---|---|---|---|---|---|---|---|
| Hamster | 16 days | 10% (Ref. 28) | E10–P3 (Ref. 29) (75%–postnatally) | First 10 postnatal days (Ref. 30) | Overlap from P4 to P6; segregate on P7 (Ref. 31) | 1% (Ref. 32) | More than 50% (Ref. 33) |
| Frog | 56 days (to complete metamorphosis) | 60% (Ref. 34) | 50%–post metamorphically (Ref. 35) | No net loss of ganglion cells (Ref. 36) | Overlap from earliest stages; remain overlapped (Ref. 37) | 2% (Ref. 38) | More than 50% (Ref. 39) |

1. Stone et al., 1973.
2. Rakic, 1975.
3. Rakic and Riley, 1983a.
4. Rakic and Riley, 1983a.
5. Stone et al., 1973.
6. Stone et al., 1973, Bunt et al., 1977.
7. Stone, 1966; Wassle 1982.
8. Walsh et al., 1983; Walsh and Polley, 1985.
9. Ng and Stone, 1982; Williams et al., 1983; Sretavan and Shatz, 1986.
10. Shatz, 1983.
11. Wassle, 1982.
12. Stone, 1966.
13. Morgan et al., 1987; J. Cucchiaro, personal communication.
14. Cucchiaro and Guillery, 1984.
15. J. Cucchiaro, personal communication.
16. Morgan et al., 1987.
17. Cowey and Perry, 1979.
18. Lam et al., 1982; Perry et al., 1983.
19. Bunt et al., 1983.
20. Polyak, 1957; Jeffery, 1984.
21. Cowey and Perry, 1979.
22. Dräger and Olsen, 1980.
23. Dräger, 1985b.
24. Young, 1984.
25. Godement et al., 1984.
26. Dräger and Olsen, 1980; Godement et al., 1980.
27. Dräger and Olsen, 1980.
28. Hsiao, 1984.
29. Sengelaub et al., 1986.
30. Sengelaub and Finlay, 1981.
31. So et al., 1978.
32. Hsiao, 1984.
33. D. Sengelaub, personal communication.
34. Hoskins and Grobstein, 1985a.
35. Hoskins and Grobstein, 1985b.
36. Dunlop and Beazley, 1984.
37. Currie and Cowan, 1974; Kennard, 1981; Hoskins and Grobstein, 1985b.
38. Hoskins and Grobstein, 1985a.
39. Hoskins and Grobstein, 1985a.

during development (Grigonis *et al.,* 1986) but are found in morphologically distinct laminae in adults (Holcombe and Guillery, 1984). In the vLGN, and IGL, ipsilateral and contralateral inputs overlap. In the superior colliculus, ipsilateral innervation is usually restricted to the rostral third of the structure, while contralateral projections extend across the entire nucleus (Holcombe and Guillery, 1984). Unilateral eye removal on the day of birth has different effects on the contralateral and ipsilateral projections to the dLGN. In the dLGN contralateral to the remaining eye of unilaterally enucleated animals, the projection is found in a larger area than normal, occupying in addition to its normal territory the medial region that would normally receive ipsilateral innervation (Grigonis *et al.,* 1986). Thus, eye removal seems to halt the process of segregation of contralateral inputs to the dLGN. The terminal zone made in the ipsilateral dLGN of monocularly enucleated animals, however, is indistinguishable from that seen in normal animals; thus, the ipsilateral fibers, in contrast to contralateral ones, do not sprout or spread into surrounding deafferented regions of the dLGN.

In the vLGN, in contrast, ipsilateral projections were altered in monocularly enucleated cases, being both denser and more extensive, while contralateral projections in vLGN in monocular enucleates were like those seen in normal rabbits. Similarly in the IGL, contralateral termination zone size and label density did not change in monocular enucleates, but ipsilateral projections increased in size. Thus, with respect to the retinothalamic projections, the contralateral projection to dLGN and ipsilateral projection to vLGN and IGL are altered in monocularly enucleated rabbits. Such a result leads to the interesting conclusion that the mechanisms which underlie the observed specific termination patterns made by retinal axons in different nuclei may differ even for groups of axons destined for the same target, albeit on different sides of the brain.

Similar analysis of projections to the superior colliculus of rabbits monocularly enucleated as neonates found ipsilateral projections in their usual territories, whether analyzed within a few days of eye removal or as adults (Ostrach *et al.,* 1986). Only the radial distribution of the projection was altered, with more SC laminae receiving ipsilateral innervation in uniocular than in normal rabbits. Thus, eye removal in rabbit has differential effects on subgroups of axons which terminate in different targets, or in ipsilateral versus contralateral terminal fields of the same targets. Cell death has not yet been examined in the rabbit; it would be interesting to know whether, for example, the different effects on projections to dLGN (contralateral projection expands, ipsilateral projection does not) in enucleates reflects a differential effect on cell survival in contralaterally and bilaterally projecting retina.

In frogs the ipsilateral projection to the thalamus develops much later than the contralateral projection; (*R. pipiens,* Currie and Cowan, 1974; *X. laevis,* Lazar, 1971; Khalil and Szekely, 1976; Kennard, 1981; Hoskins and Grobstein, 1985b). The first ipsilateral projections are not seen until approximately 3 weeks of tadpole development have elapsed, while the first contralateral projections appear during the first few days of development (Holt and Harris, 1983). This contrasts with developing mammalian systems, in which the delay is one of hours or days (reviewed in Shatz and Sretavan, 1986). An additional substantial difference between the development of ipsilateral projections in frogs and mammals is that in contrast to the mammalian case, at no time do the ipsilateral projections of frogs occupy regions of the thalamus from which they will subsequently withdraw. Instead, the ipsilateral projection is first detected at tadpole stage 55 (Hoskins and Grobstein, 1985b) as a small "dusting" of terminal labeling in the region of the nucleus of Bellonci, the largest ipsilateral terminal zone (Levine, 1980). During subsequent tadpole and postmetamorphic frog stages, the projection continues to increase in density and to occupy a larger proportion of the thalamus, until it acquires the adultlike morphology. Whether one eye is removed before axons exit the retina, just as the first axons grow out, or after the contralateral projections are well established but before any ipsilateral projections are seen, the ipsilateral projections assessed after metamorphosis are normal. Fibers originate from the usual retinal regions and do not occupy more territory than usual in the ipsilateral thalamus (Kennard, 1981; Hoskins and Grobstein, 1985a). In addition, fibers are able to form terminal zones of appropriate morphology even in the absence of the contralateral terminal fields with which they usually partially overlap, and which have usually been established well before ipsilateral projections begin to develop. This is in contrast to the situation in mammals, where in many species expanded ipsilateral projections are seen in enucleates, apparently representing an arrest in the normal developmental "pruning back" of initially overlapped ipsilateral and contralateral projections from the two eyes. The results of eye removal experiments in frogs are more consistent with a mechanism involving pre-programming of the fibers for interaction with a particular identifiable target than with mechanisms involving competition for synaptic space at the target, cell death, or interactions in the chiasm.

It is not clear whether cell death plays an important role in frog retinal development. Ganglion cell number in *Xenopus* increases postmetamorphically and into adulthood (Dunlop and Beazley, 1984), indicating that if it does occur, cell death does not produce any net decrease in the number of ganglion cells. In rodents, at early stages a small but significant number of ipsilaterally projecting cells are found outside the temporal crescent; in normal animals most of these are eliminated by cell death (Sengelaub and Finlay, 1981; Insausti *et al.*, 1984), but as described above, in one-eyed animals they persist. In *Xenopus*, in contrast, in both the normal and the unilaterally enucleated cases, few if any cells in ipsilateral nasodorsal retina can be labeled by retrograde transport of HRP, and lesions of the nasodorsal retinal periphery do not produce detectable degeneration in target regions of the ipsilateral LGN (Hoskins and Grobstein, 1980, 1985a; Kennard, 1981). That is, as in unilaterally anophthalmic mice, in one-eyed frogs it does not appear that fibers make errors in crossing choice at the optic chiasm. It is notable that in *Xenopus* this result is obtained whether the enucleation is done extremely early, before a chiasm has been established (Kennard, 1981), or in the tadpole, when substantial contralateral projections have been formed and the visual system is functional (Hoskins and Grobstein, 1985b). In the former case the nerve from one eye must form a bilateral projection in the complete absence of any fibers from the other eye, while in the latter case the chiasm forms normally during embryogenesis but the later eye removal may leave a substantial amount of debris from the severed nerve of the removed eye in the chiasm region. If such debris persists until the stage when ipsilaterally destined fibers from the remaining eye begin to grow out, it might be expected to influence fiber trajectories in the optic chiasm. This possibility has been raised in other species discussed earlier as a possible explanation for differing results seen after early versus late eye removals. In *Xenopus*, in contrast, neither early nor late enucleations alter the retinal origins of the ipsilateral projection. Thus, in *Xenopus*, the ability of a fiber to project ipsilaterally does not depend on a normally structured chiasm.

It is worth noting that the absence of anomalous ipsilateral retinothalamic projections in frogs, and their presence in mammals, after unilateral enucleation does not seem to indicate a simple species difference, or the absence in frogs of some mechanism which is present in higher vertebrates. Under some conditions, anomalous ipsilateral projections *can* be produced in *Xenopus*. If one eye is removed just after the stage at which the first ipsilateral projections are seen, an aberrant direct ipsilateral projection to the tectum is formed, from peripheral temporoventral retina (Fraser, 1979). It is tempting to speculate that fibers heading for the contralateral tectum become misrouted to the ipsilateral side because they follow the newly established path taken by fibers growing from temporoventral retina to the ipsilateral thalamus. In any case, it is notable that eye removal causes aberrant behavior of contralateral retinotectal but not of ipsilateral retinothalamic axons in the visual system of *Xenopus*.

In general then, ipsilateral projections are made primarily by regions of retina which receive input from binocular visual field. In addition, a small number of neurons scattered throughout the rest of the retina projects ipsilaterally. The development of ipsilateral projections is delayed relative to that of contralateral projections, and this delay is prolonged for *Xenopus*. In mammals, ipsilateral and contralateral projections initially overlap at their CNS targets; this overlap may be partial or complete. The adultlike pattern of segregated or partially overlapping ipsilateral and contralateral terminal fields is achieved by cell death and/or fiber retraction and rearrangement. Cell death may be partially regulated by a competition for synaptic space or trophic substances at the targets, although this is not always the case. Depending on the stage at which an eye is removed, enucleation experiments may produce retinas with increased numbers of ipsilaterally projecting ganglion cells in regions that do not view binocular space, or may produce retinas indistinguishable in terms of projection laterality from those of normal two-eyed animals. For late enucleations, the presence of degenerating fibers at the optic chiasm may be an important variable. Eye removal brings about expanded terminal fields and what were once considered "unusually located" ipsilaterally projecting cells not by inducing errors in crossing choice at the chiasm, or bilateral branching of axons, but rather by altering, to different extents in different species, two normal developmental events, the elimination of a proportion of retinal ganglion cells by naturally occurring cell death and the partial retraction of an initially widespread developing ipsilateral projection.

### 6.3.1.2. Genetic Effects on Developing Ipsilateral Projections

In the visual system of albino mammals, too many fibers cross in the optic chiasm, and the ipsilateral projection is correspondingly reduced (Guillery, 1969, 1982; see Chapter 4 in this volume). In the Siamese cat, a temperature-sensitive albino, the line of decussation for the projection to the LGN is shifted temporally (Cooper and Pettigrew, 1979a,b) and there is an abnormal projection from a portion of temporal retina to the contralateral dLGN. In all albino mammals studied there is a correlation between the number of misrouted optic fibers and the amount of melanin pigment missing from the retinal pigment epithelium (Sanderson *et al.*, 1974; Creel *et al.*, 1982), suggesting that melanin may in some way influence axonal trajectories. It appears unlikely that melanin could label retinal ganglion cells directly, since in mice at least, the absence of pigment in the immediately adjacent pigment epithelium does not correlate with the accuracy of the trajectory taken at the optic chiasm by an axon from a given cell (Guillery *et al.*, 1973). Instead, the degree of misrouting of the entire population of axons correlates with the severity of the melanin deficit. Melanin is produced in the eyecup early in development and some melanin-containing cells are found in the embryonic eyestalk, possibly in a position from which they could influence the trajectories of developing axons (Silver and Sapiro, 1981; Strongin and Guillery, 1981; Dräger, 1985b). Studies of developing axons in Siamese cat fetuses indicate that the ipsilateral projection shows abnormalities from very early developmental stages (Kliot and Shatz, 1985), supportive of the idea that the effect of the mutation and the abnormalities seen in adults are due to an alteration at the level of the optic chiasm, rather than a secondary effect mediated by changes at the target (Guillery, 1969).

Unilaterally microophthalmic mice have one normally sized eye and one tiny eye; thus, the environment encountered by developing fibers at the optic chiasm is "unbalanced" and atypical but no debris is present. If one eye is removed from normal or microophthalmic mice on day 1, ipsilaterally projecting cells are subsequently found, as in normal mice, to be concentrated in the temporal crescent, with a few found in nontemporal retina (Dräger, 1985b). In the microophthalmic mice, in contrast to enucleated normal mice, throughout embryonic as well as postnatal development axons from the larger eye have had access to more central target space, yet the number of ipsilaterally projecting cells is not greater than in normal animals. Thus, in this system, interocular competition does not seem to play a major role in determining the laterality of projection of ipsilaterally and contralaterally projecting retinal ganglion cells. Developing retinal ganglion cell axons in mouse appear intrinsically capable of making appropriate laterality choices in the optic chiasm, whether or not any fibers from the other eye are present.

### 6.3.1.3. Temporal Perturbations of Developing Ipsilateral Projections

Since ipsilateral projections normally develop at a characteristic time, timing might be critical for accurate routing of fibers. One way to test such a possibility is to challenge the projection to develop earlier or later than usual. In the amphibian visual system, such a situation can be brought about by severing one optic nerve in adults. Four to eight weeks later, after regeneration, the distribution of ipsilaterally projecting cells in the regenerated retina as well as the terminal fields made by the regenerated projection can be examined. In such an experiment, regenerating ipsilaterally projecting retinothalamic axons are challenged to reestablish in adulthood a pattern that was originally set up premetamorphically. The distributions of ipsilaterally projecting cells in regenerates are like those of normal animals, with most cells found in temporoventral and fewest in nasodorsal retina, and anterograde fills of the optic nerve show that the usual ipsilateral terminal fields are reestablished (Hoskins and Grobstein, 1985a). Although axons regenerate in a different environment from that in which they originally developed, and may seek axonal debris at the targets rather than the target itself, it is nevertheless clear that the laterality "choice" can be made accurately in adulthood, as well as during development. Interestingly, in the optic tract regenerated fibers run along the lateral wall of the diencephalon, following a different path than did the developing fibers and suggesting that the pathway taken between the chiasm and the target is not critical for guidance of ipsilaterally projecting fibers to their terminal fields (Gaze and Grant, 1978).

A second temporal perturbation involves creating an age disparity between developing ipsilaterally projecting fibers and the environment into which they grow. This can be accomplished in *Xenopus* by inducing metamorphic growth in the eye but maintaining the tadpole in a premetamorphic state. Thus, a "postmetamorphic eye" is challenged to project into a premetamorphic target, in the presence of the projection from the other, premetamorphic eye. The ipsilateral projection which develops in such animals innervates the usual CNS targets and forms terminal zones of appropriate morphology (Hoskins and Grobstein, 1984, 1985c), suggesting that the ability to form such terminal zones reflects intrinsic properties of the developing ipsilateral fibers and/or the thalamic neuropil on which they synapse, rather than a necessary interaction between developing fibers from one eye and an equivalently aged population of developing fibers from the other eye, either in the chiasm or at the target.

Therefore, both of these manipulations support the idea of intrinsic programming of ipsilaterally projecting fibers. In the regeneration study, one optic nerve regenerates an ipsilateral projection in the presence of a stable projection from the intact other eye. In the induced projection case, one optic nerve forms an ipsilateral projection in the presence of a developmentally "younger" and completely crossed projection from the untreated eye. These results, and the results of eye removal experiments presented above, suggest that accurate routing of the ipsilaterally projecting subpopulation of optic nerve fibers does not depend critically on interactions between fibers from the two optic nerves in the optic chiasm.

An alternative possibility is that interactions among fibers within a nerve could be critical, perhaps by bringing about selective fasciculation of ipsilaterally destined axons, which then might run as a contiguous bundle through the optic chiasm. Such fasciculation is seen, for example, in optic nerves of the goldfish, where axons from an annulus of ganglion cells born at the same time run together in the nerve (Easter *et al.*, 1981). This possibility was investigated in *Xenopus* by retrogradely labeling ipsilaterally projecting fibers of young postmetamorphic frogs with HRP and examining the distribution of labeled fibers in the chiasm and nerve (Hoskins, 1987). HRP-labeled ipsilaterally projecting fibers were seen to run at the nerve periphery, interspersed with unlabeled contralaterally projecting fibers. In the chiasm the fibers made turns of diverse trajectories, rarely made contact with each other, and did not appear to interact with any obvious glial structure in the chiasm. Thus, the ipsilaterally projecting fibers do not fasciculate either in the nerve or in the tract and appear to make their projection laterality choice at the chiasm as individuals.

### 6.3.1.4. Spatial Perturbations of Developing Ipsilateral Projections

Early work in the retinotectal system established the lack of left–right "laterality labels" in the developing amphibian retina. The retinotectal projection in *Xenopus* is completely crossed, with the right eye projecting to the left tectum and the left eye to the right tectum. If during embryogenesis a right eye anlage is transplanted to the left side of the brain, it still projects to the *contralateral* tectum during subsequent development, indicating that the "program" or "label" in tectally destined fibers has to do with a decision to "cross or not cross" in the chiasm rather than with a choice of "right versus left side" of the brain (Sperry, 1945; Beazley, 1975a,b). Although crossed projections to both the tectum and thalamus originate from all retinal regions, fibers from temporoventral retina may cross if projecting to contralateral thalamus or tectum or may run ipsilaterally to the thalamus.

Two lines of evidence argue strongly that the determination of whether *Xenopus* neurons project ipsilaterally occurs very early in development. First, rotation of the embryonic eye rudiment by 180° at stage 32 (about 1.5 days of development; Nieuwkoop and Faber, 1967), long before ipsilaterally projecting axons begin to develop, shows that laterality specification occurs during embryogenesis. In a series of such eye-rotated animals, lesions were made after metamorphosis, either in the temporoventral (originally nasodorsal) or nasodorsal (originally temporoventral) region of the retinal periphery, and degeneration products were looked for in the ipsilateral thalamic terminal zones. Degeneration was seen in ipsilateral thalamic terminal zones only if the lesion was in the region of retina which had been temporoventral in the stage 32 embryo (Kennard, 1981). A similar result is seen if "double temporal" eyes are created at stage 32 by replacing a nasal half eye with a temporal half eye. Both the undisturbed and the grafted temporal half eyes form ipsilateral projections during metamorphosis (Straznicky and Tay, 1977). These results are interesting in several respects. The results indicate that the "labeling" of retinal regions occurs well before the projection begins to develop, since there is a delay of several weeks between the time of eye rotation or grafting and the appearance of the first ipsilaterally projecting fibers. In addition, they suggest that axons can make an appropriate crossing choice in the optic chiasm even if they initiate growth from an unusual position. Formerly temporoventral axons in rotated eyes enter the optic nerve from a dorsonasal position, yet still end up in appropriate terminal zones. Finally, since other work indicates that ipsilaterally projecting cells do not begin to be produced until stage 54/55 (Hoskins and Grobstein, 1982, 1985b), the polarity "labels" must be passed along through several generations of cells during the 2.5 weeks which elapse between stage 32, when the retina is already regionally labeled, and stage 55, when axons can first be detected in ipsilateral terminal zones.

Second, the burst of cell division during which the ipsilaterally projecting neurons are born is itself determined early in embryogenesis. In *Xenopus*, retinas grow symmetrically until metamorphosis begins, at which time ventral retina begins to add more cells than does dorsal retina, creating a ventral growth bias which persists into adulthood (Hollyfield, 1971; Straznicky and Gaze, 1971; Jacobson, 1976; Beach and Jacobson, 1979a). Cell cycle times in dorsal and ventral retina remain the same during the "ventral growth burst," suggesting that an increased fraction of dividing cells in ventral retina underlies the asymmetry (Beach and Jacobson, 1979a). The asymmetry appears just after the stage when thyroxine appears in the system and can be induced at earlier, symmetrically growing stages by injection of thyroxine into the eye (Beach and Jacobson, 1979a,b). Interestingly, although cells normally do not respond to thyroxine until stage 55, the polarization of the retina for thyroxine-regulated growth occurs very early in development. If a compound eye is created at embryonic stage 32 by grafting a ventral half eye into the position normally occupied by the dorsal half eye, *both* regions undergo increased growth beginning at stage 55 (Straznicky and Tay, 1977; Beach and Jacobson, 1979c). Such a result indicates that the embryonic retina is "programmed" during embryogenesis for an eventual differential mitotic response to a signal which its descendant cells will encounter 3 weeks later, and that the programming is retained by cells of the retinal anlage even when they are relocated to unusual positions.

### 6.3.2. Hormonal Control of Ipsilateral Projections

As outlined above, three events occur at stage 55 of *Xenopus:* growth in ventral retina shifts from a symmetric to an asymmetric pattern, with more cells added in ventral than in dorsal retina (Hollyfield, 1971; Jacobson, 1976;

Beach and Jacobson, 1979a); ipsilaterally projecting neurons begin to be born (Hoskins and Grobstein, 1985b); and thyroxine appears in the circulation (Leloup and Buscaglia, 1977). Furthermore, asymmetric growth can be induced precociously in the symmetrically growing eyes of younger tadpoles using intraocular injection of thyroxine (Beach and Jacobson, 1979b). These observations coupled with the correlation in time between the first stage at which ipsilateral projections are seen and the time thyroxine normally appears in the developing system suggested that the appearance of thyroid hormone in the system and the development of ipsilaterally projecting fibers might be linked. Metamorphosis in *Xenopus* can be blocked reversibly by immersion of tadpoles in 0.01% propylthiouracil (PTU) (MacLean and Turner, 1976), which prevents the iodination of tyrosine necessary for thyroxine synthesis (Green, 1978). PTU-reared tadpoles reach stage 54 along the usual time course but do not progress further morphologically, although they continue to grow (Hoskins and Grobstein, 1984). The retinofugal projections of PTU-reared tadpoles are like those of normal stage 54 tadpoles, and their eyes grow by symmetrical addition of cells to the periphery. If animals are transferred from PTU medium to normal rearing solution, they resume metamorphic development and acquire normal ipsilateral projections, indicating that the development of such projections is thyroxine dependent at some level. Injection of thyroid hormone to one eye of PTU-reared tadpoles causes the development of an ipsilateral projection from that eye (Hoskins and Grobstein, 1984). Induced ipsilateral projections run only to the proper ipsilateral targets and form terminal fields of typical morphology. The ipsilateral projection can be induced by doses of hormone which do not affect retinal growth or ipsilateral projections of the untreated eye. This suggests both that thyroxine acts in this system at the level of the eye and that induction of ipsilateral projections does not involve thyroxine-induced changes at the chiasm or target, as such changes would be expected to effect projections from both eyes (see Hoskins, 1986, for discussion). In addition, in this experiment an "unbalanced" situation has been created at the optic chiasm, with many more fibers present in the optic nerve from the thyroxine-treated eye than are in the optic nerve from the untreated eye. Nevertheless, fibers are able to make an ipsilateral crossing choice and innervate the usual targets, forming terminal zones of appropriate morphology.

In the visual system of *Xenopus,* although crossed projections begin to develop within a few days of fertilization, the production of ipsilaterally projecting ganglion cells and the development of ipsilateral axonal projections begins midway through premetamorphic development, under the influence of thyroxine. The role of thyroxine is unclear. The retrograde labeling and birthdate studies in adults, as well as the anterograde studies in tadpoles, indicate that thyroxine does not act by inducing ipsilaterally projecting branches from postmitotic cells produced before stage 55. It remains a possibility that some of the ganglion cells produced during metamorphosis have branching axons, although in other systems such axons are rare (Hsiao, 1984). Since throughout development the eye grows by adding new neurons from a germinal zone at the periphery, this situation suggests that there must be a change in cell lineage at the retinal periphery during metamorphosis. The Beach–Jacobson analysis of cell cycle time during the period of asymmetric retinal growth indicated that ventral retina did not grow faster than dorsal retina during metamorphosis, but rather that the ventral germinal zone behaved as if it contained more precursor cells than did the dorsal germinal zone (Beach and Jacobson, 1979a). It is not clear at this point whether ipsilaterally projecting cells are derived from cell lineages which also produce contralaterally projecting cells, or whether the precursors of the ipsilaterally projecting population constitute a unique class. Thyroxine's role in the system may be to alter the pattern of cell production from a given lineage, to produce a new type of precursor cell (see Chapter 2 in this volume), or to stimulate division of an existing precursor cell in the peripheral germinal zone (see Hoskins, 1986, for discussion).

Overall, these results indicate that the development of the ipsilateral retinothalamic projection in *Xenopus* is regulated by thyroid hormone, as evidenced by the absence of ipsilateral projections in PTU-reared animals and by their development when thyroxine is restored to the system. The ability of thyroid hormone treatment of the eye to bring about ipsilateral projections characteristic of postmetamorphic frogs in the brains of premetamorphic tadpoles argues strongly that the normal development of the ipsilateral retinothalamic projection during metamorphosis is an event mediated by the action of thyroxine within the eye. Axons that develop in thyroxine-treated PTU-blocked animals must grow toward the chiasm along a nerve composed of previously developed contralaterally projecting axons, must diverge from those axons in the optic chiasm, and must stop growing and make synapses in the appropriate regions of the ipsilateral thalamus. The ability of thyroxine-induced axons to execute this program in the brains of premetamorphic tadpoles suggests that if environmental cues influence the growing axons, these cues must either be present by stage 54/55 or inducible by thyroxine concentrations below those required to produce detectable growth in the eyes or changes in external morphological features. Alternatively, the results may indicate that once the particular population of ipsilaterally destined neurons is produced, they can execute their "ipsilateral program" in the absence of environmental cues. While the distribution of thyroxine-induced axons in optic nerves of PTU-reared hormone-treated tadpoles has not been examined, the lack of fasciculation of ipsilaterally projecting axons in nerves of normal animals and the divergent trajectories of such axons in the optic chiasm (Hoskins, 1987) suggest that the axons may be intrinsically "programmed" for ipsilateral target sites and consequently able to find their way whether singly or in a group, whether or not they meet their corresponding axons from the other eye at the chiasm, and whether or not they are growing into a target of the appropriate age. That is, the terminal field morphology seen in the thalamic terminal zones of PTU-reared thyroxine-treated tadpoles resembles that seen in normal postmetamorphic frogs, though the hormonally induced projections superimpose on the "scaffolding" provided by a younger, stage 54 contralateral projection from the untreated eye to the same thalamic region. A similar conclusion regarding intrinsic "pathfinding ability" for a different group of developing retinofugal axons was reached by Harris (1986) in experiments in which optic vesicles were grafted to unusual locations but retinotectal fibers still succeeded in finding their targets. The most parsimonious explanation for the behavior of ipsilateral fibers may be to assume that thyroxine affects gene expression in the cells born during metamorphosis, and that thyroxine-induced cells produce axons that read or respond to existing environmental signals in a different way than did their predecessors. In this context it should be pointed out that the target cell(s) for thyroid hormone action in the eye have not yet been identified. In addition to affecting retinal ganglion cells, thyroxine is known to affect synapse organization in the inner plexiform layer (Fisher, 1972) and photopigment type (Wilt, 1959). Thyroxine also has been implicated in the control of axonal and dendritic growth and branching in other regions of the nervous system (reviewed in Kollros, 1981; Hoskins, 1986). The thyroid hormone receptor has recently been cloned and characterized as a DNA-binding protein (Weinberger *et al.,* 1986), and one well-characterized example of the role of thyroxine in gene regulation is its control of growth hormone production in a pituitary cell line (Samuels *et al.,* 1983). It is thus possible that the effects of thyroxine treatment on the development of the ipsilateral retinothalamic projection may be mediated by effects of one thyroxine-sensitive retinal cell type on another, or by thyroxine's regulation of the local production of another growth factor. Further work in the system is needed to resolve questions of lineage and of the molecular basis of the effects of this hormone on cell production and axon outgrowth.

## 6.4. CONCLUDING REMARKS

It is clear that cells in the ganglion cell layer of the vertebrate retina are heterogeneous. Experimental manipulations and descriptive anatomical analyses have revealed that ganglion cells may differ from their neighbors in size, birthdate, physiological properties, and target. In some species, such as the monkey and megachiropteran bat, cell position in the retina is precisely correlated with laterality of projection, but in other species, while the retina is regionally specified for contralaterally and bilaterally projecting regions, individual cells may bear a laterality "label" different from that of their neighboring cells. The ipsilaterally projecting ganglion cell population clearly plays an important role in the development and maintenance of binocular vision, and visual abnormalities result if ipsilateral projections are misrouted, as in albino mammals (Guillery, 1969; see Chapter 4 in this volume). Little information is available regarding the biochemical differentiation of ipsilaterally projecting ganglion cells, and the way in which the cells might be cytochemically distinguishable from other cells, or "labeled," in the sense of Sperry. In particular, one would like to know more about the axons of ipsilaterally projecting cells, as the suggestion from a number of studies is that these axons are either intrinsically "programmed" to take a particular route through the optic chiasm or are able to read and respond to environmental signals or trophic factors differently than do contralaterally projecting axons. One possibility is that subgroups of axons behave differently because their surfaces are different or because their growth cones are sensitive to different molecules in the environment. Alternatively, or possibly in addition, it may be that differential placement of glial elements in the pathway of developing retinal axons serves a guidance function (Maggs and Scholes, 1986; Guillery and Walsh, 1987). While a review of biochemical aspects of retinal differentiation is beyond the scope of this chapter, it is worth noting that a variety of rapidly evolving techniques hold great promise for revealing the molecular basis of differences between small groups of ganglion cells, differences which at this point are usually inferred or defined operationally after fairly gross perturbations of the system. Existing data in several systems offer hints into aspects of uniqueness of ipsilaterally projecting cells.

Biochemical markers are needed for ipsilaterally projecting cells, particularly in cases where such cells lie interspersed in the retina with contralaterally projecting cells. The possibility that ipsilaterally projecting cells differ metabolically from contralaterally projecting cells is suggested by existing data on cytochrome oxidase activity in adult rat retina. Cytochrome oxidase activity, as assessed by histochemical staining, is moderate in most regions, but a subpopulation of darkly stained ganglion cells is seen predominantly in temporal retina. While in rats temporal retina contains contralaterally as well as ipsilaterally projecting cells, the additional finding that regions of dLGN receiving uncrossed input stain more intensely than regions receiving only crossed input suggests that axons of the ipsilaterally projecting population may be metabolically dissimilar from axons of the crossed projection (Land, 1987). In developing hamster, a similar increase in cytochrome oxidase activity in temporal retina is detectable at postnatal day 1 (Wikler *et al.*, 1985), before substantial ipsilateral projections have been established (So *et al.*, 1978), suggesting that metabolic and perhaps functional differences may exist between different populations of axons during visual system development as well, at stages when axons are growing to their targets and laterality "decisions" are being made at the optic chiasm.

Metabolic markers may allow ipsilaterally projecting cells to be distinguished based on criteria other than projection laterality *per se,* and such markers would be useful for analyses of early differentiation of the cells, before axons have been elaborated and the projection laterality choice has been made. In some species it appears that ipsilaterally projecting cells can be distinguished from contralaterally projecting ones on the basis of size (Wakakuwa *et al.*, 1985a; Dräger and Olson, 1980) or amount of microfilament (Dräger and Hofbauer, 1984), but these criteria are not specific enough to identify ipsilaterally projecting cells uniquely. The apparent dependence of some retinal ganglion cells on melanin in the pigment epithelium may offer a clue as to molecules of importance during the early differentiation of ipsilaterally projecting ganglion cells, as well as suggesting that the critical site at which laterality information can be altered is in the developing eyecup rather than along the pathway of the developing optic nerve. In this light, the recent demonstration that retinal pigment epithelia of normal and albino mice differ substantially with regard to ability to sequester calcium (Dräger, 1985a) is of considerable interest.

Progress in identifying unique aspects of ipsilaterally projecting cells, which could be related to the projection laterality of their axons, will require new tools for study of the differentiating retina. For example, one problem with existing analyses of developing ipsilateral projections is that in nonprimates, where retinal position does not correlate absolutely with laterality of projection, ipsilaterally projecting cells are usually defined operationally, by using an anterograde or retrograde tracer after the axons have grown past the optic chiasm. To study the early differentiation of ipsilaterally projecting cells, and in particular the behavior of ipsilaterally projecting axons beginning at very early stages, as they begin to be elaborated in the ganglion cell layer, it will be necessary to have markers which identify such axons uniquely. Monoclonal antibody technology has been used in a variety of systems to generate antibodies specific for subpopulations of neurons (Hockfield and McKay, 1985; Hishinuma and Hildebrand, 1987). In the retina, cell-type-specific monoclonal antibodies allow unambiguous delineation of different retinal neuron classes (Akagawa and Barnstable, 1986); however, as yet no antibodies have been found to distinguish subpopulations within a cell class. Recent improvements in the technique have increased the probability of generating antibodies against rare cells (Hockfield, 1987) and will be useful for many species in which the ipsilaterally projecting cells constitute only a small fraction of all ganglion cells. Another newly devised approach, retroviral-mediated lineage tracing (Sanes *et al.*, 1986; Turner and Cepko, 1987), in which a histochemically detectable marker linked to a replication-defective retrovirus is used to infect precursor cells and define the clones derived from particular precursors, should allow the question of common versus specific precursors for ipsilaterally and contralaterally projecting ganglion cells to be resolved cleanly. For example, retrovirus could be injected into the developing eye during the period of ganglion cell production, and the laterality of projection of individual ganglion cells in labeled clones could be assessed using retrograde transport of a histochemically detectable marker from an ipsilateral or contralateral terminal zone. A finding that ipsilaterally projecting cells are clonally related only to other ipsilaterally projecting cells might suggest a genetic rather than environmental determination of axonal laterality choice.

At this point, the role of differential gene expression in specifying retinal ganglion cell differentiation or pathway choice is unclear. The correlation of albinism in mammals with misrouting of ipsilaterally targeted fibers suggests that presence of normal amounts of pigment in the embryonic eyecup are necessary for proper fiber growth, but the relationship between pigment and laterality choice is still obscure. The monoclonal antibody approach may succeed in defining molecules that distinguish ipsilaterally and contralaterally projecting cells; such molecules could then be studied using established techniques. Given the small percentage of ipsilaterally projecting cells in many systems, it may be unduly optimistic to expect such molecules to be defined directly in the near future. In the interim, it may be useful to attempt to pursue the existing phenomenology to the molecular level, using systems in which the laterality choice can be manipulated experimentally. For example, in *Xenopus,* where the production of ipsilaterally projecting cells and the development of the ipsilateral retinothalamic projection can be controlled by administration of thyroid hormone, it will be important to determine the mechanism by which the hormone acts. Specifically, does thyroxine act directly on ganglion cells or are its effects mediated by action on another retinal cell type or through induction of a locally produced growth factor? Does thyroxine change the lineage pattern of existing ganglion cell precursors at the retinal periphery, so that they begin to produce ipsilaterally projecting as well as contralaterally projecting cells, or does it cause the genesis of a new precursor whose descendants all project ipsilaterally? Is thyroxine required for the entire process of metamorphic retinal growth and development of ipsilateral projections, or once ipsilaterally projecting cells are born, will they make the novel projection choice at the chiasm whether or not thyroxine is still present? That is, is thyroxine a trigger or is it required continuously? Answering these and other questions will require examining the role of thyroid hormone in both cell production and in the laterality choice.

Molecular approaches will be needed to answer these and other questions. Since thyroid hormone receptors have recently been cloned and characterized (Weinberger *et al.,* 1986; Thompson *et al.,* 1987), it will be possible to approach the question of direct versus indirect effects of thyroxine on ganglion cells by using *in situ* hybridization or antibodies, when they become available, to localize receptor mRNA or protein. The finding that thyroxine receptor-specific mRNA is expressed in retinal ganglion cells would be consistent with the possibility that thyroxine acts on these cells directly. Interestingly, one form of thyroxine receptor is expressed at high levels in brain and not in the liver (Thompson *et al.,* 1987), while another form is expressed in both tissues (Weinberger *et al.,* 1986), raising the possibility that effects of thyroid hormone action on homeostatic regulation (reviewed in Gilbert and Frieden, 1981) and CNS development (reviewed in Lauder, 1983; Hoskins, 1986) may be mediated by activation of distinct sets of genes.

Overall, whether or not thyroid hormone and melanin act directly on retinal ganglion cells and their axons, or indirectly by influencing expression of other factors, both molecules have clear effects on the laterality of projection expressed by subgroups of retinal ganglion cell axons. Coupling molecular approaches which may identify, for example, thyroxine-regulated molecules expressed in the eye, with anatomical advances such as retrovirally mediated lineage tracing will allow a finer-grained analysis of the factors involved in axonal laterality choice than has been possible to date. Ultimately, since much anatomical and physiological work seems to confirm their existence, it will be important to seek and define the cytochemical nametags of Sperry. Clearly, much work will be required to build a case for an involvement of putative thyroxine-regulated molecules or pigment in laterality determination, and there is no reason *a priori* to expect all labels to be proteins. Nevertheless, the next few years are likely to be exciting ones in retinal cell biology, as new approaches using molecular techniques are applied to long standing questions of neuronal differentiation and projection laterality, in the retina and throughout the developing nervous system. It is not unreasonable to predict that decoding of Sperry's identification tags may be an achievable goal.

ACKNOWLEDGMENTS. Supported by grant #1 R29 NS25042 from the National Institutes of Health. My thanks to Drs. Barbara Finlay and Dale Sengelaub for organizing the meeting "Development of the Vertebrate Retina," and to Drs. Héctor Cornejo, Darcy Kelley, and Josh Wallman for reading and commenting on the manuscript.

## 6.5. REFERENCES

Akagawa and Barnstable, C., 1986, Identification and characterization of cell types in monolayer cultures of rat retina using monoclonal antibodies, *Brain Res.* **353:**110–120.

Arora, H. L., and Sperry, R. W., 1962, Optic nerve regeneration after surgical cross-union of medial and lateral optic tracts, *Am. Zool.* **2:**61.

Attardi, D. G., and Sperry, R. W., 1963, Preferential selection of central pathways by regenerating optic fibers, *Exp. Neurol.* **7:**46–64.

Bate, M., 1976, Pioneer neurons in an insect embryo, *Nature (London)* **260:**54–56.

Beach, D. H., and Jacobson, M., 1979a, Patterns of cell proliferation in the retina of the clawed frog during development, *J. Comp. Neurol.* **183:**603–615.

Beach, D. H., and Jacobson, M., 1979b, Influences of thyroxine on cell proliferation in the retina of the clawed frog at different ages, *J. Comp. Neurol.* **183:**615–624.

Beach, D. H., and Jacobson, M., 1979c, Patterns of cell proliferation in the developing retina of the clawed frog in relation to blood supply and position of the choroidal fissure, *J. Comp. Neurol.* **183:**624–632.

Beazley, L. D., 1975a, Factors determining decussation at the optic chiasma by developing retinotectal fibres in *Xenopus, Exp. Brain Res.* **23:**491–504.

Beazley, L. D., 1975b, Development of intertectal neuronal connections in *Xenopus:* The effects of contralateral transposition of the eye and of eye removal, *Exp. Brain Res.* **23:**505–518.

Bravo, H., and Pettigrew, J. D., 1981, The distribution of neurons projecting from the retina and visual cortex to the thalamus and tectum opticum of the barn owl *Tyto alba* and the burrowing owl *Speotyto cunicularia, J. Comp. Neurol.* **199:**419–441.

Bunt, A. H., Minckler, D. S., and Johanson, G. W., 1977, Demonstration of bilateral projection of the central retina of the monkey with horseradish peroxidase neuronography, *J. Comp. Neurol.* **171:**619–630.

Bunt, S. M., Lund, R. D., and Land, P. W., 1983, Prenatal development of the optic projection in albino and hooded rats, *Dev. Brain Res.* **6:**149–168.

Cooper, M. L., and Pettigrew, J. D., 1979a, The decussation of the retinothalamic pathway in the cat, with a note on the major meridians of the cat's eye, *J. Comp. Neurol.* **187:**285–312.

Cooper, M. L., and Pettigrew, J. D., 1979b, The retinothalamic pathways in Siamese cats, *J. Comp. Neurol.* **187:**313–348.

Cowey, A., and Perry, V. H., 1979, The projection of the temporal retina in rats, studied by retrograde transport of horseradish peroxidase, *Exp. Brain Res.* **35:**457–464.

Creel, D., Hendrickson, A. E., and Leventhal, A., 1982, Retinal projections in tyrosinase-negative albino cats, *J. Neurosci.* **2:**907–911.

Cucchiaro, J., and Guillery, R. W., 1984, The development of the retinogeniculate pathways in normal and albino ferrets, *Proc. R. Soc. London Ser. B* **223:**141–164.

Currie, J., and Cowan, W. M., 1974, Evidence for the late development of the uncrossed retinothalamic projections in the frog, *Rana pipiens. Brain Res.* **71:**133–139.

Dräger, U. C., 1985a, Calcium binding in pigmented and albino mice, *Soc. Neurosci. Abstr.* **11:**14.

Dräger, U., 1985b, Birthdates of retinal ganglion cells giving rise to the crossed and uncrossed optic projections in the mouse, *Proc. R. Soc. London Ser. B* **224:**57–77.

Dräger, U., and Hofbauer, A., 1984, Antibodies to heavy neurofilament subunit detect a subpopulation of damaged ganglion cells in retina, *Nature (London)* **309:**624–626.

Dräger, U. C., and Olsen, J. F., 1980, Origins of crossed and uncrossed retinal projections in pigmented and albino mice, *J. Comp. Neurol.* **191:**383–412.

Dreher, B., Sefton, A. J., Ni, S. Y. K., and Nisbett, G., 1985, The morphology, number, distribution and central projections of class I retinal ganglion cells in albino and hooded rats, *Brain Behav. Evol.* **26:**10–48.

Dunlop, S. A., and Beazley, L. D., 1984, A morphometric study of the retinal ganglion cell layer and optic nerve from metamorphosis in *Xenopus laevis, Vision Res.* **24:**417–427.

Easter, S. S., Rusoff, A. C., and Kish, P. E., 1981, The growth and organization of the optic nerve and tract in juvenile and adult goldfish, *J. Neurosci.* **1:**793–786.

Feldon, S., Feldon, P., and Kruger, L., 1970, Topography of the retinal projection upon the superior colliculus of the cat, *Vision Res.* **10:**135–143.

Fisher, L., 1972, Changes during maturation and metamorphosis in the synaptic organization in the tadpole retina inner plexiform layer, *Nature (London)* **235:**391–393.

Fraser, S. E., 1979, Late LEO: A new system for the study of neuroplasticity in *Xenopus,* in *Developmental Neurobiology of Vision* (R. D. Freeman, ed.), p. 319–330, Plenum Press, New York.

Fraser, S. E., and Hunt, R. K., 1980, Retinotectal specificity, *Annu. Rev. Neurosci.* **3:**319–335.

Fritzsch, B., Himstedt, W., and Crapon de Caprona, M. D., 1985, Visual projections in larval *Ichthyophis kohtaoensis, Dev. Brain Res.* **23:**201–210.

Frost, D., and Schneider, G., 1979, Plasticity of retinofugal projections after partial lesions of the retina in Syrian hamsters, *J. Comp. Neurol.* **185:**517–568.

Gaze, R. M., 1978, The problem of specificity in the formation of nerve connections, in *Specificity of Embryological Interactions* (D. Garard, ed.), pp. 51–96, Chapman and Hall, London.

Gaze, R. M., and Grant, P., 1978, The diencephalic course of regenerating retino tectal fibres in *Xenopus* tadpoles, *J. Embryol. Exp. Morphol.* **22:**201–216.

Gilbert, L., and Frieden, E., 1981, *Metamorphosis: A problem in developmental biology,* Plenum Press, New York.

Giolli, R. A., and Guthrie, M. D., 1969, The primary optic projections in the rabbit. An experimental degeneration study, *J. Comp. Neurol.* **136:**99–126.

Godement, P., Saillour, P., and Imbert, M., 1980, The ipsilateral optic pathway to the dorsal lateral geniculate nucleus and superior colliculus in mice with prenatal or postnatal loss of one eye, *J. Comp. Neurol.* **190:**611–626.

Godement, P., Salaun, J., and Imbert, M., 1984, Prenatal and postnatal development of retinogeniculate and retinocollicular projections in the mouse, *J. Comp. Neurol.* **230:**552–575.

Godement, P., Salaun, J., and Metin, C., 1987, Fate of uncrossed retinal projections following early or late prenatal monocular enucleation in the mouse, *J. Comp. Neurol.* **255:**97–109.

Green, W. L., 1978, Mechanism of action of antithyroid components, in *The Thyroid* (S. C. Werner and S. H. Ingbar, eds.), pp. 41–51, Harper & Row, New York.

Grigonis, A. M., Pearson, H. E., and Murphy, E. H., 1986, The effects of neonatal monocular enucleation on the organization of ipsilateral and contralateral retinothalamic projections in the rabbit, *Dev. Brain Res.* **29:**9–19.

Grobstein, P., and Comer, C., 1977, Post-metamorphic eye migration in *Rana* and *Xenopus, Nature (London)* **269:**54–56.

Guillery, R. W., 1969, An abnormal retinogeniculate projection in Siamese cats, *Brain Res.* **14:**739–741.

Guillery, R. W., 1982, The optic chiasm of the vertebrate brain, *Contrib. Sensory Physiol.* **7:**39–73.

Guillery, R. W., and Walsh, C., 1987, Changing glial organization relates to changing fiber order in the developing optic nerve of ferrets, *J. Comp. Neurol.* **265:**203–217.

Guillery, R. W., Scott, B., Cattnach, M., and Deol, M., 1973, Genetic mechanisms determining the central visual pathways of mice, *Science* **179:**1014–1016.

Guillery, R. W., Casagrande, V. A., and Oberdorfer, M. D., 1974, Congenitally abnormal vision in Siamese cats, *Nature (London)* **252:**195–199.

Harris, W. A., 1986, Homing behavior of axons in the embryonic vertebrate brain, *Nature (London)* **320:**266–269.

Harting, J. K., and Guillery, R. W., 1976, Organization of retinocollicular pathways in the cat, *J. Comp. Neurol.* **166:**133–144.

Himstedt, W., and Manteuffel, G., 1985, Retinal projections in the caecilian *Ichthyophis kohtaoensis, Cell Tissue Res.* **239:**689–692.

Hishinuma, A., and Hildebrand, J. G., 1987, Brain organization of *Manduca sexta* studied with monoclonal antibodies, *Soc. Neurosci. Abstr.* **13:**1515.

Hockfield, S., and MacKay, R., 1985, Identification of major cell classes in the developing mammalian nervous system, *J. Neuroscience* **5:**3310–3328.

Hockfield, S., 1987, A monoclonal antibody to a unique cerebellar neuron generated by immunosuppression and rapid immunization, *Science* **237:**67–70.

Holcombe, V., and Guillery, R. W., 1984, The organization of retinal maps within the dorsal and ventral lateral geniculate nuclei of the rabbit, *J. Comp. Neurol.* **225:**469–491.

Hollyday, M., and Grobstein, P., 1981, Of limbs and eyes and neuronal connectivity, in *Studies in Developmental Neurobiology* (W. M. Cowan, ed.), Oxford University Press, New York.

Hollyfield, J. G., 1971, Differential growth of the neural retina of *Xenopus laevis* larvae, *Dev. Biol.* **24:**264–286.

Holt, C. E., and Harris, W. A., 1983, Order in the initial retinotectal map in *Xenopus:* A new technique for labeling growing nerve fibers, *Nature (London)* **301:**150–151.

Hoskins, S. G., 1986, Control of the development of the ipsilateral retinothalamic projection in *Xenopus laevis* by thyroxine: Results and speculation, *J. Neurobiol.* **17:**203–229.

Hoskins, S. G., 1987, Distribution of ipsilaterally-projecting axons in the optic nerve of *Xenopus, Soc. Neurosci. Abstr.* **13:**949.

Hoskins, S. G., and Grobstein, P., 1980, Distribution of ipsilaterally and contralaterally projecting retinal ganglion cells in *Xenopus* and its re-establishment following optic nerve section, *Soc. Neurosci. Abstr.* **6:**647.

Hoskins, S. G., and Grobstein, P., 1982, Retinal histogenesis and the development of the ipsilateral retinothalamic projection in *Xenopus, Soc. Neurosci. Abstr.* **8:**513.

Hoskins, S. G., and Grobstein, P., 1984, Induction of the ipsilateral retinothalamic projection in *X. laevis* by thyroxine, *Nature (London)* **307:**730–733.

Hoskins, S. G., and Grobstein, P., 1985a, Development of the ipsilateral retinothalamic projection in the frog *Xenopus laevis:* I. Retinal distribution of ipsilaterally projecting cells in normal and experimentally manipulated frogs, *J. Neurosci.* **5:**911–919.

Hoskins, S. G., and Grobstein, P., 1985b, Development of the ipsilateral retinothalamic projection in the frog *Xenopus laevis:* II. Ingrowth of optic nerve fibers and production of ipsilaterally projecting retinal ganglion cells, *J. Neurosci.* **5:**920–929.

Hoskins, S. G., and Grobstein, P., 1985c, Development of the ipsilateral retinothalamic projection in the frog *Xenopus laevis:* III. Role of thyroxine, *J. Neurosci.* **5:**930–940.

Hsiao, K., 1984, Bilateral branching contributes minimally to the enhanced ipsilateral projection in monocular Syrian golden hamsters, *J. Neurosci.* **4:**368–373.

Hughes, A., 1977, The topography of vision in mammals of contrasting life style: Comparative optics and retinal organization, in *Handbook of Sensory Physiology*, Vol. II, Part 5. (F. Crescitelli, ed.), pp. 613–756, Springer-Verlag, New York.

Insausti, R., Blakemore, C., and Cowan, W. M., 1984, Ganglion cell death during development of ipsilateral retino-collicular projection in golden hamster, *Nature (London)* **308:**362–365.

Jacobson, M., 1976, Histogenesis of the retina of the clawed frog with implications for the pattern of development of retinotectal connections, *Brain Res.* **127:**55–67.

Jeffery, G., 1984, Retinal ganglion cell death and terminal field retraction in the developing rodent visual system, *Dev. Brain Res.* **13**:81–96.

Jeffery, G., and Perry, V. H., 1982, Evidence for ganglion cell death during development of the ipsilateral retinal projection in the rat, *Dev. Brain Res.* **2**:176–180.

Kahn, A. J., 1973, Ganglion cell formation in the chick neural retina, *Brain Res.* **63**:285–290.

Kennard, C., 1981, Factors involved in the development of ipsilateral retinothalamic projections in *Xenopus laevis, J. Embryol. Exp. Morphol.* **65**:199–217.

Khalil, E., and Szekely, G., 1976, The development of the ipsilateral retinothalamic projections in the *Xenopus* toad, *Acta Biol. Acad. Sci. Hung.* **27**:253–260.

Kliot, M., and Shatz, C. J., 1985, Abnormal development of the retinogeniculate projection in Siamese cats, *J. Neurosci.* **5**:2641–2653.

Kollros, J., 1981, Transitions in the nervous system during amphibian metamorphosis, in *Metamorphosis: A Problem in Developmental Biology* (L. Gilbert and E. Frieden, eds.), p. 445–459, Plenum Press, New York.

Lam, K., Sefton, A. H., and Bennett, M. R., 1982, Loss of axons from the optic nerve of the rat during early postnatal development, *Dev. Brain Res.* **3**:487–491.

Land, P. W., 1987, Dependence of cytochrome oxidase activity in the rat lateral geniculate nucleus on retinal innervation, *J. Comp. Neurol.* **262**:78–89.

Land, P. W., and Lund, R. D., 1979, Development of the rat's uncrossed retinotectal pathway and its relation to plasticity studies, *Science* **205**:698–699.

Lauder, J. M., 1983, Hormonal and humoral influences on brain development, *Psychoneuroendocrinology* **8**:121–155.

Lazar, G., 1971, The projection of the retinal quadrants on the optic centres in the frog, *Acta Morphol. Acad. Sci. Hung.* **19**:325–384.

Leloup, J., and Buscaglia, M., 1977, La triiodothyronine, hormone de la metamorphose des amphibiens, *C. R. Acad. Sci. Paris* **284**:2261–2263.

Levine, R., 1980, An autoradiographic study of the retinal projection in *Xenopus laevis,* with comparisons to *Rana, J. Comp. Neurol.* **189**:1–29.

Lund, R. D., 1978, *Development and Plasticity of the Brain,* Oxford University Press, New York.

Lund, R. D., Cunningham, T. J., and Lund, J. S., 1973, Modified optic projections after unilateral eye removal in young rats, *Brain Behav. Evol.* **8**:51–72.

Lund, R. D., Lund, J. S., and Wise, R. P., 1974, The organization of the retinal projection to the dorsal lateral geniculate nucleus in pigmented and albino rats, *J. Comp. Neurol.* **158**:383–404.

MacLean, N., and Turner, M., 1976, Adult hemoglobin in developmentally retarded tadpoles of *Xenopus laevis, J. Embryol. Exp. Morphol.* **35**:261–266.

Maggs, A., and Scholes, J., 1986, Glial domains and nerve fiber patterns in the fish retinotectal pathway, *J. Neurosci.* **6**:424–438.

Morgan, J. E., Henderson, Z., and Thompson, I. D., 1987, Retinal decussation patterns in pigmented and albino ferrets, *Neuroscience* **20**:519–535.

Ng, A. Y. K., and Stone, J., 1982, The optic nerve of the cat: Appearance and loss of axons during normal development, *Dev. Brain Res.* **5**:263–271.

Nieuwkoop, P. D., and Faber, J., 1967, *Normal Table of Xenopus laevis (daudin),* North-Holland, Amsterdam.

O'Leary, D. D. M., Gerfen, C. R., and Cowan, W. M., 1983, The development and restriction of the ipsilateral retinofugal projection in the chick, *Dev. Brain Res.* **10**:93–109.

Ostrach, L. H., Crabtree, J. W., and Chow, K. L., 1986, The ipsilateral retinocollicular projection in the rabbit: An autoradiographic study of postnatal development and effects of unilateral enucleation, *J. Comp. Neurol.* **254**:369–381.

Perry, V. H., Henderson, Z., and Linden, R., 1983, Postnatal changes in retinal ganglion cell and optic axon populations in the pigmented rat, *J. Comp. Neurol.* **219**:356–368.

Pettigrew, J. D., 1986, Flying primates? Megabats have the advanced pathway from eye to midbrain, *Science* **231**:1304–1306.

Polyak, S., 1957, *The Vertebrate Visual System* (H. Kluver, ed.), University of Chicago Press, Chicago.

Provis, J. M., and Watson, C. R., 1981, The distribution of ipsilaterally and contralaterally projecting ganglion cells in the retina of the pigmented rabbit, *Exp. Brain Res.* **44:**82–92.

Rakic, P., 1975, Prenatal genesis of connections subserving ocular dominance in the rhesus monkey, *Nature (London)* **261:**467–471.

Rakic, P., 1977, Prenatal development of the visual system in the rhesus monkey, *Philos. Trans. R. Soc. London Ser. B* **278:**245–260.

Rakic, P., and Riley, K. P., 1983a, Overproduction and elimination of retinal axons in the fetal rhesus monkey, *Science* **219:**1441–1444.

Rakic, P., and Riley, K. P., 1983b, Regulation of axon number in primate optic nerve by prenatal binocular competition, *Nature (London)* **305:**135–137.

Reese, B. E., and Jeffery, G., 1983, Crossed and uncrossed visual topography in dorsal lateral geniculate nucleus of the pigmented rat, *J. Neurophysiol.* **49:**877–885.

Rettig, G., and Roth, G., 1982, Afferent visual projections in three species of lungless salamanders (family *Plethodontidae*), *Neurosci. Lett.* **31:**221–224.

Rettig, G., Fritzsch, B., and Himstedt, W., 1981, Development of retinofugal areas in the brain of the alpine newt, *Triturus alpestris, Anat. Embryol.* **162:**163–171.

Rodieck, R. W., 1979, Visual pathways, *Annu. Rev. Neurosci.* **2:**193–225.

Samuels, H., Perlman, A., Raaka, B., and Stanley, F., 1983, Thyroid hormone receptor synthesis and degradation and interaction with chromatin components, in *Molecular Basis of Thyroid Hormone Action* (J. Oppenheimer and H. Samuels, eds.), Chap. 4, pp. 100–136, Academic Press, New York.

Sanderson, K. J., 1971, The projection of the visual field to the lateral geniculate and medial interlaminar nuclei in the cat, *J. Comp. Neurol.* **143:**101–118.

Sanderson, K. J., Guillery, R. W., and Shackelford, R. M., 1974, Congenitally abnormal visual pathways in mink (*Mustela vison*) with reduced retinal pigment, *J. Comp. Neurol.* **154:**225–248.

Sanes, J., Rubenstein, J. L., and Nicolas, J. F., 1986, Use of a recombinant retrovirus to study post-implantation cell lineage in mouse embryos, *EMBO J.* **5:**3133–3142.

Schmidt, J. T., 1978, Retinal fibers alter tectal positional markers during the expansion of the half retinal projection in goldfish, *J. Comp. Neurol.* **177:**279–300.

Sengelaub, D. R., and Finlay, B. L., 1981, Early removal of one eye reduces normally occurring cell death in the remaining eye, *Science* **213:**573–574.

Sengelaub, D. R., Dolan, R. P., and Finlay, B. L., 1986, Cell generation, death and retinal growth in the development of the hamster retinal ganglion cell layer, *J. Comp. Neurol.* **246:**527–543.

Shatz, C. J., 1983, The prenatal development of the cat's retinogeniculate pathway, *J. Neurosci.* **3:**482–499.

Shatz, C. J., and Sretavan, D. W., 1986, Interactions between retinal ganglion cells during the development of the mammalian visual system, *Annu. Rev. Neurosci.* **9:**171–207.

Shaw, G., and Bray, D., 1977, Movement and extension of isolated growth cones, *Exp. Cell Res.* **104:**55–62.

Silver, J., 1984, Studies on the factors that govern directionality of axonal growth in the embryonic optic nerve and at the optic chiasm of mice, *J. Comp. Neurol.* **223:**238–251.

Silver, J., and Sidman, R. L., 1980, A mechanism for the guidance and topographic patterning of retinal ganglion cell axons, *J. Comp. Neurol.* **189:**101–111.

Silver, J., and Sapiro, J., 1981, Axonal guidance during development of the optic nerve: The role of pigmented epithelia and other extrinsic factors, *J. Comp. Neurol.* **202:**521–538.

So, K.-F., Schneider, G. E., and Frost, D. O., 1978, Postnatal development of retinal projections to the lateral geniculate body in Syrian hamsters, *Brain Res.* **142:**343–352.

Sperry, R. W., 1944, Optic nerve regeneration with recovery of vision in anurans, *J. Neurophysiol.* **7:**57–69.

Sperry, R. W., 1945, Restoration of vision after crossing of optic nerves and after contralateral transplantation of eye, *J. Neurophysiol.* **8:**15–28.

Sperry, R. W., 1963, Chemoaffinity in the orderly growth of nerve fiber patterns and connections, *Proc. Natl. Acad. Sci. U.S.A.* **50:**703–710.

Sretavan, D. W., and Shatz, C. J., 1986, Prenatal development of cat retinogeniculate axon arbors in the absence of binocular interactions, *J. Neurosci.* **6:**990–1003.

Stone, J., 1966, The naso-temporal division of the cat's retina, *J. Comp. Neurol.* **124:**337–352.

Stone, J., and Fukuda, Y., 1974, The naso-temporal division of the cat's retina re-examined in terms of Y-, X-, and W-cells, *J. Comp. Neurol.* **155:**377–394.

Stone, J., Leicester, J., and Sherman, S. M., 1973, The naso-temporal division of the monkey's retina, *J. Comp. Neurol.* **150:**333–348.

Stone, J., Rowe, M. H., and Campion, J. H., 1978, The nasotemporal division of retina in the Siamese cat, *J. Comp. Neurol.* **180:**783–798.

Stone, J., Dreher, B., and Rapaport, D. H., 1984, Development of visual pathways in mammals, *Neurol. Neurobiol.* **9.**

Straznicky, K., and Gaze, R. M., 1971, The growth of the retina of *Xenopus laevis.* An autoradiographic study, *J. Embryol. Exp. Morphol.* **26:**67–79.

Straznicky, K., and Tay, D., 1977, Retinal growth in double dorsal and double ventral eyes in *Xenopus, J. Embryol. Exp. Morphol.* **26:**175–185.

Strongin, A. C., and Guillery, R. W., 1981, The distribution of melanin in the developing optic stalk and its relation to cellular degeneration, *J. Neurosci.* **1:**1193–1209.

Thompson, I. D., 1979, Changes in the uncrossed retinotectal projection after removal of the other eye at birth, *Nature (London)* **279:**63–66.

Thompson, C. C., Weinberger, C., Lebo, R., and Evans, R. M., 1987, Identification of a novel thyroid hormone receptor expressed in the mammalian central nervous system, *Science* **237:**1610–1614.

Turner, D., and Cepko, C., 1987, A common progenitor for neurons and glia persists in rat retina late in development, *Nature* **328:**131–136.

Varon, S., Manthorpe, M., and Williams, L., 1984, Neuronotrophic and neurite-promoting factors and their clinical potentials, *Dev. Neurosci.* **6:**73–100.

Wakakuwa, K., Washida, A., and Fukuda, Y., 1985a, Ipsilaterally projecting retinal ganglion cells in the eastern chipmunk (*Tamias sibiricus asiaticus*), *Neurosci. Lett.* **55:**219–224.

Wakakuwa, K., Washida, A., and Fukuda, Y., 1985b, Distribution and soma size of ganglion cells in the retina of the eastern chipmunk (*Tamias sibiricus asiaticus*), *Vision Res.* **25:**877–885.

Walsh, C., and Polley, E., 1985, The topography of ganglion cell production in the cat's retina, *J. Neurosci.* **5:**741–750.

Walsh, C., Polley, E., Hickey, T., Guillery, R. W., 1983, Generation of cat retinal ganglion cells in relation to central pathways, *Nature (London)* **302:**611–614.

Wassle, H., 1982, Morphological types and central projections of ganglion cells in the cat retina, in *Progress in Retinal Research* (N. Osborne and G. Chandler, eds.), pp. 125–152, Pergamon Press, Oxford.

Weinberger, C., Thompson, C. C., Ong, E., Lebo, R., Gruol, D., and Evans, R. J., 1986, The c-*erb*-A gene encodes a thyroid hormone receptor, *Nature (London)* **324:**641–646.

Weiss, P. A., 1960, in *Analysis of Development* (B. Willier, P. A. Weiss, and V. Hamburger, eds.), pp. 346–401, W. B. Saunders, Philadelphia.

Wikler, K. C., Raabe, J. I., and Finlay, B. L., 1985, Temporal retina is preferentially represented in the early retinotectal projection in the hamster, *Dev. Brain Res.* **21:**152–155.

Williams, R. W., Bastiani, M. J., and Chalupa, L. M., 1983, Loss of axons in the cat optic nerve following fetal unilateral enucleation: An electron microscopic analysis, *J. Neurosci.* **3:**133–144.

Wilt, F. H., 1959, The organ specific action of thyroxine in visual pigment differentiation, *J. Embryol. Exp. Morphol.* **7:**556–563.

Young, R. W., 1984, Cell death during differentiation of the retina in the mouse, *J. Comp. Neurol.* **229:**362–373.

# Dendritic Interactions between Cell Populations in the Developing Retina

7

## V. Hugh Perry

### 7.1. INTRODUCTION

It has been known for some time that the cones of different types in the retinas of fish and birds are distributed in a very precise fashion, forming an almost crystal-like pattern. The first to draw attention to the regular distribution of cells in the inner layers of the mammalian retina was Wässle and Reimann (1978). They showed from an analysis of the distribution of the nearest-neighboring cells that the cell bodies of cat alpha ganglion cells and A-type horizontal cells were each arranged in nonrandom patterns. Subsequently, there have been a number of studies demonstrating that other cell types are arranged in nonrandom distributions (e.g., Wässle *et al.*, 1981c; Tauchi and Masland, 1984; Vaney, 1986). The regular distribution of the cell bodies appears to go hand in hand with the relatively uniform coverage of the retina by the dendritic territories of a particular cell type (Wässle *et al.*, 1981a,b; Tauchi and Masland, 1984). The term coverage, the number of cells overlapping any given point on the retina, is readily computed as the product of the dendritic area and local density of a given cell type (Cleland *et al.*, 1975). The functional significance of the regular spacing of the cell bodies and the uniform coverage of the retina is clear. If the retina is to sample the visual world faithfully and convey the information to the brain, cells dealing with different aspects of the visual scene should be distributed so as to leave no holes in our perceptual world. The regular spacing of the cell bodies and relatively uniform coverage ensure this in a most economic fashion.

From a developmental point of view the problem is conceptually simple. What are the processes or mechanisms operating during development which produce nonrandom distributions of cell bodies and the uniform coverage of

**V. Hugh Perry** • Department of Experimental Psychology, University of Oxford, Oxford OX1 3UD, England.

the retina by the dendritic territories of these cells? The answer to this problem is likely to be complex, involving many different cellular interactions.

### 7.1.1. Regulation of Cellular Distribution

The nonrandom distributions of the cell bodies of different cell types could all be specified in the genome. For example, a single or some small number of progenitor cell types (Turner and Cepko, 1987), which give rise to all the various cell populations in the retina, may be distributed in a nonrandom pattern at the ventricular surface. Since cells in the retina only migrate in the radial direction and there is no evidence for lateral migration (Sidman, 1961; Turner and Cepko, 1987), the mosaic arises as a consequence of the progenitor distribution. However, in many parts of the nervous system the adult cell numbers and patterns of connectivity develop from an original number of cells which is much greater than that found in the adult, and the retina is no exception. Cell death is a major developmental event in the ganglion cell and inner nuclear layer populations (Lam *et al.,* 1982; Potts *et al.,* 1982; Perry *et al.,* 1983; Williams *et al.,* 1986; Young, 1986; Beazley *et al.,* 1987). From this larger immature population the regular patterns are generated and thus it seems unlikely that the distributions found in the adult are entirely predetermined. One of the factors that regulates cell death is the size of the available target (Oppenheim, 1981). It has been argued elsewhere that the interactions between axon terminals of ganglion cells at the subcortical visual centers are unlikely to be responsible for the generation of ganglion cell mosaics (Perry, 1984) since the axons lie outside the retina. We shall examine the evidence that an intraretinal mechanism operates during development to regulate the naturally occurring degeneration of ganglion cells. This intraretinal mechanism involves interactions between the dendrites of neighboring cells.

### 7.1.2. Regulation of Cellular Morphology

The second part of our problem concerns the factors that regulate cell shape and size. There exists in the retina a great variety of different cell forms and it is very probable that a large component of the form is intrinsically specified. It is difficult to imagine how an intrinsic program might regulate the precise size and shape of each cell so that the surface of the retina is covered in a uniform fashion by a given cell type. This is of course particularly problematic if some cells degenerate during development as a consequence of extrinsic interactions regulating cell numbers. We shall examine the evidence that the dendrites of neighboring cells in the retina interact so as to regulate the size and shape of the dendritic tree. The evidence shows that local events in the environment of a developing dendrite have a major influence on its subsequent growth. The dendritic tree is not simply a passive recipient of its afferent input but interactions between the dendrites of neighboring cells play an important part in the generation of a complex neural network such as the retina.

## 7.2. DENDRITIC COMPETITION: THE PHENOMENON

Studies on the distribution of alpha ganglion cells in the cat retina demonstrated that their cell bodies were arranged in a nonrandom pattern. Furthermore, the dendritic territories of the alpha cells were arranged so that the entire retina is covered by this cell type (Wässle *et al.*, 1981a,b). The authors suggested that there might be some sort of interaction between neighboring cells that would regulate dendritic growth. Evidence that neighboring ganglion cells influence each other's territories during development came from experiments in which a small lesion was made in the retina of newborn rats close to the optic disk. This lesion, which transects the axons of ganglion cells, results in the rapid retrograde degeneration of ganglion cells in a sector of the retina peripheral to the lesion. It was believed that despite the loss of ganglion cells the inner nuclear layer, the origin of the ganglion cell afferents, was left largely intact (see below). When the animals were mature, the remaining ganglion cells were retrogradely filled with horseradish peroxidase to reveal the cell body and a substantial portion of the dendritic tree. Those ganglion cells surrounding the region devoid of ganglion cells were now found to have their dendrites directed into the ganglion-cell-free region. Thus, it was clear that the direction of growth of dendrites was influenced by neighboring cells (Perry and Linden, 1982). We pointed out that these observations could be a result of either an attraction of the dendrites toward the depleted area and/or repulsion by neighbors on the other side and suggested that during development ganglion cell dendrites "compete" for their afferents. This competition could be for synaptic contacts or for trophic factor(s) released by the afferents. It is important to note that "dendritic competition" is a description of the phenomenology and not a mechanism. We have little idea as to the actual cell biology underlying the change in the orientation of the dendrites.

Similar experiments have been performed on cat retina. A lesion made to the retina of a kitten at 20 days of age or less results in a dramatic rearrangement of ganglion cell dendritic trees similar to that described above (Eysel *et al.*, 1985). The major difference between this study and that in the rat was that by using a reduced silver stain to identify the alpha cells and stain the whole dendritic tree, the authors assessed whether the rearrangement was accompanied by an increase in the size of the dendritic field area. No change was found in the dendritic field area and the authors suggested that the size of a ganglion cell was intrinsically specified but this now seems unlikely (Kirby and Chalupa, 1986; see Chapter 8 in this volume).

These experiments show that the shape of the dendritic tree is influenced by neighboring cells but they do not shed any light on how the regular spacing of the cell bodies might develop. In a separate series of experiments, it was shown that cell death was influenced by the number of neighboring cells in a situation where these cells did not share a common terminal field (Linden and Perry, 1982).

In adult rats the ipsilateral projection from the retina to the brain arises in the crescent-shaped region in the temporal retina, the temporal crescent (Cowey and Perry, 1979; see Chapter 6 in this volume). In newborn rats the

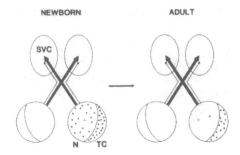

**Figure 7.1.** A schematic figure to show how the distribution of the ipsilaterally projecting ganglion cells (•) in the retina differ in newborn and adult rats. TC, temporal crescent; N, nasal retina; SVC, subcortical visual centers.

ipsilateral projection comes from a greater number of ganglion cells than are present in the adult. These cells lie not only in the temporal crescent but also in the nasal retina. Many of the immature ipsilaterally projecting cells degenerate over the first five postnatal days (Fig. 7.1; Jeffery, 1984). Removal of one eye on the day of birth prevents the normal degeneration (rescues) of many of the immature ipsilaterally projecting ganglion cells from the remaining eye (Fig. 7.2A; Sengelaub and Finlay, 1981; Jeffery and Perry, 1982). This has been interpreted as evidence for competition between axon terminals from the two eyes for postsynaptic sites in subcortical visual centers. Transection of one optic tract in a newborn rodent results in the rapid degeneration of ganglion cells in the eye contralateral to the lesion. Following such a lesion it was found that the number of ganglion cells projecting ipsilaterally from the eye, contralateral to the lesion, was greater than that found in the normal adult, particularly in the nasal retina (Fig. 7.2B). This partial preservation of

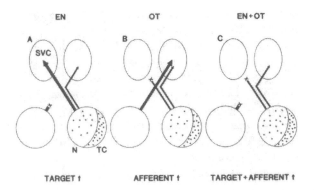

**Figure 7.2.** Schematic figures to show how different manipulations performed on the day of birth in a rat influence the final distribution of ipsilaterally projecting ganglion cells found at adulthood. (A) Monocular enucleation, EN. (B) Unilateral optic tract section, OT. (C) Combined monocular enucleation and unilateral optic tract section, EN + OT. Other symbols as in Fig. 7.1. See text for further details.

the immature pattern occurred despite the fact that the projection from the other eye was intact and there was no increase in the available synaptic space in the regions of the terminals (Linden and Perry, 1982). We suggested that some cells had survived the normal cell death process. A similar result was described by Jacobs *et al.* (1984) in the cat after neonatal optic tract section.

We interpreted our results in the following way. In a normally developing retina there is a larger number of ganglion cells when compared to the adult. As the dendritic tree of a ganglion cell grows it has to gain access to its afferent supply, amacrine and bipolar cell contacts, in a sufficient quantity (obviously unknown) such that its survival is assured. Cells that fail to gain access degenerate. We suggested that there was a "competitive" interaction between the dendrites of neighboring ganglion cells for their afferents and the outcome of this competition would determine cell survival. An optic tract lesion removes a large number of cells from the contralateral retina at a time when the dendritic trees are rapidly growing (Perry and Walker, 1980; Maslim *et al.*, 1986). The ipsilaterally projecting cells in the nasal retina of a newborn rat now find themselves in a region of the retina where they have few neighbors and thus have access to a large territory in the retina and presumably a larger number of afferents. Cells that would normally have degenerated survive due to the loss of competition for available afferents. We do not know whether the interaction with the afferents actually involves contacts or uptake of a locally diffusible growth factor. It should be remembered again that "dendritic competition" in this instance is little more than a convenient shorthand for describing the observation that cell death is regulated by some mechanism(s) that involves interactions between dendrites of neighboring cells.

## 7.3. DIFFERENTIAL EFFECTS OF AFFERENTS AND TARGETS ON CELL SURVIVAL

The idea that afferents play a role in cell survival during the period of normal cell death was suggested by Cunningham (1982) and evidence that this is indeed the case is growing (Cunningham *et al.*, 1979; Linden and Perry, 1982; Okado and Oppenheim, 1984; Clarke, 1985; Finlay *et al.*, 1986; Furber *et al.*, 1987). It is well established that the target plays an important role in the survival of developing neurons (Oppenheim, 1981) and the ganglion cells are no exception. If, however, the afferents and targets affect ganglion cell death in different ways, then it should be possible to demonstrate this.

Linden and Serfaty (1985) have shown that when the removal of one eye is combined with an optic tract lesion on the same side of the brain, the ipsilateral projection from the remaining eye to the intact hemisphere is larger than following either lesion alone (Fig. 7.2C). Increasing the target and the afferent supply was found to be additive but not in a simple linear fashion. They also showed that the increase in afferent supply following an optic tract lesion rescued a different population of cells from that following an increase in the target by unilateral enucleation. Both lesions rescued cells in nasal retina with small soma sizes, but in optic tract sectioned animals there was also a population of cells with large cell bodies, type 1 ganglion cells (Perry, 1979), rescued. These experiments suggest that different mechanisms are operating after the two different manipulations. They also raise the question as to why an increase in the target prevents the cell death of small ganglion cells but an increase in afferent supply prevents the death of both small and large ganglion cells.

We have suggested that there may be a temporal sequence of cell death for each class of ganglion cell, which occurs in the same sequence in which the ganglion cells are generated (Jacobs *et al.*, 1984). In the studies of Okado and Oppenheim (1984) and Clarke (1985) on the survival of motoneurons and isthmo-optic nucleus neurons, respectively, the effect of afferent depletion was to increase cell loss during the latter part of the normal cell death period. Thus, we might speculate that the period of cell death for a particular class of ganglion cell is not only related to its birthdate but also has two phases, a target-dependent phase and an afferent-dependent phase. We can interpret the results of Linden and Serfaty (1985) in the following way. Monocular enucleation on the day of birth increases the target field at a time when the later born small cells (Sidman, 1961; Dräger, 1985) are still in a target-dependent stage; there is no increase in the available afferents and only small cells survive. On the day of birth the earlier generated large cells (Sidman, 1961; Dräger, 1985) are in their afferent-dependent stage and are only rescued by a manipulation that increases the afferent supply, namely, by reducing the number of neighboring cells. The fact that small cells are also rescued by the latter manipulation may reflect overlap in birthdates of different cell types since there is not a perfect segregation of the birthdates of large and small cells. If this hypothesis is correct, we would expect to find evidence that the large ganglion cells are not rescued by a tract lesion made at 3 days after birth, and this appears to be the case (V. H. Perry and J. Duxbury, unpublished observations). It would be of some interest to look at the age-dependent effects of these two lesions in a species where development is more protracted than in a rodent and the different phases may be more clearly delineated.

The idea that ganglion cell survival depends on two different phases—a target phase and an afferent phase—is consistent with the results of Johnson *et al.* (1986), who have shown that ganglion cells isolated from embryonic rats can be maintained in culture with purified brain-derived neurotrophic factor (BDNF) for at least 4 days. In contrast, ganglion cells from newborn animals can only be transiently sustained by BDNF. The possibility that ganglion cells require a growth factor derived from both their targets and afferents would be consistent with the evidence that two different factors, one derived from the central target and the other from the periphery, regulate the survival of sensory neurons (Davies *et al.*, 1986). A recent report by Armson *et al.* (1987) suggests that the situation in the retina may be more complex. They have shown that Müller cell conditioned media supports the survival of retinal ganglion cells from 17 day embryos more effectively than superior colliculus extracts. However, by postnatal day 12 the superior colliculus extracts promote cell survival better than the Müller cell conditioned media. The interpretation of this experiment is complicated by the fact that the time of differentiation of the Müller cells in rat is unknown but many are generated postnatally (Young, 1985; Turner and Cepko, 1987).

Another interesting and rather different approach to the importance of afferents in ganglion cell survival comes from studies on the role of electrical activity for ganglion cell survival *in vitro* (Lipton, 1986). In cultures of retinal cells, those ganglion cells that form clusters with other cell types show spontaneous electrical activity while those growing in isolation rarely if ever have any spontaneous activity. This spontaneous activity requires functional synaptic contact with afferents since it is abolished in culture medium containing low calcium and high magnesium concentrations. Following the addition of tetrodotoxin (TTX) to the cultures (which blocks the sodium channels and thus the spontaneous action potentials), virtually all the spontaneously active ganglion cells in clusters died within 24 hr. The electrically inactive ganglion cells growing in isolation were protected from the effects of TTX. TTX-induced cell death was found in clustered ganglion cells taken from animals from 2–10 days of age but not from retinas from animals 11–13 days of age. The cell death which followed the addition of TTX to the cultures could be overridden by culturing the ganglion cells in conditioned medium. The conditioned medium was taken from the second day in culture of retinal cells from 7 and 8 day old animals. These results suggest that afferent-dependent electrical activity is necessary for the survival of some ganglion cells during a critical period of postnatal development. Furthermore, during this period an intraretinal trophic factor is produced in the presence of electrical activity. This trophic factor can promote ganglion cell survival in the absence of a normal postsynaptic target.

A final piece of evidence that suggests that afferents in part mediate the survival of ganglion cells comes from experiments in which embryonic retinas were transplanted to a site distant from their normal targets. Retinas from embryonic day 12 or 13 mice were transplanted to the cortex of newborn rats. When examined up to 6 weeks later, well past the period of naturally occurring cell death for the mouse ganglion cells (Young, 1986), the retina had formed a relatively normal laminar pattern and there were a number of ganglion cells surviving. We found that there were no axons leaving the retina and growing into the host cortex. Thus, in the absence of their target but with their afferent supply relatively intact, the ganglion cells had survived, consistent with the *in vitro* results of Lipton (1986). It could of course be argued that there is a growth factor in the cortex which is the same as that found in subcortical visual structures and this promotes the survival of the ganglion cells, but if that is so then we would have expected the ganglion cells to grow out toward it. In other experiments where pieces of tectum were cotransplanted with the retina, the ganglion cell axons leave the transplant, traverse the cortex, and innervate the transplanted tectum (Sefton *et al.*, 1987), thus showing that ganglion cell axons will indeed grow toward a target source.

These experiments further demonstrate that ganglion cells are not destined to die at a particular time if they do not find their target and can survive for long periods with only their afferent supply intact. The fact that ganglion cells degenerate in larger numbers than normal when their major target, the superior colliculus, is removed after they have innervated it (Perry and Cowey, 1982; Carpenter *et al.*, 1986) suggests that they develop a dependency on the target once they have met it, but if never confronted with it they do not develop this dependency.

Thus, there is good evidence that not only does the postsynaptic target play a role in ganglion cell survival but so too do the afferents. Parallel studies in other systems support this. During the appropriate period of development, an increase in the available afferents prevents cell death (Cunningham *et al.*, 1979; Linden and Perry, 1982) and a decrease enhances cell death (Okado and Oppenheim, 1984; Clarke, 1985; Furber *et al.*, 1987). Combining increases in both the target and afferents rescues more cells than either manipulation alone (Linden and Serfaty, 1985) and removal of the target and afferents induces more cell death than removal of either alone (Furber *et al.*, 1987). A working hypothesis is that the ganglion cells become dependent on their afferents during the latter part of the cell death period.

## 7.4. THE AFFERENTS IN THE INNER NUCLEAR LAYER

Before we further examine the factors that regulate the survival and form of the ganglion cells, we should take a more detailed look at our system and one of the underlying assumptions. We have argued that changes in the shape of dendritic territories and the survival of ganglion cells are related to interactions between dendrites as they grow and gain access to their potential afferent supply. These manipulations have largely involved the depletion of ganglion cells. We need to know how the afferent populations, the amacrine and bipolar cells, might be changed by the loss of ganglion cells. It is quite possible that the afferents have also been depleted as a result of retrograde transneuronal degeneration such as happens with ganglion cells following visual cortex lesions in primates and cats (Van Buren, 1963; Cowey, 1974; Tong *et al.*, 1982). The retrograde transneuronal effects are particularly prominent in neonates (Dineen and Hendrickson, 1981; Tong *et al.*, 1982; Payne *et al.*, 1984). The loss of a substantial number of afferents from the inner nuclear layer would alter the interpretation of our results.

To discover whether the development of cells in the inner nuclear layer shows the same sort of target dependency as has been described for ganglion cells, we have examined how optic nerve section on the day of birth affects the subsequent development of the other retinal cells (Perry, 1981; Osborne and Perry, 1985; Beazley *et al.*, 1987). Optic nerve section on the day of birth results in the rapid degeneration of the ganglion cells over the next 48 hr (Miller and Oberdorfer, 1981; Beazley *et al.*, 1987), and this is prior to the time that significant numbers of synapses are formed in the inner plexiform layer (Wiedman and Kuwabara, 1968).

Eayrs (1952) noted that optic nerve transection in neonatal rats had little effect on the development of the retinal lamination and the width of the inner nuclear layer. In a subsequent analysis of the displaced amacrine cell population in the rat, it was found that the final number of displaced amacrine cells was not influenced by optic nerve transection on the day of birth. The morphology of several classes of amacrine cell found in the ganglion cell layer was similar to that found in normal animals (Perry, 1981). In the normal retina, cell death proceeds across the retina in an inner to outer sequence. The maximal number of dying cells in the ganglion cell layer is found on the day of birth. The maximal cell death for the amacrine and bipolar cells in the inner nuclear layer is found on the seventh and tenth postnatal days, respectively. In the inner nuclear layer there is a weak center–periphery gradient. Very few dying cells were found in the outer nuclear layer at any age. It was previously suggested that the wave of cell death passing from the inner to outer retina might reflect a matching of the number of inner nuclear layer cells to the ganglion cell population (Perry, 1984). This is clearly wrong. Following optic nerve transection, the time course and amount of cell death are not significantly different from normal.

To assess further the effects of optic nerve transection, we examined several populations of amacrine cells. Through the use of immunocytochemistry, the amacrine cells containing tyrosine-hydroxylase, choline acetyltransferase, and substance P were labeled. Qualitatively, there was no difference in the distribution of the cell bodies and the levels of dendritic branching when allowances were made for the shrinkage of the inner plexiform layer. A biochemical analysis of the choline acetyltransferase and glutamate decarboxylase content of the retina after optic nerve section revealed a reduction of about 10% when compared to normal and the dopamine content was reduced by about 15% (Osborne and Perry, 1985).

The evidence reviewed above shows that ganglion cells have little influence on several aspects of inner nuclear layer development. In the rat, ganglion cell dendrites represent about 30% of the postsynaptic targets for amacrine cells and also provide a major postsynaptic target for bipolar cells (Sosula and Glow, 1970). While we found that the number of cells in the inner nuclear layer and the distribution of several classes of amacrine cell are apparently normal when viewed in section, we were interested to learn whether the regular spacing of a class of amacrine cell was affected. The cholinergic amacrine cells in rat, as in other species, have their cell bodies in both the inner nuclear and ganglion cell layers where they form nonrandom distributions (Voight, 1986). After transection of one optic nerve on the day of birth, rats were left to adulthood before the retinas were prepared as whole mounts and stained using a monoclonal antibody to choline acetyltransferase to identify the cholinergic amacrine cells. As we might have expected from our previous results, there was no difference in the number and distribution of the cholinergic amacrine cells between operated and unoperated retinas (V. H. Perry and C. Herbert, unpublished observations). The precision of the mosaic was also statistically indistinguishable between the normal and ganglion cell depleted retinas (Fig. 7.3).

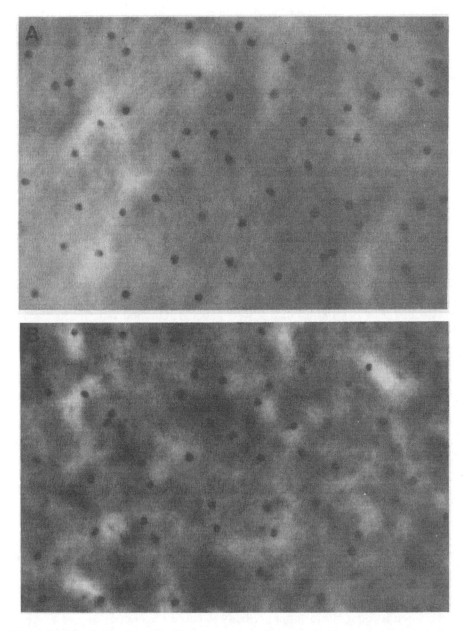

**Figure 7.3.** The distribution of cholinergic amacrine cells revealed with a monoclonal antibody to choline acetyltransferase (a generous gift from Dr. B. Weiner) in the inner nuclear layer of a normal retina (A) and following optic nerve transection on the day of birth (B).

The fact that the inner retina appears to be resistant to retrograde trans-neuronal degeneration is not only useful from an experimental point of view but also raises some interesting questions as to what regulates the numbers of local circuit neurons in development. We (Beazley *et al.*, 1987) calculated that over the first 15 days of postnatal life the amacrine and bipolar cell population is reduced by as much as 66%, a comparable loss to that found for the ganglion cell population. However, unlike ganglion cells they do not seem to be target dependent in the same way. It has also been shown (Blanks and Bok, 1977) that the degeneration of photoreceptors in the rd/rd mutant mouse, which begins on the seventh postnatal day and is virtually complete by the twentieth postnatal day, has little effect on the number of inner nuclear layer cells up to 41 days postnatally, well after the normal cell death phase for inner nuclear layer cells (Young, 1986). Despite the almost complete loss of af-ferents to the inner nuclear layer, normal cell death is not greatly affected. Other evidence also points to a lack of transneuronal effects on the inner nuclear layer. In chick the loss of 50% of the ganglion cells has no significant effect on inner nuclear layer numbers (Clarke, 1985). Following the local depletion of ganglion cells from kitten retina, the underlying horizontal cell population is not reduced in number (Eysel *et al.*, 1985).

The inescapable conclusion is that during development many inner nu-clear layer cells die for reasons other than the availability of the afferents or target. It is worth remembering that in the majority of studies in which ma-nipulations have been made, which prevent cell death of long axon neurons, only a fraction of the excess cells is rescued. The cells in the inner nuclear layer differ in an important respect from cell populations examined in most other studies; namely, these cells are in the main local circuit neuron. It is too early to say whether all such populations of local circuit neurons behave in this way. It is tempting to believe that inner nuclear layer cells are an example of intrinsically programmed cell death similar to that described in invertebrates (Horvitz *et al.*, 1982), but the present data do not allow such a conclusion. The cells of the inner nuclear layer make diverse synaptic connections and thus their developmental dependencies may also be diverse and able to regulate in spite of the loss of a major target or afferent. These cells also have a distinctive mode of development in that they are at all times in close contact with many cells with which they will ultimately make their adult synaptic connections. The possibilities for the regulation of cell numbers operating through these direct cell–cell contacts abound and remain to be explored.

## 7.5. FINDINGS CONSISTENT WITH THE DENDRITIC COMPETITION HYPOTHESIS

In its broadest sense "dendritic competition" says that the dendrites of neighboring ganglion cells exert some influence on each other: they are not simply passive, predetermined recipients of their inputs. The survival, size, and shape of a ganglion cell will be influenced by its neighbors. We would predict that experimentally reducing the ganglion cell density would allow the remaining ganglion cells to grow larger. Conversely, increasing the ganglion cell density would produce cells with smaller dendritic fields.

In the retinas of many mammals there is a conspicuous density gradient across the retina, with the density falling with increasing eccentricity (see Chapters 9 and 10 in this volume). As the density declines so the sizes of the dendritic territories of some ganglion cells increase in a proportional fashion so that the coverage remains fairly constant (Wässle *et al.*, 1981b; Leventhal, 1982; Perry *et al.*, 1984). This sort of observation suggests that local cell density determines territorial size, and in the primate the change in density can account for about 90% of the variance of dendritic field size (Perry *et al.*, 1984). However, this may be nothing more than the operation of an intrinsic program of ganglion cell form, if size were genetically predetermined. Experimental modification of the cell density is required to answer this.

Monocular enucleation in the embryonic cat increases the density of ganglion cells in the remaining eye by about 30% (Chalupa *et al.*, 1984), as a result of a reduction in the amount of cell death found in the embryonic cat (Williams *et al.*, 1986). Analysis at maturity of the dendritic field sizes of alpha cells in the remaining eye shows that the size of the dendritic territory is reduced by an amount proportional to the increase in alpha cell density (Kirby and Chalupa, 1986). In these retinas it was also found that despite the increase in cell density the mosaic of alpha cell somas was comparable to that found in a normal retina, the mosaic regulating to accommodate the excess cells. The potential postsynaptic target for ganglion cells in a monocular animal is larger than that of a normal animal but the cell bodies of alpha cells in embryonic enucleated animals were significantly smaller than normal. The intraretinal interactions, not the target size, determine the size of the cell body and the dendritic field.

The converse experiment of reducing the ganglion cell density has been done by Leventhal and his colleagues (see Chapter 8 in this volume). They have shown that a reduction in cell density allows ganglion cells to grow considerably larger than normal, although they do not grow to fill all the available space. Thus, the first prediction of a dendritic competition model is fulfilled, namely, that the dendritic field size of ganglion cells is not intrinsically specified but is regulated by the local cell density. In conditions where there is an overabundant afferent supply, the dendritic territory is limited by other factors, possibly the metabolic limit of the cell, or cessation of growth once some critical number of synapses is reached.

A second prediction from the dendritic competition type of model is that we would expect to find evidence that local interactions regulate the shape of the dendritic tree. When examined electrophysiologically, ganglion cells in cat retina have ovoid receptive fields (Levick and Thibos, 1982). The morphological basis of this is to be found in the shape of the dendritic trees. The dendritic territories of the majority of ganglion cells are ovoid in shape and the long axis of the dendritic field is directed toward the optic disk. The cells have a preferred orientation (Leventhal and Schall, 1983). In addition, the

**Figure 7.4.** Schematic diagram to illustrate how the elliptical dendritic trees of ganglion cells might be generated in a part of the cat retina with the center of the dendritic territory peripheral to the cell body. In the shaded portion of the retina centered on the area centralis (AC), the ganglion cell bodies are present and the dendrites of these cells occupy the available space. Three cells, which were generated later, have begun to grow their dendrites. The dendritic interactions described in the text will produce maximal dendritic growth away from the area centralis.

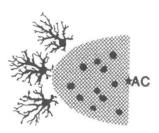

weighted centers of the dendritic territories are invariably located peripheral to the cell body (Schall and Leventhal, 1987). A simpleminded model of dendritic interactions that would account for these observations is shown in Fig. 7.4. As the retina develops, it is the central region that first produces a trophic factor for ganglion cells (Lipton, 1986) and this region becomes rapidly occupied by the dendrites of ganglion cells. At a slightly later stage the retina has further matured and the trophic factor is now produced by inner nuclear layer cells situated some distance from the area centralis. The interactions of neighboring ganglion cells are determined by the local availability of this trophic factor and possibly, in addition, a substrate on which the dendrites grow. As is shown in Fig. 7.4, the dendrites will grow preferentially away from the central region and away from the cell body toward the periphery. In addition, they will be elongated because neighboring cells on either side prevent access to the appropriate growth substrate or use up the local trophic factor. The idea that the trophic factor is first produced at the area centralis is consistent with results showing that the central retina matures first and then the maturation of the retina spreads peripherally (Rappaport and Stone, 1982). However, if this sequence of events were the case, we might expect the most central ganglion cells to have their dendrites directed toward the center of the area centralis: this appears to be the case (Rowe and Dreher, 1982). This idea is similar to the notion of a "wave of maturation" spreading across the developing retina (Schall and Leventhal, 1987), but here the wave of maturation is suggested to be the production of a trophic factor by cells of the inner nuclear layer. In normal development the shape of the dendritic territory as well as the size is influenced by dendritic interactions between neighboring cells.

## 7.6. DENDRITIC COMPETITION: COMPETITION FOR WHAT?

The evidence from observations on normal retina and following experimental manipulations are all consistent with the idea of dendritic interactions of some sort. The problem now is to discover the nature of these interactions. The abnormal dendritic trees formed by ganglion cells along the border of a region of the retina depleted of ganglion cells is surely the exaggeration of a normal developmental event. If we can understand more about this phenomenon, then possibly this will give us insights into the underlying mechanisms of normal development and guide future experiments directed toward a more molecular approach.

In the course of a series of experiments looking at the age-related effects of retinal lesions on the abnormal dendritic fields, we found in newborn operated rats that the axons of cells bordering the depleted area were typically directed away from the depleted area. This effect is shown schematically in Fig. 7.5. The same lesion to an animal older than 20 days of age produced no asymmetry in the dendritic tree and also the axons were not directed away from the depleted area (Perry and Maffei, 1988). The reason for this polarization of the cell body, where the axon arises from one side of the cell and the dendrites from the other (axodendritic polarity), seems obvious. The cells with the axons passing through the lesion area or with axons joining these fascicles will be transected and those cells will die (the dotted cells in Fig. 7.5). Why the same effect is not found in the adult is unclear.

### 7.6.1. Does Synapse Availability Regulate Competition?

Two questions are raised by these observations. First, is there evidence for axodendritic polarity in the normal ganglion cell population when viewed in the horizontal plane (as in a whole mount)? We know that for the vast majority of ganglion cells the axon arises from the vitread side of the cell body and the dendrites from the sclerad side, but it is not known whether ganglion cells show a polarity in the horizontal dimension as when viewing a whole mount. Second, is the asymmetry of the dendritic field present before the onset of synaptogenesis or does it require synapse formation for the asymmetry to develop? In brief, we have found for the largest class of ganglion cell in rat, cat, and monkey that most cells are weakly polarized in the horizontal plane. There are more primary dendrites on the side of the cell body opposite the axon initial segment and the vector describing their direction is invariably pointed away from the initial segment (Maffei and Perry, 1988). These results

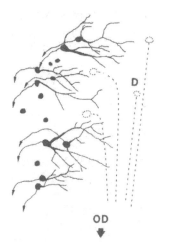

**Figure 7.5.** Horseradish peroxidase filled cells in an adult rat bordering a region of the retina depleted of ganglion cells (D) after a small retinal lesion was made close to the optic disk (OD) on the day of birth. The cells along the border have their dendrites directed into the depleted area and their axons (arrowheads) directed away from this region. These cells show an exaggerated axodendritic polarity. The dotted axons and cells are the presumed positions of some of the cells that degenerated as a consequence of the lesion. See text for further details.

are not entirely unexpected in view of the following observations. Schall and Leventhal (1987) have shown that the center of the dendritic tree of cat ganglion cells is invariably located peripheral to the cell body and in the cat the axon initial segment is commonly located on the side of the cell body closer to the optic disk (Maffei and Perry, 1988). Thus, a lesion in a newborn rat selects for or produces a population of cells with their axons directed away from the lesion and in addition these cells will have an intrinsic bias to have more dendrites pointing toward the depleted area.

We then examined how rapidly the asymmetry of the dendritic tree developed beyond that seen in the normal ganglion cell population. At the earliest time examined, 5 days after a lesion to the retina of a 1 day old rat, there was already a conspicuous asymmetry, greater than that seen for the normal population. In the newborn rat there are very few synapses in the inner plexiform layer until 10 days after birth (Weidman and Kuwabara, 1968) and in the mouse, where we have more quantitative data, the number of conventional synapses is only 50% of the total 10 days after birth and ribbon synapses are not yet formed (Fisher, 1979). The conclusion must be that large numbers of synapses are not a critical element in the generation of the abnormal dendritic field. Indeed, we have found that as lesions are made at successively later postnatal ages the magnitude of the abnormal dendritic asymmetry decreases hand-in-hand with the predicted increase in the number of synapses. Thus, it is the increasing number of synaptic contacts that is likely to limit the modifiability and size of the dendritic field. The relative paucity of synapses found in the early stages of dendritic growth does not mean that they are unimportant in the context of ganglion cell survival or interactions between dendrites. The large numbers of synapses present on the adult ganglion cell may be necessary for the construction of the visual receptive field, but small numbers of synapses or tight junctions may be all that is necessary for the transfer of important molecules. Studies on growth factors in other contexts show that only a small fraction of the receptors on a target cell need be occupied for the full biological effect to be observed (De Meyts, 1976; Sutter *et al.*, 1984). Also, a small number of synapses may be important in the regulation of the shape of the dendritic tree in the early stages of development, for if dendrites were to "compete" for synaptic contacts, then a small number would provide a more restrictive environment than a large excess.

### 7.6.2. Does Dendritic Contact Regulate Competition?

As a possible explanation for the territorial behavior of ganglion cell dendrites, it has been suggested that some sort of contact inhibition operates between the dendrites of neighboring cells (Wässle *et al.*, 1981a; Eysel *et al.*, 1985). We might envisage that the contact inhibition is of the sort that has been described for neurite growth in culture (Dunn, 1971). There are, however, a number of problems with a contact inhibition type model. In the first instance, it is hard to see how such an interaction would explain the overlap of the dendritic trees. While it might be argued that the dendrites have to reach a particular density before the inhibition would operate, it is clear from the analysis of the dendritic trees of alpha cells in a patch of retina that the density of the dendrites is not uniform but shows marked variance (Schall *et al.*, 1986). In the area centralis of the cat, the dendritic coverage (the number of cells with dendrites covering each point on the retina) for the beta cell population increases to a value of about 11, whereas in the rest of the retina it is about 4 (Leventhal, 1982). Our observations on the dendritic trees of ganglion cells bordering a depleted region also suggest that the loss of contact inhibition is not the whole explanation for the formation of abnormal dendritic trees. It is not the case that the dendritic trees bordering a region depleted of ganglion cells fill the available space around them; the dendritic trees appear to lack processes running parallel to the border, and this is also clear for the beta cell in the cat (see Fig. 2 in Eysel *et al.*, 1985). We have also found that reducing the density across the whole retina by about 30% prior to making a lesion in the neonatal rat retina has little effect on the amount of the dendritic tree directed into the depleted area (Perry and Maffei, 1988).

It is also difficult to envisage how a contact inhibition type model can account for a coverage factor much greater than 1, and this problem is further exaggerated when we consider some of the amacrine cell populations. The coverage factor of the cholinergic amacrine cells in the rabbit retina has been estimated to be between 30 and 70, depending on the area of the retina studied (Tauchi and Masland, 1984, 1985; Vaney 1984). The dendrites of these cells form a plexus in the inner plexiform layer with spaces surrounded by bundles of dendrites. Tauchi and Masland (1985) suggest three possibilities to account for this behavior. The dendrites of cholinergic amacrine cells might be attracted to each other, they might run along surfaces of greater adhesiveness or through channels, or they might be attracted to a common target. At present, it is not possible to distinguish between these alternatives. The problem of high coverage factors possibly reaches its pinnacle with one type of serotoninergic amacrine cell studied by Vaney (1986) in the cat retina, which has an estimated coverage of up to 900 in the peripheral retina. We should note in this context that coverage is a relatively crude measure of how the dendrites cover the retina, since it is the density of the processes that may be the important measure from both a functional and developmental point of view (Tauchi and Masland, 1984).

### 7.6.3. Do Chemotrophic Factors Regulate Competition?

The hypothesis that we favor to account for our observations following retinal lesions is that the dendritic trees of ganglion cells bordering a depleted region are attracted toward the region devoid of ganglion cells (Perry and Maffei, 1988). We envisage that following the depletion of the ganglion cells from a small portion of the retina the cells of the inner nuclear layer continue to develop normally (Beazley *et al.*, 1987) and synthesize a chemotrophic factor to which the dendrites of cells bordering the depleted areas are attracted. As the dendrites growing into the depleted area hypertrophy, this is at the expense of the dendrites on the other side of the cell body, which now mature less well than normal, and the cell develops a highly asymmetric dendritic territory. The dendrites do not grow across the entire depleted area for two reasons: first, the cell has an upper limit on its size, presumably because of some metabolic limits; and second, once a cell has attained a critical but unknown number of synaptic connections, further growth is prevented. If the depleted area acts as a source of a chemotrophic substance, then we would expect that dendritic asymmetry would be present for cells not only adjacent to the border but also some distance away and the asymmetry would decrease with distance. This is exactly what we find (Perry and Linden, 1982; Perry and Maffei, 1988).

Therefore, we can think of the development of a ganglion cell in the following way. The ganglion cell is generated and has an intrinsic program for a particular dendritic form. The growth of the dendritic tree is driven by growth factors supplied by both the target and the afferent neurons. As the dendritic tree matures it becomes anchored at some critical sites of greatest adhesion, possibly immature synapses or tight junctions. Since the sites of adhesion are themselves likely to be the afferents supplying the growth factor, further growth at the dendritic tips requires an ever increasing amount of growth factor to produce further growth. [In this context it is interesting to note that the primary dendrites of beta ganglion cells lack synaptic inputs and the majority of the synapses are found on lower-order dendrites with increasingly fewer on higher-order dendrites (McGuire *et al.*, 1986)]. In normal development the availability of the growth factor and of contacts depends on the number and distribution of neighboring cells of the same type, since neighboring cells will use up the growth factor and occupy sites. As far as the occupancy of sites is concerned, we do not have to invoke contact inhibition, but this can be thought of as operating on a "first-come-first-served" basis. Not only will neighboring cells influence cell survival during the period of cell death but also the shape of the dendritic territory will be influenced by these cells. It may be thought that from this hypothesis we would require a growth factor for each type of ganglion cell. However, a single or small number of growth factors could be produced by the afferents, and the precision of the connectivity and local interactions between the neighboring cells of the same type is a result of specificity at the adhesion sites or at the synaptic connections.

It is clear that there is a variety of different ganglion cell forms in the retina of any given species, and these differences in morphology reflect differences in connectivity, and in turn different receptive field properties. It was suggested in the introduction that the differences in form are likely to reflect intrinsic predetermined programs. The different types of ganglion cell can be distinguished prior to the formation of a large number of synaptic contacts in the inner plexiform layer (Perry and Walker, 1980; Maslim *et al.*, 1986). It remains an unanswered question as to what extent the form of a ganglion cell is intrinsically specified and what the nature of this specification might be. We need to establish the relative weight that is given to the intrinsically and extrinsically determined components.

## 7.7. INTERACTIONS BETWEEN CELL TYPES

One question that we have not touched on is whether the nonrandom distributions of the different cell types are related. Are all the dendritic interactions homotypic or are there also heterotypic interactions? The simple answer is that we do not know. In the best studied retina, that of the cat, several cell types which have been examined all show a density gradient from center to periphery (see Vaney, 1985, Fig. 30, for a summary diagram). An interesting example as an exception to this trend is to be found in the monkey, where the dopaminergic amacrine cells have a density distribution that parallels the rod photoreceptor distribution and has a peak density in a ring of radius 3 mm around the fovea (Mariani *et al.*, 1984). On a more local scale there is no evidence either way for the mosaic of one class of cell being yoked to the mosiac of another, for example, a class of amacrine cell showing a direct interaction with another.

On the other hand, it has been shown that alpha and beta ganglion cells of the cat each branch at two different levels in the inner plexiform layer and that this difference in the level of branching reflects whether the cell is an ON- or OFF-center cell (Nelson *et al.*, 1978; Saito, 1983). ON- and OFF-center cells branch in the middle and outer parts of the inner plexiform layer, respectively. Thus, inspection of the level of branching of the dendrites has allowed the separation of alpha and beta cells into their two different subpopulations of ON and OFF cells. Analysis of the mosaic formed by the total population of alpha cells shows that the distribution is nonrandom but not very precise. If the ON or OFF cells are considered separately, the precision of the mosaic is much improved for both types. It has been suggested that this may reflect the independent development of the ON- and OFF-cell types (Wässle *et al.*, 1981b). The same analysis has been performed for the beta population and the same inferences made (Wässle *et al.*, 1981c).

While it may indeed be the case that the ON and OFF subpopulations are generated independently, the present data do not allow such a conclusion. We can look at the problem in a slightly different way, particularly in the light of the work of Maslim *et al.* (1986), who have shown that ganglion cells in the early stages of development do not respect the divisions of the inner plexiform layer seen in the adult. The ganglion cell subpopulations are not predetermined to ramify in a particular subdivision of the inner plexiform layer. In the published distributions of alpha or beta cells, where the cells have been separated into ON or OFF cells (Wässle *et al.*, 1981b,c; Kirby and Chalupa, 1986), it is clear that the nearest neighbor to an ON cell has a high probability of being an OFF cell and vice versa. This is formally shown in the contingency tables in Fig. 7.6A–D. If we take any two populations of elements, where the two populations are of the same density and similar precision of the mosaic, and superimpose them we are likely to get essentially the same result, the nearest neighbor of one type is going to be a cell of the other type. This can be demonstrated by taking the mirror image of the ON-center populations illustrated in Fig. 4 of Kirby and Chalupa (1986) and overlaying it on the OFF

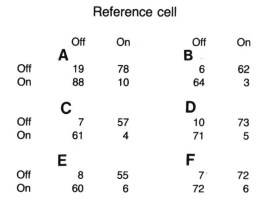

Reference cell

|   | Off | On |   |   | Off | On |
|---|---|---|---|---|---|---|
|   | **A** |   |   |   | **B** |   |
| Off | 19 | 78 |   | Off | 6 | 62 |
| On | 88 | 10 |   | On | 64 | 3 |
|   | **C** |   |   |   | **D** |   |
| Off | 7 | 57 |   | Off | 10 | 73 |
| On | 61 | 4 |   | On | 71 | 5 |
|   | **E** |   |   |   | **F** |   |
| Off | 8 | 55 |   | Off | 7 | 72 |
| On | 60 | 6 |   | On | 72 | 6 |

**Figure 7.6.** Contingency tables to show the relationship between ON and OFF cells and their nearest neighbors. (A) A population of alpha cells in cat retina (Wässle *et al.,* 1981b). (B) A population of beta cells in cat retina (Wässle *et al.,* 1981c). (C) and (D) Alpha cells from cat retina (Kirby and Chalupa, 1986). (C) is from a normal cat, (D) from the remaining eye of an animal from which one eye was removed in embryogenesis; note that for (C) and (D) the samples are taken from the same eccentricity but there are more cells in (D) than (C). It is clear that the nearest neighbor is likely to be of the opposite sign. In (E) the OFF cells and the mirror image of the ON cells that were analyzed in (C) were superimposed to give a pseudomosaic. The same procedure was adopted in (F) for the cells analyzed in (D). (C) and (E), and (D) and (F), are very similar. The pairing of ON and OFF cells could be dependent or independent.

population. The contingency table gives the same result as before (Fig. 7.6E and F) and this is essentially true of any of the inversions tried. The data cannot tell us about dependence or independence of the two subpopulations. Maslim *et al.* (1986) have suggested that the segregation of the ON and OFF cells may be related to the development of the bipolar cells, since the development of bipolar-like synapses and inner plexiform layer sublamination appear at about the same time.

The data of Kirby and Chalupa (1986) provide a possible insight into the generation of the mosaic of alpha cells in cat. Following monocular prenatal enucleation, which increases the ganglion cell density in the remaining eye by about 30%, alpha cells are not only more numerous and smaller but the precision of the mosaic is preserved. However, the cells that have been prevented from degenerating are located in the mosaic such that ON- and OFF-center cells are paired together in similar ratios as in a normal retina (Fig. 7.6C and D). We could argue that cells of the opposite sign attract each other, but a more parsimonious explanation may be that the distribution of cells is actually dependent. The following hypothesis is suggested. The dendrites of alpha cells, for example, grow and branch in both sublaminae. As the cells mature, some cells begin to ramify more in one particular division of the inner plexiform layer. The neighbors of these cells that lag slightly behind in their development are now less readily able to ramify in this layer and develop their dendrites in the other sublamina. Thus, it is a dendritic interaction that produces the nearest neighbors of the opposite type. We also propose that it is the

ON-center cells that mature first and occupy the central portion of the inner plexiform layer. The idea that the ON-center cells mature first is not entirely without foundation. Polyak (1941) noted and we have confirmed (Perry and Silveira, 1988) that the type described as "midget ganglion cells" by Polyak could be divided into two subtypes: those that branch in about the middle of the inner plexiform layer and those that terminate in the outer part of the inner plexiform layer. The cell bodies of those that branch in the middle of the inner plexiform layer have a strong bias to lie more vitread than the others. Since the retina is generated in an inner to outer sequence (Sidman, 1961), it is likely that the ON-center cells are born prior to the OFF-center cells. If the hypothesis is correct, the level of branching in the inner plexiform layer is another example of how dendritic interactions might influence the form of ganglion cell dendrites. One prediction of this hypothesis is that if the alpha cell population were depleted during the period that the ganglion cells were first developing their dendritic trees and prior to a commitment to branch in one particular lamina, we would expect that the majority of the remaining cells would branch in the inner part of the IPL. This remains to be tested.

## 7.8. SUMMARY

We began with two questions: How is the regular spacing of cell bodies generated in the retina and how is the size and shape of the dendritic territory of retinal cells regulated so as to produce a uniform coverage of the retinal surface? The evidence shows that interactions between the dendrites of neighboring ganglion cells play an important part in the regulation of ganglion cell number and thus are likely to play a role in the generation of ganglion cell mosaics. Interactions between the dendrites of ganglion cells in part determine the size and shape of the dendritic territory, and it is suggested that such interactions may also play a role in determining the level of branching within the inner plexiform layer. The nature of the dendritic interactions are unclear but the evidence points to a trophic factor being produced by cells of the inner nuclear layer which is important for the survival and growth of ganglion cells and which synaptic contacts serve to limit the size and overlap of their dendritic territories. A scenario incorporating these ideas is proposed to explain the generation of regularly spaced cells and the uniform coverage of the retina.

The evidence from these studies on the retina shows that interactions between dendrites of neighboring cells play an important role in the development of a complex neural network. It seems likely that the same interactions will operate in other parts of the nervous system where the regular sampling of the afferent input is essential.

ACKNOWLEDGMENTS. This work was supported by the Medical Research Council, U.K. and the Wellcome Trust. V.H.P. is a Wellcome Senior Research

Fellow. I thank Drs. M. Brown and K. Martin for their critical comments on the manuscript and Drs. J. Heath and J. Krebs for many helpful discussions.

## 7.9. REFERENCES

Armson, P. F., Bennett, M. R., and Raju, T. R., 1987, Retinal ganglion cell survival and neurite regeneration requirements: The change from Müller cell dependence to superior colliculi dependence during development, *Dev. Brain Res.* **32:**207–216.

Beazley, L. D., Perry, V. H., Baker, B., and Darby, J. E., 1987, An investigation into the role of ganglion cells in the regulation of division and death of other retinal cells, *Dev. Brain Res.* **33:**179–184.

Blanks, J. C., and Bok, D., 1977, An autoradiographic analysis of postnatal cell proliferation in the normal and degenerative mouse retina, *J. Comp. Neurol.* **174:**317–328.

Carpenter, P., Sefton, A. J., Dreher, B., and Lim, W.-L., 1986, The role of the target tissue in regulating the development of retinal ganglion cells in the albino rat: Effects of kianate lesions in the superior colliculus, *J. Comp. Neurol.* **251:**240–259.

Chalupa, L. M., Williams, R. W., and Hendrickson, Z., 1984, Binocular interactions in the fetal cat regulate the size of the ganglion cell population, *Neuroscience* **12:**1139–1146.

Clarke, P. G. H., 1985, Neuronal death during development in the isthmo-optic nucleus of the chick: Sustaining role of afferents from the tectum, *J. Comp. Neurol.* **234:**365–379.

Cleland, B. G., Levick, W. R., and Wassle, H., 1975, Physiological identification of a morphological class of cat retinal ganglion cells, *J. Physiol. (London)* **248:**151–171.

Cowey, A., 1974, Atrophy of retinal ganglion cells after removal of striate cortex in a rhesus monkey, *Perception* **3:**257–260.

Cowey, A., and Perry, V. H., 1979, The projection of the temporal retina in rats, studied by the retrograde transport of horseradish peroxidase, *Exp. Brain Res.* **35:**457–464.

Cunningham, T. J., 1982, Naturally occurring cell death and its regulation by developing neural pathways, *Int. Rev. Cytol.* **74:**163–186.

Cunningham, T. J., Huddleston, C., and Murray, M., 1979, Modification of neuron numbers in the visual system of the rat, *J. Comp. Neurol.* **184:**423–434.

Davies, A. M., Thoenen, H., and Barde, Y.-A., 1986, Different factors from the central nervous system and the periphery regulate the survival of sensory neurons, *Nature (London)* **319:**497–499.

De Meyts, P., 1976, Cooperative properties of hormone receptors in cell membranes, *J. Supramol. Struct.* **4:**241–258.

Dineen, J., and Hendrickson, A. E., 1981, Age-correlated differences in the amount of retinal degeneration after striate cortex lesions in monkeys, *Invest. Ophthal. Vis. Sci.* **21:**749–752.

Dräger, U. C., 1985, Birth dates of retinal ganglion cells giving rise to the crossed and uncrossed optic projections in the mouse, *Proc. R. Soc. London Ser. B* **224:**57–77.

Dunn, G. A., 1971, Mutual contact inhibition of extension of chick sensory fibres in vitro, *J. Comp. Neurol.* **143:**491–508.

Eayrs, J. T., 1952, Relationship between the ganglion cell layer of the retina and the optic nerve in the rat, *Br. J. Ophthalmol.* **36:**453–459.

Eysel, U., Peichl, L., and Wassle, H., 1985, Dendritic plasticity in the early postnatal feline retina: Quantitative characteristics and sensitive period, *J. Comp. Neurol.* **242:**134–145.

Finlay, B. L., Sengelaub, D. R., and Berian, C. A., 1986, Control of cell number in the developing visual system. I. Effects of monocular enucleation, *Dev. Brain Res.* **28:**1–10.

Fisher, L. J., 1979, Development of synaptic arrays in the inner plexiform layer of the neonatal mouse retina, *J. Comp. Neurol.* **147:**359–372.

Furber, S., Oppenheim, R. W., and Prevette, D., 1987, Naturally occurring neuron death in the ciliary ganglion of the chick embryo following removal of preganglionic input: Evidence for the role of afferents in ganglion cell survival, *J. Neurosci.* **7:**1816–1832.

Horvitz, H. R., Ellis, H. M., and Sternberg, P. W., 1982, Programmed cell death in nematode development, *Neurosci. Comment* **1:**56–65.

Jacobs, D. S., Perry, V. H., and Hawken, M. J., 1984, The postnatal reduction of the uncrossed projection from the nasal retina in the cat, *J. Neurosci.* **4:**2425–2433.

Jeffery, G., 1984, Retinal ganglion cell death and terminal field retraction in the developing rodent visual system, *Dev. Brain Res.* **13:**81–96.

Jeffery, G., and Perry, V. H., 1982, Evidence for ganglion cell death during the development of the ipsilateral retinal projection in the rat, *Dev. Brain Res.* **2:**176–180.

Johnson, J. E., Barde, Y.-A., Schwab, M., and Thoenen, H., 1986, Brain-derived neurotrophic factor supports the survival of cultured rat retinal ganglion cells, *J. Neurosci.* **6:**3038.

Kirby, M. A., and Chalupa, L. M., 1986, Retinal crowding alters the morphology of alpha ganglion cells, *J. Comp. Neurol.* **251:**532–541.

Lam, K., Sefton, A. J., and Bennett, M. R., 1982, Loss of axons from the optic nerve of the rat during early postnatal development, *Dev. Brain Res.* **3:**487–491.

Leventhal, A. G., 1982, Morphology and distribution of retinal ganglion cells projecting to different layers of the dorsal lateral geniculate nucleus in normal and Siamese cats, *J. Neurosci.* **2:**4–1042.

Leventhal, A. G., and Schall, J. D., 1983, Structural basis of orientation sensitivity of cat retinal ganglion cells, *J. Comp. Neurol.* **220:**465–475.

Levick, W. R., and Thibos, L. N., 1982, Analysis of orientation bias in cat retina, *J. Physiol. (London)* **329:**243–261.

Linden, R., and Perry, V. H., 1982, Ganglion cell death within the developing retina: A regulatory role for ganglion cell dendrites? *Neuroscience* **11:**2813–2827.

Linden, R., and Serfaty, C. A., 1985, Evidence for differential effects of terminal and dendritic competition upon developmental neuronal death in the retina, *Neuroscience* **15:**853–868.

Lipton, S. A., 1986, Blockade of electrical activity promotes the death of mammalian retinal ganglion cells in culture, *Proc. Natl. Acad. Sci. U.S.A.* **83:**9774–9778.

Maffei, L., and Perry, V. H., 1988, The axon initial segment as a possible determinant of ganglion cell morphology, *Dev. Brain Res.* **41:**185–194.

Mariani, A. P., Kolb, H., and Nelson, R., 1984, Dopamine containing amacrine cells of the rhesus monkey retina parallels rods in spatial distribution, *Brain Res.* **322:**1–7.

Maslim, J., Webster, M., and Stone, J., 1986, Stages in the structural differentiation of retinal ganglion cells, *J. Comp. Neurol.* **254:**382–402.

McGuire, B. A., Stevens, J. K., and Sterling, P., 1986, Microcircuitry of beta ganglion cells in cat retina, *J. Neurosci.* **4:**2920–2938.

Miller, N. M., and Oberdorfer, M., 1981, Neuronal and neuroglial responses following retinal lesions in the neonatal rat, *J. Comp. Neurol.* **202:**493–504.

Nelson, R., Famiglietti, E. V., and Kolb, H., 1978, Intracellular staining reveals different levels of stratification for on- and off-center ganglion cells in cat retina, *J. Neurophysiol.* **41:**472–483.

Okado, N., and Oppenheim, R. W., 1984, Cell death of motoneurons in the chick embryo spinal cord. IX. The loss of motoneurons following removal of afferent inputs, *J. Neurosci.* **4:**1639–1652.

Oppenheim, R. W., 1981, Neuronal death and some related regressive phenomena during neurogenesis: A selective historical review and progress report, in *Studies in Developmental Neurobiology: Essays in Honour of Victor Hamburger* (W. M. Cowan, ed.), pp. 74–133, Oxford University Press, New York.

Osborne, N. N., and Perry, V. H., 1985, Effects of optic nerve transection on some classes of amacrine cells in the rat retina, *Brain Res.* **343:**230–235.

Payne, B. R., Pearson, H. E., and Cornwell, P., 1984, Transneuronal degeneration of beta retinal ganglion cells in the cat, *Proc. R. Soc. London Ser. B* **222:**15–32.

Perry, V. H., 1979, The ganglion cell layer of the rat retina: A Golgi study, *Proc. R. Soc. London Ser. B* **204:**363–375.

Perry, V. H., 1981, Evidence for an amacrine cell system in the ganglion cell layer of the rat retina, *Neuroscience* **5:**931–944.

Perry, V. H., 1984, The development of ganglion cell mosaics, in *Development of Visual Pathways in Mammals* (J. Stone, B. Dreher, and D. H. Rappaport, eds.), pp. 57 – 73, Alan Liss, New York.

Perry, V. H., and Cowey, A., 1982, A sensitive period for ganglion cell degeneration and the formation of aberrant retinofugal connections following tectal lesions in rats, *Neuroscience* **7:**583–594.

Perry, V. H., and Linden, R., 1982, Evidence for dendritic competition in the developing retina, *Nature (London)* **297:**683–685.

Perry, V. H., and Maffei, L., 1988, Dendritic competition: Competition for what? *Dev. Brain Res.* **41:**195–208.

Perry, V. H., and Walker, M., 1980, Morphology of cells in the ganglion cell layer during development of the rat retina, *Proc. R. Soc. London Ser. B* **208:**433–455.

Perry, V. H., and Silveira, L. C. L., 1988, Functional lamination in the ganglion cell layer of the macaque's retina, *Neuroscience* **25:**217–223.

Perry, V. H., Linden, R., and Henderson, Z., 1983, Postnatal changes in retinal ganglion cell and optic axon populations in the pigmented rat, *J. Comp. Neurol.* **219:**356–368.

Perry, V. H., Oehler, R., and Cowey, A., 1984, Retinal ganglion cells that project to the dorsal lateral geniculate nucleus in the macaque monkey, *Neuroscience* **12:**1101–1123.

Polyak, S. L., 1941, *The Retina*, University of Chicago Press, Chicago.

Potts, R. A., Dreher, B., and Bennett, M. R., 1982, The loss of ganglion cells in the developing retina of the rat, *Dev. Brain Res.* **3:**481–486.

Rappaport, D. H., and Stone, J., 1982, The site of commencement of maturation in mammalian retina: Observations in the cat, *Dev. Brain Res.* **5:**273–279.

Rowe, M. H., and Dreher, B., 1982, Functional morphology of beta cells in the area centralis of cat's retina: A model for the evolution of central specializations, *Brain Behav. Evol.* **21:**1–23.

Saito, H. A., 1983, Morphology of physiologically identified X-, Y-, and W-type retinal ganglion cells of the cat, *J. Comp. Neurol.* **221:**279–288.

Schall, J. D., and Leventhal, A. G., 1987, Relationships between ganglion cell dendritic structure and retinal topography in the cat, *J. Comp. Neurol.* **257:**149–159.

Schall, J. D., Vitek, D. J., and Leventhal, A. G., 1986, Retinal constraints on orientation specificity in cat visual cortex, *J. Neurosci.* **6:**823–836.

Sengelaub, D. R., and Finlay, B. L., 1981, Early removal of one eye reduces naturally occurring cell death in the remaining eye, *Science* **213:**573–574.

Sefton, A. J., Lund, R. D., and Perry, V. H., 1987, Target regions enhance the outgrowth and survival of ganglion cells in embryonic retina transplanted to cerebral cortex in neonatal rats, *Dev. Brain Res.* **33:**145–149.

Sidman, R. L., 1961, Histogenesis of mouse retina studied with thymidine-$^3$H, in *The Structure of the Eye* (G. K. Smelser, ed.), Academic Press, Orlando, FL.

Sosula, L., and Glow, P. H., 1970, A quantitative ultrastructural study of the inner plexiform layer of the rat retina, *J. Comp. Neurol.* **140:**439–478.

Sutter, A., Hosang, M., Vale, R. D., and Shooter, E. M., 1984, The interaction with nerve growth factor with its specific receptors, in *Cellular and Molecular Biology of Neuronal Development* (I. B. Black, ed.), pp. 201–214, Plenum Press, New York.

Tauchi, M., and Masland, R. H., 1984, The shape and arrangement of the cholinergic neurons in the rabbit retina, *Proc. R. Soc. London Ser. B* **223:**101–119.

Tauchi, M., and Masland, R. H., 1985, Local order among the dendrites of an amacrine cell population, *J. Neurosci.* **9:**2494–2501.

Tong, L., Spear, P. D., Kalil, R. E., and Callahan, E. C., 1982, Loss of retinal X-cells in cats with neonate or adult visual cortex damage, *Science* **217:**72–75.

Turner, D. L., and Cepko, C. L., 1987, A common progenitor for neurons and glia persists in rat retina late in development, *Nature (London)* **328:**131–136.

Van Buren, K. M., 1963, *The Retinal Ganglion Cell Layer*, Charles C. Thomas, Springfield, IL.

Vaney, D. I., 1984, "Coronate" cells in the rabbit retina have the "starburst" dendritic morphology, *Proc. R. Soc. London Ser. B* **220:**501–508.

Vaney, D. I., 1985, The morphology and topographic distribution of AII amacrine cells in the cat retina, *Proc. R. Soc. London Ser. B* **224:**475–488.

Vaney, D. I., 1986, Morphological identification of serotonin accumulating neurons in the living retina, *Science* **223:**444–446.

Voigt, T., 1986, Cholinergic amacrine cells in the rat retina, *J. Comp. Neurol.* **248:**19–35.

Wässle, H., and Riemann, H. J., 1978, The mosaic of nerve cells in the mammalian retina, *Proc. R. Soc. London Ser. B* **200:**441–461.

Wässle, H., Peichl, L., and Boycott, B. B., 1981a, Dendritic territories of cat retinal ganglion cells, *Nature (London* **292:**344–345.

Wässle, H., Peichl, L., and Boycott, B. B., 1981b, Morphology and topography of on- and off-alpha cells in the cat retina, *Proc. R. Soc. London Ser. B* **212:**157–175.

Wässle, H., Boycott, B. B., and Illing, R.-B., 1981c, Morphology and mosaic of on- and off-beta cells in the cat retina and some functional considerations, *Proc. R. Soc. London Ser. B* **212:**177–195.

Weidman, T. A., and Kuwabara, T., 1968, Postnatal development of the rat retina, *Arch. Ophthalmol.* **79:**470–484.

Williams, R. W., Bastiani, M. J., Lia, B., and Chalupa, L. M., 1986, Growth cones, dying axons, and developmental fluctuations in the fiber population in the cat optic nerve, *J. Comp. Neurol.* **246:**32–69.

Young, R. W., 1985, Cell differentiation in the retina of the mouse, *Anat. Rec.* **212:**199–205.

Young, R. W., 1986, Cell death during differentiation of the retina, *J. Comp. Neurol.* **229:**362–373.

# Extrinsic Determinants of Retinal Ganglion Cell Development in Cats and Monkeys

8

Audie G. Leventhal and Jeffrey D. Schall

## 8.1. INTRODUCTION

Vision in vertebrates is subserved by a number of specialized channels that function in parallel in order to extract simultaneously a variety of features of the environment (reviewed by Stone *et al.*, 1979). As with all things visual, these parallel pathways begin in the retina. The mammalian retina contains a variety of ganglion cell types which have different forms, functional properties, and central projections (Enroth-Cugell and Robson, 1966; Boycott and Wässle, 1974; Cleland and Levick, 1974a,b; Stone and Fukuda, 1974). The developmental processes through which neurons in general and ganglion cells in particular differentiate into distinct types have recently begun to be explored (Eysel *et al.*, 1985; Kirby and Chalupa, 1986; Maslim *et al.*, 1986; Ramoa *et al.*, 1987; Leventhal *et al.*, 1988a,b). Both intrinsic (genetic) factors and extrinsic influences of the cell's environment appear to mediate the determination of adult cell type.

This chapter describes studies of the factors controlling the development of the different classes of retinal ganglion cells in cats and monkeys. Specifically, a number of recent studies indicate that experimental alterations in the density of ganglion cells in the immature retina profoundly affect development of retinal ganglion cell structure (Linden and Perry, 1982; Ault *et al.*, 1985; Eysel *et al.*, 1985; Leventhal *et al.*, 1988a). In the experiments described here, retinal ganglion cell density was manipulated experimentally in the neonate either by unilateral optic tract section or by pinpoint retinal lesions

**Audie G. Leventhal** • Department of Anatomy, University of Utah, School of Medicine, Salt Lake City, Utah 84132.     **Jeffrey D. Schall** • Department of Brain and Cognitive Science, Massachusetts Institute of Technology, Cambridge, Massachusetts 02139.

around the perimeter of the optic disk in cats and new-world (*Saimiri scirueus*) and old-world (*Macaca fascicularis*) monkeys. The procedures used for the surgery, extracellular single unit recordings, electrophoretic injection of horseradish peroxidase (HRP), histology, histochemistry, and computer-aided morphometric analysis are now standard and have been described in detail previously (Leventhal, 1982; Leventhal *et al.*, 1981, 1985, 1988a,b; Schall and Leventhal, 1987). These experiments show that competition among retinal ganglion cells for afferents as well as direct interactions among neighboring cells in the immature retina contribute to the development of both the structure and retinal distribution of ganglion cells in mammals.

## 8.2. EXPERIMENTAL OBSERVATIONS

### 8.2.1. Retinal Ganglion Cell Development in the Cat

#### 8.2.1.1. Retinal Ganglion Cell Classes in the Cat

Cat retinal ganglion cells can be classified as alpha, beta, epsilon, and other types of cells. Alpha cells have large cell bodies, large dendritic fields, and coarse axons (Boycott and Wässle, 1974). Beta cells have medium size cell bodies, small dendritic fields, and medium gauge axons (Boycott and Wässle, 1974). Epsilon cells have medium size cell bodies, large dendritic fields, and thin axons (Leventhal *et al.*, 1980). All remaining ganglion cells are referred to as "other types, or g1 and g2 cells." These cells have large dendritic fields, small cell bodies, and very thin axons. Cells with these characteristics were termed gamma cells by Boycott and Wässle (1974). A complete description of the morphologies and central projections of cat retinal ganglion cells can be found in Leventhal *et al.* (1985).

#### 8.2.1.2. Manipulation of Retinal Ganglion Cell Density in the Neonate

In the cat most but not all ganglion cells in temporal retina project ipsilaterally, while virtually all ganglion cells in nasal retina project contralaterally (Stone, 1966) (see Chapter 6, Fig. 6.1). Unilaterally sectioning the optic tract in the neonate severely depletes the temporal retina ipsilateral to the section of ganglion cells and virtually eliminates the ganglion cells in the nasal retina contralateral to the section (Leventhal *et al.*, 1988b). We have analyzed quantitatively the morphology of about 1000 ganglion cells which survived in retinas depleted of ganglion cells. Some of the cells studied were sparsely distributed within regions of reduced ganglion cell density and thus were isolated. Most of these cells were in the temporal retina contralateral to the intact hemisphere, but a few were in the nasal retina ipsilateral to the intact hemisphere. Other cells studied were located on the borders between ganglion cell-dense and ganglion cell-sparse zones. These cells thus had many neighbors on one side and very few on the other. The morphological attributes of all cells studied were related quantitatively to the density and distribution of the cell's neighboring neurons. The structures of more than 1000 ganglion cells in normal retinas, which have been analyzed previously (Leventhal and Schall, 1983; Schall and Leventhal, 1987), provided a normal data base. The results described below show that profound changes in the structure of some but not all types of cat retinal ganglion cell result from alterations in the distribution of neighboring ganglion cells during development.

### 8.2.1.3. Relationship between the Ganglion Cell Density Gradient and Dendritic Displacement

In our experiments we analyzed the dendritic structure of 200 well-filled cells in temporal retina on the border between the normal and depleted regions of the retina contralateral to the optic tract section. These cells were filled with HRP injected into the intact LGNd and thus projected ipsilaterally. In optic tract sectioned animals, very few cells in nasal retina project ipsilaterally (Jacobs *et al.*, 1984; Leventhal *et al.*, 1988b) (see Chapter 6, Fig. 6.1, in this volume). In our material the sharpest border, that is, the steepest ganglion cell density gradient, is found along the nasotemporal division in the retina contralateral to the optic tract lesion (Leventhal *et al.*, 1988b). It is for this reason that the contralateral retina was chosen for this analysis.

In agreement with the observations of Linden and Perry (1982), Perry and Linden (1982), and Eysel *et al.* (1985), we find (Ault *et al.*, 1985; Leventhal *et al.*, 1988a) that on the border between normal regions and regions depleted of ganglion cells the dendritic fields of all types of cell extend preferentially into the depleted area (Fig. 8.1). This is actually an exaggeration of the normal situation. In normal regions of cat retina, ganglion cell dendrites are also directed down the ganglion cell density gradient (Schall and Leventhal, 1987).

### 8.2.1.4. Relationship between Ganglion Cell Density and Cell Body Size

Ganglion cells that develop in areas of retina depleted of other ganglion cells are much larger than their normal counterparts (Ault *et al.*, 1985; Leventhal *et al.*, 1988a). The magnitude of the increase in size varies with retinal position and across cell types as indicated in Fig. 8.2. The cell body sizes of beta cells in the part of the area centralis depleted of ganglion cells are significantly larger than beta cells in corresponding regions of normal density (Figs. 8.2 and 8.3). Similarly, isolated central alpha cells are much larger than their normal counterparts (Figs. 8.2 and 8.3). Such differences are not as pronounced for alpha and beta cells in peripheral retina. It is important to note that while both alpha and beta cells increase dramatically in size, the relative differences between the two cell types are always preserved. In other words, even though alpha and beta cells in cell-poor regions are both abnormally large, alpha cells in cell-poor regions are always still far larger than adjacent beta cells.

To demonstrate that the relationship between cell body size and the

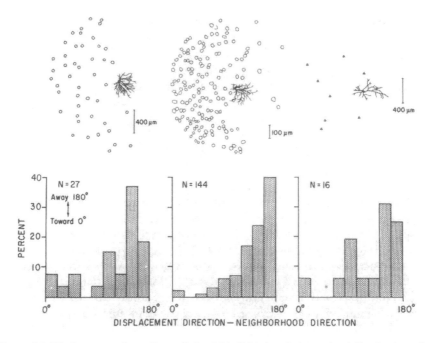

**Figure 8.1.** Displacement of ganglion cell dendritic fields in relation to local distribution of ganglion cells of the same morphological type. The top row shows the structure of representative alpha, beta, and epsilon ganglion cells from left to right; in each panel only the cell bodies of cells of like type are outlined. Below each cell, histograms illustrate the distribution of angle differences between the direction of dendritic displacement and the direction of peak density for the different cell types. Displacement direction is the angle of the center of the dendritic field from the center of the cell body. Neighborhood direction is the angle from the center of the cell body of the cell under investigation to the point of peak local density. An angle difference of 0° indicates that the dendritic field is displaced toward the densest point, and an angle difference of 180° indicates dendritic displacement away from local density. For alpha, beta, and epsilon cells it is evident that the dendritic fields tend to be displaced away from neighboring cells.

density of neighboring cells was truly independent of retinal position, we related the cell body size of beta cells within a restricted region near the area centralis to the local density of neighboring beta cells. Figure 8.4 illustrates this result. We find that within 1 mm of the center of the area centralis in normal and cell-poor regions of the retina, the soma size of beta cells is related to the decrease in local cell density. It is clear that beta cells in areas of reduced cell density within 1 mm of the area centralis have cell bodies as large as their counterparts in far peripheral areas of retina where similar numbers of neighboring beta cells are normally found. This indicates that cell density and not retinal position per se is important in determining the size of the cell body.

In marked contrast to the results described above for alpha and beta cells, a reduction in ganglion cell density does not affect the cell bodies of other types of ganglion cell; epsilon, g1, and g2 (Fig. 8.2) in areas of reduced ganglion cell density are not larger than normal.

It has been reported that about 50% of gamma (g1, g2) cells in temporal

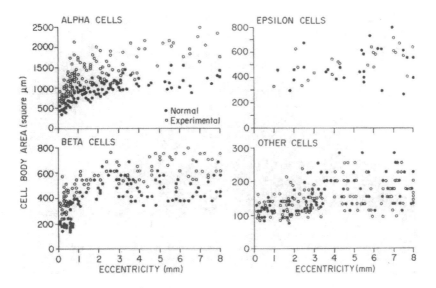

**Figure 8.2.** Cell body area versus eccentricity. Each point represents the measure for an individual cell; all cells were sampled in the temporal retina contralateral to the intact hemisphere. Solid points show normal cells, and open points show experimentally affected cells. It is evident that the cell bodies of alpha and beta cells which develop in areas of reduced cell density are larger than normal at all eccentricities but that the relative difference decreases with eccentricity. In contrast, the cell bodies of epsilon and other cell types which develop in areas of reduced cell density are no different from normal.

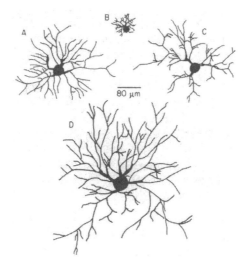

**Figure 8.3.** Camera lucida drawings of HRP-filled ganglion cells in corresponding retinal areas of normal and reduced ganglion cell density. (A) Normal alpha cell; (B) normal beta cell; (C) beta cell from area of reduced density; (D) alpha cell from area of reduced density. Each of these cells was approximately 1 mm from the center of the area centralis. Alpha and beta cells that developed in areas of reduced density have significantly larger dendritic fields than normal. Also note the morphological similarity between the normal central alpha cell (A) and the isolated central beta cell (C).

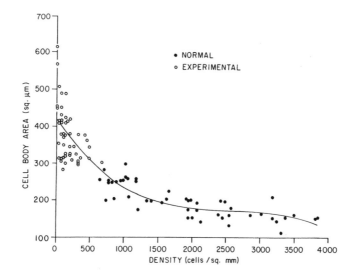

**Figure 8.4.** Cell body area versus cell density for beta cells in normal and experimentally depleted regions within 1 mm of the area centralis. The line passing through the data points is a third-order polynomial regression. It is clear that soma size increases dramatically as the density of neighboring beta cells decreases within a restricted region of central retina. These data indicate that cell density and not retinal position per se determines cell size.

retina project contralaterally in the normal cat (Wässle and Illing, 1980). Consequently, neonatal optic tract section should not reduce the density of this cell type to the same extent as it does the other cell types. For this reason all the retinas studied in our experiments were counterstained and we only analyzed g1 and g2 cells in regions in which the density of these cells was truly reduced. In addition, to make certain that the sizes of these cells did not increase in areas of reduced density, we also analyzed the sizes of the ipsilaterally projecting g1 and g2 cells in nasal retina which survived optic tract section (Jacobs *et al.*, 1984). These were the most isolated cells in our sample; we only found 20–40 labeled cells in the ipsilateral nasal retina following injections in the LGNd of normal and OTX cats (Leventhal *et al.*, 1988b). As illustrated in Fig. 8.5, the size of these cells does not change with retinal eccentricity and ipsilaterally projected cells in nasal retina of OTX cats are not significantly larger than normal.

### 8.2.1.5. Relationship between Ganglion Cell Density and Dendritic Field Structure

The effects of reduced ganglion cell density on dendritic field size are demonstrated in Fig. 8.6. In this figure alpha and beta cells from corresponding points of retina in regions of normal and experimentally reduced ganglion cell density are compared. It is evident that the dendritic fields of alpha and beta cells in regions of reduced ganglion cell density are far larger than

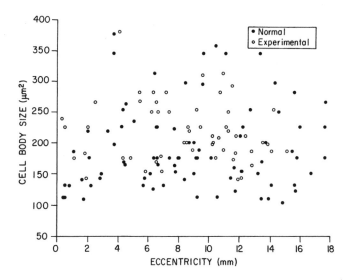

**Figure 8.5.** Cell body area versus eccentricity for isolated ipsilaterally projecting g1 and g2 cells in nasal retina of normal and OTX cats. Conventions are as in Fig. 8.3. The cell bodies of g1 and g2 cells which develop in areas of reduced cell density are not significantly larger than normal.

those of their normal counterparts. We also analyzed the dendritic fields of epsilon, g1, and g2 cells in regions of reduced cell density. We found no evidence that the dendritic fields of these cells were larger than normal.

Since the dendritic fields of alpha and beta cells in depleted areas of retina were so much larger than normal, the branching patterns of the dendritic arbors of these cells were analyzed in detail. The results indicate that the abnormally large alpha and beta cells in areas of reduced density have both longer and more dendritic segments comprising their dendritic fields (Leventhal *et al.*, 1988a).

### 8.2.2. Retinal Ganglion Cell Development in Primates

#### 8.2.2.1. Retinal Ganglion Cell Classes in the Primate

Retinal ganglion cells in both new-world monkeys (*S. sciureus*) and old-world monkeys (*M. fascicularis*) can be classified as A cells (p alpha), B (p beta) cells, C cells (p gamma), and E cells (p epsilon) (Leventhal *et al.*, 1981; Perry and Cowry, 1984; Perry *et al.*, 1984; A. G. Leventhal *et al.*, unpublished observations). In both species, A cells project heavily to the magnocellular laminae of the LGNd, have large cell bodies, large dendritic trees, and coarse axons. B cells project to the parvocellular laminae of the LGNd, have small cell bodies, very small dendritic fields and medium gauge axons. Within central retina, B cells are "midget" ganglion cells (Polyak, 1941; Leventhal *et al.*, 1981; Perry and Cowey, 1984; Perry *et al.*, 1984; Rodieck *et al.*, 1985). C cells project to the superior colliculus and pretectum and constitute a hetero-

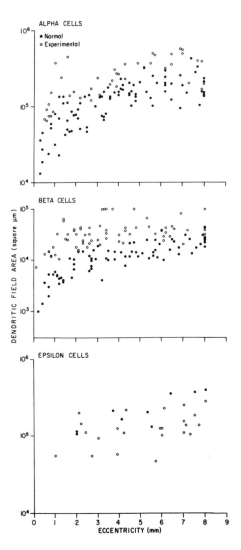

**Figure 8.6.** Dendritic field area versus eccentricity for alpha, beta, and epsilon cells. Note that alpha and beta cell dendritic fields that develop in areas of reduced ganglion cell density are larger than normal and that the relative increase in size decreases with eccentricity. Epsilon cell dendritic fields that develop in areas depleted of ganglion cells are not larger than normal.

geneous group of cells with small to medium sized cell bodies, large dendritic fields, and fine axons. In both species, the cell bodies and dendritic fields of A and B cells, but not other cells, are larger in peripheral than in central regions of retina. However, within the central 2 mm of retina, the sizes of B cells (midget ganglion cells) do not vary significantly (Perry and Cowey, 1984; A. G. Leventhal *et al.*, unpublished observations). In both *M. fascicularis* and *S. sciureus,* A and B cells in temporal retina tend to be larger than those in nasal retina (Perry *et al.*, 1984). The one notable difference between the two species is that A and B cells in *M. fascicularis* are somewhat larger than their counterparts in *S. sciureus*. This difference (see below) may reflect the fact that retinal ganglion cell density is greater in *S. sciureus* than *M. fascicularis* (Stone and Johnston, 1981).

### 8.2.2.2. Effects of Neuronal Density on Foveal Development

The spatial distribution of ganglion cells is not uniform across the retina; the retinas of higher mammals are characterized by a central area containing a high density of ganglion cells (Mann, 1964; Stone, 1965; Rolls and Cowey, 1970; Hendrickson and Kupfer, 1976; Stone and Johnston, 1981; Abramov *et al.*, 1982; Hendrickson and Yuodelis, 1984). In the cat, the central area or area centralis mediates high-resolution vision and consists of a large number of small cells crowded into the central millimeter of retina (Stone, 1965). Advanced primates have 8–10 times as many ganglion cells as the cat (Rolls and Cowey, 1970; Stone and Johnston, 1981). Primate central retina is characterized by a foveal region consisting of a roughly circular area devoid of ganglion cells (the foveal pit) surrounded by a multilayered, annular region of very small, densely packed ganglion cells (the foveal slope) (Mann, 1964; Rolls and Cowey, 1970; Hendrickson and Kupfer, 1976; Abramov *et al.*, 1982; Hendrickson and Yuodelis, 1984). While in the cat the area centralis is the first region to mature and the ganglion cells there are the first to differentiate (Rapaport and Stone, 1983; Maslim *et al.*, 1986), in monkeys the fovea resembles an area centralis prenatally; the foveal pit begins to develop later in gestation and continues to develop during the first months of life (Hendrickson and Kupfer, 1976). In humans, foveal development continues for years postnatally as the neurons in central retina progressively mature and redistribute. During this period the principal dendrites of retinal ganglion cells normally elongate as their cell bodies migrate away from the center of the foveal pit (Hendrickson and Yuodelis, 1984). Foveal development is in large part responsible for the dramatic improvement in visual function which normally occurs during early childhood, since in primates the fovea mediates both high resolution and color vision. We have studied foveal development in monkeys in which central retina was experimentally depleted of cells in the neonate.

We find that in monkeys in which central retina is depleted of cells at birth, foveal development is abnormal; an abnormally large number of ganglion cells projecting to the LGNd are found throughout and around the foveal pit and the area of the foveal pit is about 40% smaller than normal (Figs. 8.7 and 8.8C). Thus, unlike in normal monkeys, these animals had no cell-free region in central retina. This is similar to what we have observed in neonates following injections of HRP into the LGNd; retinal ganglion cells projecting to the LGNd are found throughout the foveal pit in newborn monkeys.

### 8.2.2.3. Effects of Neuronal Density on the Structure of Primate Retinal Ganglion Cells

In these studies, small retinal lesions were made close to the optic disc in newborn monkeys. These lesions cut the axons of retinal ganglion cells as they exited the eye. The affected ganglion cells died, resulting in small strips of retina which were devoid of ganglion cells (Perry and Linden, 1982) (Fig. 8.7B). The central projections, morphologies, and retinal distributions of

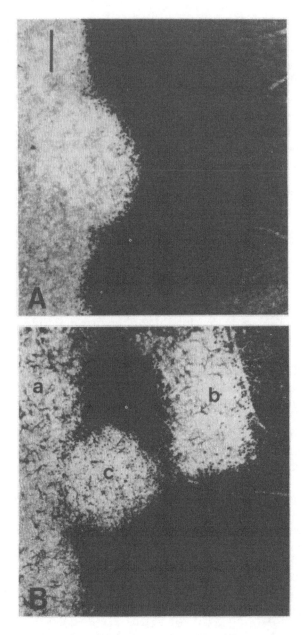

**Figure 8.7.** Retinal ganglion cells labeled by large HRP injections into one LGNd of a 3 year old monkey (*S. sciureus*). The regions in (A) and (B) are contralateral and ipsilateral, respectively, to the injection. At 1 day of age, this animal received a small lesion in one retina just temporal to the optic disc. This created a cell-free region adjacent to the foveal pit [shown in (B)]. In (B) the cells indicated by the letters a, b, and c are magnified and shown in parts (A), (B), and (C) of Fig. 8.8, respectively. In animals in which central retina has been depleted of cells, the foveal pit is abnormally small and contains an abnormally large number of B (midget) ganglion cells. (A) and (B) are at the same magnification; the scale bar in (A) = 200 μm.

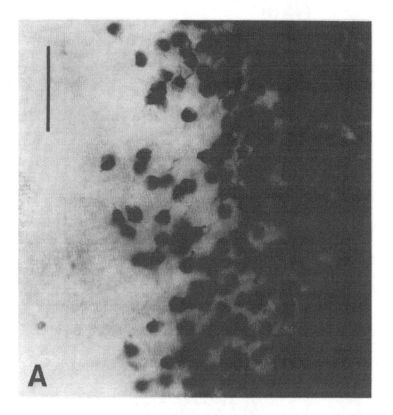

**Figure 8.8.** Higher-power photomicrographs of the cells in regions a, b, and c in Fig. 8.7. The cells in (A) and (B) are in regions of central retina approximately equidistant from the center of the foveal pit. The cells in (A) are in a region of normal density. The cells in (B) are within the experimentally induced cell-poor region and are much larger than normal. The cells shown in (C) are within the foveal pit and are also abnormally large. All cells shown in (A)–(C) are B (midget) ganglion cells. (A)–(C) are at the same magnification; the scale bar in (A) = 50 μm.

ganglion cells that developed in normal regions of monkey retina and in regions depleted of ganglion cells were compared.

In primate central retina the cell bodies and dendritic fields of retinal ganglion cells which survive within and on the borders of regions depleted of ganglion cells are much larger than normal. Changes in size are relatively greatest close to the foveal pit, where the relative decrease in ganglion cell density is greatest (Figs. 8.8–8.10). In fact, the cell bodies of B midget ganglion cells that develop in cell-poor regions of the fovea are almost as large as normal peripheral A cells and thus are actually significantly larger than normal central A cells as well as normal peripheral B cells. The cell bodies of isolated central cells are over 10 times their normal volume; their dendritic fields are about 10 times their normal area (Figs. 8.9 and 8.10). Despite these impressive increases in size, isolated central B cells are still morphologically similar to midget ganglion cells (Fig. 8.9). That is, they have a spherical cell

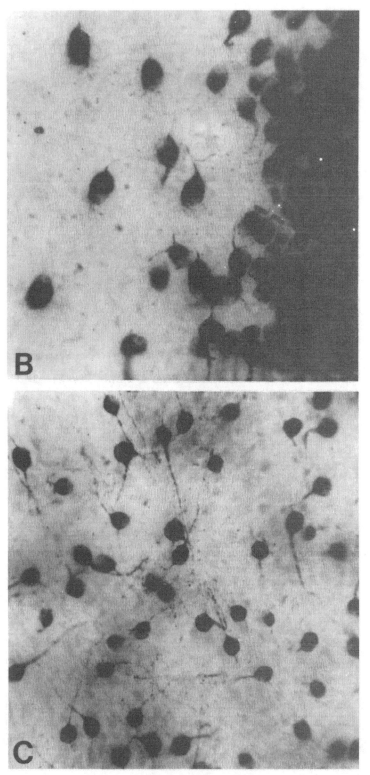

**Figure 8.8.** (*Continued*)

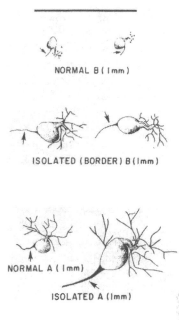

**Figure 8.9.** Camera lucida drawings of A- and B-type ganglion cells in normal and cell-poor regions of primate (*S. sciureus*) retina. This monkey's retina was lesioned close to the optic disk at 1 day of age; the animal received HRP injections into one LGNd as an adult. Note that the cell bodies and dendritic fields of isolated central A and B cells are much larger than those of their normal counterparts. The scale bar = 80 μm.

body and a single principal dendrite that branches profusely in a single sublamina of the inner plexiform layer.

Cells in paracentral regions (2–5 mm from the fovea) of reduced density are only slightly larger than normal and closely resemble normal paracentral A and B cells. Cells in peripheral regions (more than 5 mm from the fovea) do not appear to be abnormally large. These cells have multiple principal dendrites and large, relatively diffuse dendritic fields. While the cell bodies of isolated cells in central and peripheral regions are similar in size, the dendritic

**Figure 8.10.** Cell body and dendritic field sizes of A and B cells in regions of normal density and of isolated A and B cells which developed in regions of the foveal slope depleted of ganglion cells. All cells included in this figure were sampled from the same retina and were equidistant from the center of the foveal pit. Note that the ordinate is logarithmic. The cross-sectional areas of the cell bodies of isolated A and B cells (ordinate) are about four times larger than normal; their dendritic field areas are over 10 times larger than normal (abscissa). Since the cell bodies of central A and B cells are normally very close to spherical, a four- to fivefold increase in cross-sectional area indicates a roughly 10-fold increase in volume.

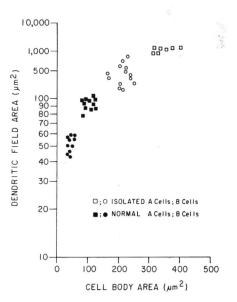

fields of isolated central cells, even though abnormally large, are substantially smaller than those of both normal and isolated peripheral cells. Thus, unlike in the cat (Leventhal *et al.,* 1988a), isolated cells in central and peripheral regions of monkey retina remain morphologically distinct.

The case illustrated in Fig. 8.7B is of special interest in that an experimentally induced cell-poor region was created just adjacent to the foveal pit. This resulted in two cell-poor regions separated by a narrow cell-rich zone. It is noteworthy that in both the cell-rich and cell-poor regions the ganglion cells are directed normally and thus their dendrites extended toward the center of the foveal pit. As a result, the dendrites of cells on the border of the experimentally induced cell-free region are actually directed toward neighboring cells and away from the cell-poor zone in regions where the foveal pit was in the same direction as the area of highest ganglion cell density. In such regions, the dendritic fields of the cells on the border of the cell-rich zone arborize in the same region of the inner plexiform layer as do those of adjacent cells within the cell-rich zone. Nevertheless, ganglion cells within the cell-rich region appear normal and are very small, while cells on the border of the ganglion cell-poor region are abnormally large.

## 8.3. SUMMARY OF EXPERIMENTAL OBSERVATIONS

This chapter describes work showing that the development of cat and monkey retinal ganglion cells is influenced by the density and distribution of neighboring ganglion cells in the immature retina. The specific results are summarized below:

1. In the cat the direction of displacement of the dendritic field from the cell body of all types of ganglion cell is determined by the distribution of neighboring cells; the center of the dendritic field is displaced down the experimentally induced density gradient, that is, away from other cells. In the primate, this is not always the case; in monkey central retina ganglion cell dendrites are directed toward the fovea even in cases where the fovea is in the direction of the highest density of neighboring cells.
2. If ganglion cell density is reduced in the neonate, then the cell bodies and dendritic fields of alpha and beta cells in the cat and A and B cells in the monkey grow to be abnormally large. The increase in size is most pronounced in central retina where the reduction in ganglion cell density is greatest. In the ganglion cell-free area centralis region of the cat, isolated beta cells bear a morphological resemblance to alpha cells in the same retina in the region of the area centralis containing normal numbers of cells. In contrast, cells in the central and peripheral regions of reduced density in primate retina remain morphologically distinct even though isolated central cells in primates can be up to 10 times larger than normal.
3. In ganglion cell-poor regions of cat retina alpha cells always have

larger cell bodies and dendritic fields than do nearby beta cells. In ganglion cell-poor regions of primate retina, differences between abnormally large A and B cells are also obvious. Thus, while the absolute sizes of cells which develop in areas of reduced density increase dramatically, the relative differences between the two major cell types in cat and monkey are always maintained.

4. The results described above cannot be generalized to cell types other than alpha and beta in the cat and A and B in the monkey. The sizes of cells of other types which develop in ganglion cell-poor regions of cat and monkey retina are not different from normal.

5. An experimentally induced reduction in ganglion cell density in immature primate central retina interferes with normal foveal development.

## 8.4. IMPLICATIONS OF EXPERIMENTAL FINDINGS

### 8.4.1. Normal Changes in Morphology Associated with Retinal Eccentricity

The results described in this chapter indicate that the normal increases in the cell body and dendritic field sizes of alpha and beta cells associated with increasing distance from the cat area centralis (Boycott and Wässle, 1974; Kolb *et al.*, 1981; Leventhal, 1982) are in large part due to the decrease in the density of ganglion cells normally associated with eccentricity. This has been suggested previously by Wässle *et al.* (1981a). Our finding that isolated beta cells in the area centralis region of cat retina resemble their normal peripheral counterparts supports this idea. It is also consistent with this hypothesis that the morphologies of cell types other than alpha and beta cells normally change little with eccentricity (Boycott and Wässle, 1974), and the cell body and dendritic field sizes of these cells are not affected significantly by a reduction in the number of neighboring cells.

In consonance with our findings, two other recent investigations have demonstrated increases in ganglion cell body size following an experimental reduction in ganglion cell density in the cat (Rapaport and Stone, 1983) and rat (Carpenter *et al.*, 1986). Results that are complementary to ours have been provided by Kirby and Chalupa (1986); they demonstrated that when ganglion cell density is increased in one eye by prenatal enucleation of the other, alpha cell somas and dendritic fields are smaller than their normal counterparts. Hence, the foregoing data, taken together, support the notion that the ganglion cell size changes normally associated with eccentricity are in large part due to neuronal interactions which occur during development of the normal retinal ganglion cell density gradient.

### 8.4.2. Implications for Retinal Ganglion Cell Classification

Boycott and Wässle (1974) were the first investigators to note the morphological similarity between alpha and beta cells. These authors reported that beta cells in far peripheral regions of cat retina closely resemble alpha cells close to the area centralis. The results reviewed here include two additional findings consistent with the idea that alpha and beta cells (cat) and A and B cells (monkey) are more closely related to each other than to other ganglion cell types. First, our results extend the observations of Boycott and Wässle (1974) and show that beta cells within a cell-poor area centralis region of cat retina bear a striking resemblance to alpha cells in the normal, high-density area centralis region of the same retina. Second, we have found that a reduction in the number of neighboring cells during development profoundly and similarly affects cell body and dendritic field sizes of alpha and beta cells (cat) and A and B cells (monkey) but does not affect the sizes of other types of cell at all. Thus, different developmental mechanisms seem to affect the morphological differentiation of alpha and beta cells and other types of cell in the cat and A and B cells and other types of cell in the monkey.

It is tempting to speculate that cat and monkey retina actually contains two supergroups of ganglion cells. Such an idea has already been put forth by Rowe and Stone (1980) for cat retinal ganglion cells. These authors postulated that X/Y cells constituted one group and W cells the other. According to this scheme, X and Y cells, which correspond to beta and alpha cells, respectively, are subtypes of the X/Y supergroup, while all other cell types are subtypes of the W cell supergroup. Our findings indicate that the mechanisms that mediate ganglion cell development are consistent with this view.

### 8.4.3. Retinal Ganglion Cell Class Differences

While the present study provides evidence that interactions between neighboring ganglion cells during development contribute to the size differences evident among ganglion cells, our results also indicate that (1) different cell types are affected differently, (2) cells in different regions of retina are affected differently, and (3) the relative differences between cell types develop independent of such interactions.

Thus, a number of questions arise. First, in ganglion cell-poor zones why do alpha cells (cat) and A cells (monkey) grow to be larger, respectively, than adjacent beta cells (cat) and B cells (monkey)? Second, in regions of peripheral retina depleted of cells, why do alpha and beta cells in the cat and A and B cells in the monkey grow to be only slightly larger than normal peripheral cells even though the density of peripheral cells in the depleted regions is much lower than in normal peripheral retina? Third, why do epsilon, g1, and g2 cells not grow to be abnormally large?

At least three answers to these questions come to mind. The first and most obvious is that some of the differences between ganglion cells are genetically determined. There are several lines of evidence which are consistent with there being genotypic distinction of the different ganglion cell types. For example, different ganglion cell types differentiate at different times (Walsh and Polley, 1985). Also, the different ganglion cell types exhibit different membrane proteins as revealed by the selective staining of the CAT-301 monoclonal antibody (Hendry *et al.*, 1984). Finally, the synaptic connections of the different ganglion cell types within the retina are different (Kolb, 1979; Stevens *et al.*, 1980).

Another possibility is that many of the differences between ganglion cell types are not genetically specified, but rather result from interactions not revealed by the present study. For example, it is possible that different cell types mature at different times and the cell's retinal environment varies during development. Thus, cells with similar genetic makeup may develop quite differently since the extrinsic factors to which they are exposed during development vary. It is also quite possible that our manipulations are done too late in the course of retinal development to affect certain aspects of ganglion cell morphology. In fact, the different classes of retinal ganglion cells are already morphologically distinct in neonatal kittens (Maslim *et al.*, 1986; Ramoa *et al.*, 1987), as well as in newborn monkeys (our observations), so class-related differences already exist at the time of our manipulations. It should be possible to test whether some or all of the relative differences in ganglion cell morphology can be eliminated by carrying out studies similar to those described here in the neonatal ferret. Ferret retina contains morphological classes of ganglion cells similar to those in the cat, but at birth ferret retina is immature and the morphological classes of ganglion cells have not yet differentiated (Vitek *et al.*, 1985).

The third possibility is that some of the differences between the cell body and dendritic morphologies of alpha and beta cells in the cat and A and B cells in the monkey result from differences in the axonal morphologies and patterns of central projection of the different cell types. Differences in the patterns of central projections of the different cell types in cats and monkeys are well documented (Kelly and Gilbert, 1975; Wässle and Illing, 1980; Illing and Wässle, 1981; Leventhal *et al.*, 1981, 1985; Leventhal, 1982; Perry and Cowey, 1984; Perry *et al.*, 1984). For example, in the cat most alpha cells branch and project to many visual nuclei including the A and C laminae of the LGNd, the medial intralaminar nucleus (MIN), the superior colliculus, and the pretectum. On the other hand, most beta cells project only to the A laminae of the LGNd (Friedlander *et al.*, 1979; Illing and Wässle, 1981; Bowling and Michael, 1984; Leventhal *et al.*, 1985). Also, in the laminae of the LGNd alpha cell axonal arbors are more extensive than beta cell arbors (Bowling and Michael, 1981). Hence, it is possible that alpha cells grow to be larger than beta cells in order to balance their more extensive axonal arborizations. It is also likely that A cells in the monkey have more extensive axonal arborizations than do B cells (Michael, 1983).

This hypothesis can be tested in the cat by unilaterally ablating the superior colliculus in the neonate. This should reduce the number of alpha cells but not beta cell axonal branches since all alpha cells and few if any beta cells project to the superior colliculus (Illing and Wässle, 1981; Leventhal *et al.*, 1985). In these animals the relative differences between alpha and beta cells should be reduced if there is a relationship between a cell's axonal morphology and its cell body and dendritic structure.

### 8.4.4. Independence of Dendritic Field Size and Displacement

Retinal ganglion cell dendrites extend preferentially into areas of reduced ganglion cell density (Linden and Perry, 1982; Perry and Linden, 1982; Ault *et al.*, 1985; Eysel *et al.*, 1985; Leventhal *et al.*, 1988a). It has also been demonstrated that in the normal cat retina ganglion cell dendrites tend to be displaced from the cell body down the normal ganglion cell density gradient (Schall and Leventhal, 1987); such a tendency is not observed, however, in the rat, which has a very shallow ganglion cell density gradient (Schall *et al.*, 1987), or in primate central retina, where the dendrites of midget ganglion cells in central retina are normally directed toward the center of the foveal pit regardless of the density of neighboring cells.

The results described in this chapter indicate that the development of cell body and dendritic field size seems to be independent of the spatial distribution of the dendritic field relative to the cell body. For example, in the cat, the dendrites of all cell types extend preferentially into ganglion cell-free regions even though the cell bodies and dendritic fields of only alpha and beta cells increase in size if cell density is reduced. More support for this view comes from our observations in the monkey. A and B cells in central primate retina grow to be abnormally large even though their dendrites can extend directly toward the region containing the greatest density of neighboring cells. Thus, it appears that more than one "competitive" mechanism is involved in normal ganglion cell development. One determines the direction of dendritic growth, and the other determines cell body size and the amount of dendrite produced.

The dual mechanism hypothesis is attractive since it suggests that uniform coverage of the retina can be achieved for all classes of cells as a result of the mechanism that normally results in the displacement of the dendrites down the ganglion cell density gradient. This mechanism is likely to be competition for afferents (described below; see Chapter 7 in this volume) and may be responsible for the development of the ganglion cell mosaic, that is, the complete and uniform coverage of ganglion cell dendritic fields over the retina (Wässle *et al.*, 1981a,b; Schall and Leventhal, 1987). The second mechanism, which may be a sort of direct interaction among neighboring cells (described below), is hypothesized to be specific for alpha and beta cells in the cat and A and B cells in the monkey. This mechanism is hypothesized to be responsible for the changes in the morphology of these cell types normally associated with eccentricity. A dual mechanism hypothesis can provide for uniform coverage of the retina by all cell types even though changes in neuronal density do not affect the overall sizes of cell types other than alpha and beta in the cat and A and B in the monkey.

### 8.4.5. Retinal Ganglion Cell Development and Competition for Afferents

It has been hypothesized that, in rats and cats, retinal ganglion cell dendrites exhibit territorial behavior (Wässle *et al.*, 1981a; Perry and Linden, 1982; Eysel *et al.*, 1985; Leventhal *et al.*, 1988a). It has been postulated that ganglion cells are competing for the same afferents and thus tend to spread their dendrites into regions of retina free of the dendrites of other cells. To date, increases in the overall sizes of retinal ganglion cell dendritic fields resulting from reductions in ganglion cell density have not been reported. In fact, it has been suggested that the overall size of a retinal ganglion cell's dendritic field may be specified genetically (Eysel *et al.*, 1985). The results summarized in this chapter are thus important since they show that, in cats and monkeys, the dendritic fields of retinal ganglion cells increase dramatically in size if the density of neighboring ganglion cells is reduced.

The results described here also indicate that, in primates, dendritic field structure varies depending on whether or not isolated cells are in central or peripheral retina. One explanation for this is that the density of retinal ganglion cell afferents is much higher in central than peripheral primate retina. Cone density, for example, decreases by 80% over the central 5° of monkey retina and 2.5° of human retina (Rolls and Cowey, 1970). Thus, isolated cells are able to "saturate" their dendritic fields with synapses over a smaller region of central than peripheral retina. The result is that central A and B cells in the monkey have smaller dendritic fields than peripheral A and B cells regardless of the density of their neighbors.

### 8.4.6. Direct Interactions among Neighboring Cells and Ganglion Cell Development

The results reviewed in this chapter demonstrate that, in cats and primates, a retinal ganglion cell's cell body increases dramatically in size if the density of neighboring neurons is reduced. In the cat, it is not possible to tell whether increases in cell body and dendritic field size are due to reduced competition for afferents (discussed above) or to some sort of direct interactions among neighboring cells; in cat retina all the abnormally large cells within and on the borders of ganglion cell-poor zones have a reduced number of cells with which to interact directly and also extend their dendrites into the ganglion cell-free region (Leventhal *et al.*, 1988a). These cells both contact very few other cells and are at a competitive advantage with respect to their afferents.

In the monkey, it is possible to differentiate between these two possibilities; our results suggest that a cell's overall size is, at least in part, limited by direct interactions among neighbors (see also Kirby and Chalupa, 1986). The best evidence for this comes from our observations in the "bifoveate" monkey. In this sort of retina, cells on the border of the strip of cells separating the foveal pit from the experimentally depleted area are abnormally large yet their dendrites extend directly into the region of highest ganglion cell density. Thus, their dendrites may not be at a competitive advantage with respect to their afferents. Finally, it should also be noted that other factors such as the amount of space available in the LGNd to the axons of retinal ganglion cells in areas of reduced density may also contribute to the changes we have observed.

## 8.4.7. Relation to Other Neuronal Systems

It is important to recognize that the mechanisms of dendritic growth revealed by our experiments on retinal ganglion cells are probably not different from the mechanisms occurring in other parts of the nervous system. Whereas in the retina it has been possible to deplete the number of neighboring ganglion cells and thus to provide ganglion cells with more afferents than normal, in other parts of the nervous system it is possible to remove the afferents to a group of cells. It is well documented that reducing the number of afferents results in fewer, shorter dendritic branches. For example, elimination of the granule cells in the cerebellum results in a reduced distal arbor of Purkinje cell dendrites (Altman and Anderson, 1972; Rakic and Sidman, 1973a,b; Sotelo and Changeux, 1974; Sotelo, 1975; Altman, 1976; Bradley and Berry, 1976; Mariani *et al.*, 1977). Elimination of the climbing fibers results in a decrement in the proximal dendrites of Purkinje cells (Bradley and Berry, 1976; Sotelo and Arsenio-Nunes, 1976). Elimination of the cochlear nerve input to brainstem auditory nuclei results in reduced dendritic growth (Parks, 1981; Deitch and Rubel, 1984).

The foregoing investigations have considered overall dendritic shape. However, there is an example in another system of the experimental induction of directed dendritic growth which may be analogous to what has been observed in the retina. Specifically, in layer 4 of the rodent somatosensory cortex the cells are arranged in barrels which correspond to the distribution of facial vibrissae (Woolsey and Van der Loos, 1970). The dendritic trees of cells on the rim of a barrel extend into the center of the barrel (Woolsey *et al.*, 1975). If a row of vibrissae is removed in the neonate, the corresponding barrels coalesce, and the dendritic fields adjust their spatial distribution accordingly (Steffan and Van der Loos, 1980; Harris and Woolsey, 1981). As discussed above, in these studies as well as the present one, dendritic field structure may be determined by "competition" for afferents or trophic substances as well as by more direct interactions with neighboring cells of like type.

## 8.4.8. Foveal Development

The present results indicate that normal foveal development is dependent on very high ganglion cell densities in immature central retina; primates having very high densities of cells in central retina develop cell-free foveal pits while nonprimate species and experimental monkeys which have relatively low densities of ganglion cells in central retina do not. It appears that a critical density of ganglion cells must exist in immature central retina if foveation is to proceed normally. Consistent with this, the foveal pit is larger in *S. sciureus* than in *M. fascicularis* (Rolls and Cowey, 1970; Stone and Johnston, 1981; our observations); the density of ganglion cells in central retina is greater in *S. sciureus* than in *M. fascicularis* (Stone and Johnston, 1981; our observations). Also, albino humans have a poorly developed fovea and impaired visual function; the density of ganglion cells is abnormally low in the central retina of albino mammals (Stone *et al.*, 1978; Leventhal, 1982; Leventhal and Creel, 1985). Future studies of the mature and developing mammalian retina should clarify the mechanisms mediating human visual development as well as those that guide the development of the nervous system in general.

## 8.5. REFERENCES

Abramov, I., Gordon, J., Hendrickson, A., Mainline, L., Dobson, V., and Labossiere, E., 1982, The retina of the newborn human infant, *Science* **217**:265–267.

Altman, J., 1976, Experimental reorganization of the cerebellar cortex VII. Effects of late X-irradiation schedules that interfere with cell acquisition after stellate cells are formed, *J. Comp. Neurol.* **165**:65–76.

Altman, J., and Anderson, W. J., 1972, Experimental reorganization of the cerebellar cortex. I. Morphological effects of elimination of all microneurons with prolonged X-irradiation started at birth, *J. Comp. Neurol.* **146**:355–406.

Ault, S. J., Schall, J. D., and Leventhal, A. G., 1985, Experimental alterations of cat retinal ganglion cell dendritic field structure, *Soc. Neurosci. Abstr.* **11**:15.

Bowling, D. B., and Michael, C. R., 1984, Terminal patterns of single, physiologically characterized optic tract fibers in the cat's lateral geniculate nucleus, *J. Neurosci.* **4**:198–216.

Boycott, B. B., and Wässle, H., 1974, The morphological types of ganglion cells of the domestic cat's retina, *J. Physiol. (London)* **240**:397–419.

Bradley, P., and Berry, M., 1976, The effects of reduced climbing and parallel fibre input on Purkinje cell dendritic growth, *Brain Res.* **109**:133–151.

Carpenter, P., Sefton, A. J., Dreher, B., and Lim, W., 1986, Role of target tissue in regulating the development of retinal ganglion cells in the albino rat: Effects of kainate lesions of the superior colliculus, *J. Comp. Neurol.* **251**:240–259.

Cleland, B. G., and Levick, W. R., 1974a, Brisk and sluggish concentrically organized ganglion cells in the cat's retina, *J. Physiol. (London)* **240**:421–456.

Cleland, B. G., and Levick, W. R., 1974b, Properties of rarely encountered types of ganglion cells in the cat's retina and an overall classification, *J. Physiol. (London)* **240**:457–492.

Deitch, J. S., and Rubel, E. W., 1984, Afferent influences on brainstem auditory nuclei of the chicken: Time course and specificity of dendritic atrophy following deafferentation, *J. Comp. Neurol.* **229**:66–79.

Enroth-Cugell, C., and Robson, J. G., 1966, The contrast sensitivity of retinal ganglion cell of the cat, *J. Physiol. (London)* **187**:517–552.

Eysel, U. T., Peichl, L., and Wässle, H., 1985, Dendritic plasticity in the early postnatal feline retina: Quantitative characteristics and sensitive period, *J. Comp. Neurol.* **242**:134–145.

Friedlander, M. J., Lin, C.-S., Stanford, L. R., and Sherman, S. M., 1979, Structure of physiologically identified X and Y cells in the cat's lateral geniculate nucleus, *Science* **204**:1114–1117.

Harris, R. M., and Woolsey, T. A., 1981, Dendritic plasticity in mouse barrel cortex following postnatal vibrissa follicle damage, *J. Comp. Neurol.* **196**:357–376.

Hendrickson, A., and Kupfer, C., 1976, The histogenesis of the fovea in the macaque monkey, *Invest. Ophthalmol.* **15**:746–756.

Hendrickson, A., and Yuodelis, C., 1984, The morphological development of the human fovea, *Ophthalmology* **91**:603–612.

Hendry, S. H. C., Hockfield, S., Jones, E. G., and McKay, R., 1984, Monoclonal antibody that identifies subsets of neurons in the central visual system of monkey and cat, *Nature (London)* **307**:267–269.

Illing, R. B., and Wässle, H., 1981, The retinal projection to the thalamus in the cat: A quantitative investigation and a comparison with the retinotectal pathway, *J. Comp. Neurol.* **202**:265–285.

Jacobs, D. S., Perry, V. H., and Hawken, M. J., 1984, The postnatal reduction of the uncrossed projection from the nasal retina in the cat, *J. Neurosci.* **4:**2425–2433.

Kelly, J. P., and Gilbert, C. D., 1975, The projections of different morphological types of ganglion cells in the cat retina, *J. Comp. Neurol.* **163:**65–80.

Kirby, M. A., and Chalupa, L., 1986, Retinal crowding alters the morphology of alpha ganglion cells, *J. Comp. Neurol.* **251:**532–541.

Kolb, H., 1979, The inner plexiform layer in the retina of the cat: Electron microscopic observations, *J. Neurocytol.* **8:**295–329.

Kolb, H., Nelson, R., and Mariani, A., 1981, Amacrine cells, bipolar cells and ganglion cells of the cat retina: A Golgi study, *Vision Res.* **221:**1081–1114.

Leventhal, A. G., 1982, Morphology and distribution of retinal ganglion cells projecting to different layers of the dorsal lateral geniculate nucleus in normal and Siamese cats, *J. Neurosci.* **2:**1024–1042.

Leventhal, A. G., and Creel, D. J., 1985, Retinal projection and functional architecture of cortical areas 17 and 18 in the Tyrosinase-negative albino cat, *J. Neurosci.* **5:**795–807.

Leventhal, A. G., and Schall, J. D., 1983, Structural basis of orientation sensitivity of cat retinal ganglion cells, *J. Comp. Neurol.* **220:**465–475.

Leventhal, A. G., Keens, J., and Törk, I., 1980, The afferent ganglion cells and cortical projections of the retinal recipient zone (RRZ) of the cat's "pulvinar complex," *J. Comp. Neurol.* **194:**535–554.

Leventhal, A. G., Rodieck, R. W., and Dreher, B., 1981, Retinal ganglion cell classes in old-world monkey: Morphology and central projections, *Science* **213:**1139–1142.

Leventhal, A. G., Rodieck, R. W., and Dreher, B., 1985, Central projections of cat retinal ganglion cells, *J. Comp. Neurol.* **237:**216–226.

Leventhal, A. G., Schall, J. D., and Ault, S. J., 1988a, Extrinsic determinants of retinal ganglion cell structure in the cat, *J. Neurosci.*, **8:**2028–2038.

Leventhal, A. G., Schall, J. D., Ault, S. J., Provis, J. M., and Vitek, D. J., 1988b, Class specific cell death during development shapes the distribution and pattern of central projection of cat retina ganglion cells, *J. Neurosci.*, **8:**2011–2027.

Linden, R., and Perry, V. H., 1982, Ganglion cell death within the developing retina: A regulatory role for retina dendrites? *Neuroscience* **7:**2813–2837.

Mann, I., 1964, *The Development of the Human Eye*, British Medical Association, London.

Mariani, J., Crepel, F., Mikoshiba, K., Changeux, J. P., and Sotelo, C., 1977, Anatomical, physiological and biochemical studies of the cerebellum from reeler mutant mouse, *Philos. Trans. R. Soc. London* **281:**1–28.

Maslim, J., Webster, J., and Stone, J., 1986, Stages in the structural differentiation of retinal ganglion cells, *J. Comp. Neurol.* **254:**382–402.

Michael, C. R., 1983, Functional classes of neurons in monkey's lateral geniculate nucleus have distinctive morphology, *Soc. Neurosci. Abstr.* **9:**1047.

Parks, T. N., 1981, Changes in the length and organization of nucleus laminaris dendrites after unilateral otocyst ablation in chick embryos, *J. Comp. Neurol.* **202:**47–57.

Perry, V. H., and Cowey, A., 1984, Retinal ganglion cells that project to the superior colliculus and pretectum in the macaque monkey, *Neuroscience* **12:**1125–1137.

Perry, V. H., and Linden, R., 1982, Evidence for dendritic competition in the developing retina, *Nature (London)* **297:**683–685.

Perry, V. H., Oehler, R., and Cowey, A., 1984, Retinal ganglion cells that project to the dorsal lateral geniculate nucleus in the macaque monkey, *Neuroscience* **12:**1101–1123.

Polyak, S., 1941, *The Retina*, University of Chicago Press, Chicago.

Rakic, P., and Sidman, R. L., 1973a, Sequence of developmental abnormalities leading to granule cell deficit in cerebellar cortex of weaver mutant mice, *J. Comp. Neurol.* **152:**103–132.

Rakic, P., and Sidman, R. L., 1973b, Organization of cerebellar cortex secondary to deficit of granule cells in weaver mutant mice, *J. Comp. Neurol.* **152:**133–162.

Ramoa, A. S., Campbell, G., and Shatz, C. J., 1987, Transient morphological features of identified ganglion cells in living fetal and neonatal retina, *Science* **237:**522–525.

Rapaport, D. H., and Stone, J., 1983, Time course of morphological differentiation of cat retinal ganglion cells: Influences on soma size, *J. Comp. Neurol.* **221**:42–52.

Rodieck, R. W., Binmoeller, K. F., and Dineen, J., 1985, Parasol and midget ganglion cells of the human retina, *J. Comp. Neurol.* **233**:115–132.

Rolls, E. T., and Cowey, A., 1970, Topography of the retina and striate cortex and its relationship to visual acuity in rhesus monkeys and squirrel monkeys, *Exp. Brain Res.* **10**:298–310.

Rowe, M. H., and Stone, J., 1980, The interpretation of variation in the classification of nerve cells, *Brain Behav. Evol.* **17**:1233–1251.

Schall, J. D., and Leventhal, A. G., 1987, Relationships between ganglion cell dendritic structure and retinal topography in the cat. *J. Comp. Neurol.* **257**:149–159.

Schall, J. D., Perry, V. H., and Leventhal, A. G., 1987, Ganglion cell dendritic structure and retinal topography in the rat, *J. Comp. Neurol.* **257**:160–165.

Sotelo, C., 1975, Anatomical, physiological and biochemical studies of the cerebellum from mutant mice. II. Morphological study of cerebellar cortical neurons and circuits in the weaver mouse, *Brain Res.* **94**:19–44.

Sotelo, C., and Arsenio-Nunes, M. L., 1976, Development of Purkinje cells in absence of climbing fibers, *Brain Res.* **111**:389–395.

Sotelo, C., and Changeux, J. P., 1974, Transsynaptic degeneration in cascade in cerebellar cortex of staggerer mutant mice, *Brain Res.* **67**:519–526.

Steffan, H., and Van der Loos, H., 1980, Early lesions of mouse vibrissal follicles: Their influence on dendritic orientation in the cortical barrelfield, *Exp. Brain Res.* **40**:419–431.

Stevens, J. K., McGuire, B. A., and Sterling, P., 1980, Toward a functional architecture of the retina: Serial reconstruction of adjacent ganglion cells, *Science* **207**:317–319.

Stone, J., 1965, A quantitative analysis of the distribution of ganglion cells in the cat's retina, *J. Comp. Neurol.* **124**:337–352.

Stone, J., 1966, The nasotemporal division of the cat's retina, *J. Comp. Neurol.* **126**:585–600.

Stone, J., and Fukuda, Y., 1974, Properties of cat retinal ganglion cells: A comparison of W-cells with X- and Y-cells, *J. Neurophysiol.* **37**:722–748.

Stone, J., and Johnston, E., 1981, The topography of primate retina: A study of the human, bushbaby, and new- and old-world monkeys, *J. Comp. Neurol.* **196**:205–223.

Stone, J., Campion, J. E., and Leicester, J., 1978, The nasotemporal division of the retina in the Siamese cat, *J. Comp. Neurol.* **180**:783–798.

Stone, J., Dreher, B., and Leventhal, A. G., 1979, Hierarchical and parallel mechanisms in the organization of the visual cortex, *Brain Res. Rev.* **1**:345–394.

Vitek, D. J., Schall, J. D., and Leventhal, A. G., 1985, Morphology, central projections and dendritic field orientation of retinal ganglion cells in the ferret, *J. Comp. Neurol.* **241**:1–11.

Walsh, C., and Polley, E. H., 1985, The topography of ganglion cell production in the cat's retina, *J. Neurosci.* **5**:741–750.

Wässle, H., and Illing, R. B., 1980, The retinal projection to the superior colliculus in the cat: A quantitative study with HRP, *J. Comp. Neurol.* **190**:333–356.

Wässle, H., Peichl, L., and Boycott, B. B., 1981a, Dendritic territories of cat retinal ganglion cells, *Nature (London)* **292**:344–345.

Wässle, H., Peichl, L., and Boycott, B. B., 1981b, Morphology and topography of on- and off-alpha cells in the cat retina, *Proc. R. Soc. London Ser. B* **212**:157–175.

Woolsey, T. A., and Van der Loos, H., 1970, The structural organization of layer IV in the somatosensory region (SI) of mouse cerebral cortex. The description of a cortical field composed of discrete cytoarchitectonic units, *Brain Res.* **17**:205–242.

Woolsey, T. A., Dierker, M. L., and Wann, D. F., 1975, Mouse Smi cortex: Qualitative and quantitative classification of Goldi-impregnated barrel neurons, *Proc. Natl. Acad. Sci. U.S.A.* **72**:2165–2169.

# Phylogenetic, Evolutionary, and Functional Aspects of Retinal Development

# II

Chapters 1–8 have discussed retinal development at a cellular level: neurogenesis, the specification of cell type and the control of cell survival, and dendritic, somal, and axonal growth. There are other levels of analysis, however, that affect our understanding of the mechanisms of cellular development. Each vertebrate eye has a phylogenetic history that constrains which cellular mechanisms can be employed ontogenetically. In most vertebrate species, the eye grows throughout life, which requires that development and ongoing visual function be constantly accommodated to each other; even in mammals and birds, where substantial cell addition is not known to occur once visual function begins, the eye grows and reconforms substantially during early vision. Visual function also directly alters the course of retinal development. This section deals with the phylogenetic, life history, and functional levels of analysis within which the development of the retina must be understood.

Ocular growth is at an unusually interesting confluence of descriptions of retinal development that are cellular versus systemic and is a theme common to many of the chapters in this volume. Whole-eye growth in teleosts, whose retinas add cells throughout life coupled with different acuity requirements for scotopic and photopic vision, requires that the neurogenesis of rods and cones be under separable control (Chapter 11). Nonuniform retinal growth is a principal mechanism that produces varying distributions of retinal cells in different species (Chapters 9 and 10). Nonuniform growth appears to be produced, at least in part, by maturational gradients that have also been shown to be important in the changing production of different cell types, process formation, the pattern of axon outgrowth, and projection laterality (Chapters 1 and 3–6). The nonuniform growth and resulting nonuniform spatial distribution of cells influences directly somal and dendritic morphology whose coverage of the retinal surface is critical for the appropriate sampling of the visual field (Chapters 7 and 8). The rate of whole-eye growth is also modifiable by visual function: activity of retinal neurons changes the rate of ocular growth to produce an eye of appropriate focal length (Chapter 12), and this growth must secondarily alter both the distribution and morphology of retinal neurons.

In this section, Beazley, Dunlop, Harman, and Coleman raise the problem of how a common feature of vertebrate eyes—nonuniform cell distribution—is produced in two classes of vertebrates, one in which retinal cell addition occurs throughout life (amphibia), and a second in which cell addition ceases embryonically (marsupials). In fish and amphibians, maintenance of an area centralis during growth is a problem whose solution involves differential cell addition. In mammals, by contrast, the process of retinal growth is employed to produce regional specializations in cell number. Wikler and Finlay further describe how retinal growth may be employed differently in various mammalian species to produce eyes of various conformations. Fernald describes the consequences of changing eye size in fish for their optical quality, visual acuity, and visual sensitivity, and how the addition of cellular elements is orchestrated to maintain good vision in the face of continual ocular growth. Finally, Howland and Schaeffel describe how retinal defocus and the accommodative state of the eye may be used to regulate early eye growth in humans and chickens to produce an eye of optimal ocular quality.

# Development of Cell Density Gradients in the Retinal Ganglion Cell Layer of Amphibians and Marsupials

9

## Two Solutions to One Problem

Lyn D. Beazley, Sarah A. Dunlop, Alison M. Harman, and Lee-Ann Coleman

## 9.1. OVERVIEW

A feature of the mature vertebrate retina is the nonuniform distribution of ganglion cells. High ganglion cell densities in the temporally situated area centralis and in the nasotemporally aligned visual streak subserve high acuity vision in the frontal field and along the horizon. We have described the different developmental strategies adopted in amphibia and in mammals to form an area centralis and visual streak from an essentially uniform cell distribution. To facilitate the mammalian studies, we chose to study marsupials since they are born at a much more immature stage than eutherians.

In frogs, cell division continues at the retinal circumference throughout life, adding cells to each retinal layer. During the formation of cell density gradients, more cells are added at the nasal and temporal poles than elsewhere, suggesting that high cell densities are generated by differential cell addition. Dying cells have not been observed during this process, suggesting

**Lyn D. Beazley, Sarah A. Dunlop, Alison M. Harman, and Lee-Ann Coleman**
• Department of Psychology, University of Western Australia, Nedlands Western Australia 6009, Australia.

that cell death does not play a part in shaping live cell density gradients. Continued areal growth reduces cell densities throughout life.

By contrast, in mammals, all ganglion cells are generated in an early phase of cell division well before the area centralis and visual streak are present. As these specializations appear, ganglion cell numbers fall by approximately one third. Dying cells are seen and are present in greater numbers in regions destined to become of low cell density than elsewhere, indicating that cell death may play a role in the formation of live cell topography. Asymmetric retinal expansion may also be important. A second phase of mitosis adds cells to the inner and outer nuclear layers at the time cell density gradients become pronounced in the ganglion cell layer. This later phase of cell generation ceases first in areas adjacent to the presumptive area centralis. Continued cell addition to peripheral retina would differentially expand the retina and may thereby reduce cell densities more in those parts of the ganglion cell layer outside the area centralis and visual streak. However, even after the completion of mitosis and cell death, ganglion cell density gradients are not as steep as in the adult. During the later stages, ganglion cell topography must therefore be accentuated by other factors of retinal growth such as changes in the size and shape of cells.

## 9.2. INTRODUCTION

Many studies of the mature vertebrate retina have shown that ganglion cells are not uniformly distributed. The adult pattern appears to reflect the animal's environment, a concept formalized as the Terrain theory (Hughes, 1977). Until recently, little was known of the developmental events leading to the establishment of these adult patterns. The only data were for primates and suggested that the fovea gradually developed from an even cell distribution (Mann, 1964; Hendrickson and Kupfer, 1976).

In many adult vertebrates, retinal ganglion cells are organized as a temporally situated area centralis of highest density embedded within a nasotemporal visual streak. For example, these specializations are found in some frogs (Jacobson, 1960, 1962; Fite, 1976; Bousfield and Pessoa, 1980; Dunlop and Beazley, 1981; Coleman *et al.*, 1984; Humphrey and Beazley, 1985; Beazley *et al.*, 1986), turtle (Peterson and Ulinski, 1979), rabbit (Provis, 1979), cat (Stone, 1965, 1978; Hughes, 1975, 1985), wallaby (Beazley and Dunlop, 1983; Wong *et al.*, 1986), and kangaroo (Dunlop *et al.*, 1987a). We have examined the events that underlie formation of the area centralis and visual streak in the ganglion cell layer of amphibians and mammals. We anticipated interesting differences between their patterns of growth, since previous studies had indicated that the retina continues to add new cells throughout life in lower vertebrates (Saxen, 1954; Straznicky and Gaze, 1971; Jacobson, 1976; Johns, 1977; Johns and Easter, 1977; Meyer, 1978; Beach and Jacobson, 1979) but not in mammals (Sidman, 1961; Kuwabara and Wiedman, 1974; Johns *et al.*, 1979; Walsh and Polley, 1985).

In our amphibian studies, we initially examined the frog *Heleioporus eyrei* (Dunlop and Beazley, 1981) and later its close relative *Limnodynastes dorsalis* (Coleman *et al.*, 1984; Dunlop and Beazley, 1985a). This switch allowed us to collect and study neurula stages since *Limnodynastes* eggs are laid in water and are readily accessible, whereas *Heleioporus* breeds in burrows.

For the mammalian studies we chose to concentrate on marsupials. In most mammals many crucial developmental events take place *in utero*. By contrast, marsupials are born at a very immature stage and therefore share with most amphibians the advantage of being readily accessible for manipulation throughout the majority of visual development (Dunlop *et al.*, 1988). To illustrate the different degrees of mammalian visual maturation in relation to birth, fibers reach the chiasm by postnatal day (P) 1 in quokka but at embryonic day (E) 23 for cat (Williams *et al.*, 1986); the area centralis is first seen between P50 and P60 (Dunlop and Beazley, 1985b) and between E40 and E50 (Stone *et al.*, 1982; Lia et al., 1987; Wong and Hughes, 1987a) for the two species, respectively. An additional advantage is the protracted development of marsupials. This feature allowed us to analyze complex sequences of events such as cell death and neurogenesis, which are described later in this chapter. However, from descriptions for cat (Dunlop and Beazley, 1984a; Wong and Hughes, 1987a,b) it is clear that retinal development follows similar principles despite taking place in postnatal life for marsupials but prenatally in eutherians.

For our marsupial studies we selected two Macropod marsupials, the smallest of the wallabies *Setonix brachyurus*, quokka, and the grey kangaroo *Macropus fuliginosus*. We chose these species since the adult ganglion cell topographies are specialized to different degrees. Comparison might reveal those aspects of development controlling changing cell distributions.

## 9.3. IDENTIFICATION OF GANGLION CELLS

Retinal ganglion cells are defined as those cells which send an axon into the optic nerve. In Nissl-stained preparations, these cells may be difficult to distinguish from displaced amacrine cells, which form the other main component of the ganglion cell layer (Perry, 1981). In the adults of some species, it is possible to distinguish tentatively between cell types in the ganglion cell layer by their morphology. Ganglion cells are usually larger than displaced amacrine cells, tend to be more irregular in shape, and contain clumped rather than wispy Nissl substance. Neuroglia represent a small proportion of cells in this layer and can tentatively be identified by their small, round somas which become deeply stained.

To confirm a morphological identification of ganglion cells, it is necessary to adopt other techniques, including retrograde labeling with a tracer such as horseradish peroxidase (HRP). However, it is possible that not all ganglion cells become labeled. For example, those cells with axons of small diameter may not transport sufficient tracer to be clearly labeled.

Therefore, in both amphibians and adult mammals it is useful to support a tentative identification of ganglion cells by determining whether their numbers match optic axon counts. An approximate 1 : 1 relationship would be expected since mature optic axons are considered not to form side branches within the optic nerve. In frogs (Dunlop *et al.*, in press) and mammals (quokka: Coleman *et al.*, 1987; rat: Linden, 1987), displaced ganglion cells are present in low numbers. For example, in the frog *Hyla moorei* and in quokka these cells represent less than 2% of the total ganglion cell population and their exclusion is unlikely to change markedly the ratio of counts for optic axons of presumed ganglion cells. Furthermore, an efferent projection from the brain to the retina appears to be absent from frogs (Scalia and Tietelbaum, 1978); the existence of such a projection remains controversial for mammals (Brooke *et al.*, 1965; Itaya, 1980; Schnyder and Kunzle, 1984).

During development of the mammalian visual system a correspondence of ganglion cell and optic axon counts cannot be assumed. Axon numbers will be raised by the transient retino-retinal projection from each eye into the opposite optic nerve (Bunt and Lund, 1981; Bunt *et al.*, 1983) although this projection is probably relatively small (Dunlop and Beazley, 1986). In addition, immature optic axons within the fetal cat retina possess short side branches (Campbell *et al.*, 1987), a feature which has also been observed in quokka (Fig. 9.1). By reconstructing serial sections, we have recently demonstrated similar structures within the optic nerve (S. A. Dunlop and L. D. Beazley, unpublished observation); such processes would appear to be separate axons in a cross section and presumably would be included in axon counts (Fig. 9.2).

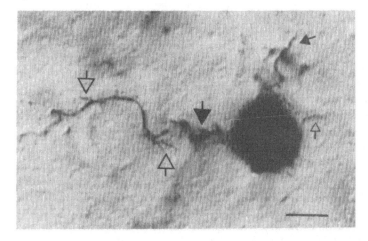

**Figure 9.1.** Montage of HRP-labeled ganglion cell photographed at 5 planes of focus from a wholemounted retina of a P6 quokka. The soma is round and bears a short somatic process (small open arrow); only one immature dendrite has formed and is shown in its entirety (small closed arrow). The optic axon (large closed arrow) has numerous short side branches (large open arrows). The eye was removed, HRP applied to the severed retrobulbar portion of the optic nerve and the eye-cup maintained in culture for several hours to allow for transport. This method (Wong and Dunlop, 1988) labels a small percentage of ganglion cells and allows individual axons and dendrites to be visualized. Nomarski optics. Scale = 5 μm.

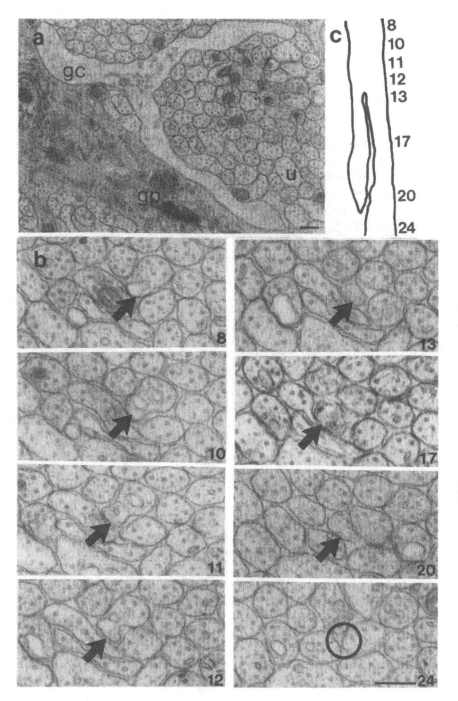

**Figure 9.2.** Transverse sections of a P50 quokka optic nerve taken immediately behind the eye. (a). A fascicule containing a growth cone. u = unmyelinated axon, gc = growth cone, gp = glial process. (b). Selected sections from a consecutive series showing a small side process (arrow) budding off an axon and terminating (open circle). (c) Reconstruction of the axon and its side branch shown in (b), indicating the levels from which micrographs were taken. Scale = 0.25 μm.

### 9.3.1. Amphibians

Labeling with HRP reveals that most ganglion cells in *Heleioporus* and *Limnodynastes* are larger than nonganglion cells. Only a minority of ganglion cells, however, can be identified by morphological criteria and they are usually those cells with the largest somas. These observations are comparable to studies of other frogs such as *Hyla moorei* (Humphrey and Beazley, 1985) and *Rana pipiens* (Beazley *et al.*, 1986).

As in other amphibians, labeling with HRP and optic axon counts (reviewed by Beazley, 1984) indicate that a substantial majority of cells in the ganglion cell layer of Leptodactylids at tadpole and later stages are ganglion cells. In some studies we therefore have examined the total cell population of the ganglion cell layer (Dunlop and Beazley, 1981; Coleman et al., 1984; Dunlop and Beazley, 1985a).

### 9.3.2. Marsupials

In adult marsupials, ganglion cells tend to be more readily distinguishable morphologically from other cells of the ganglion cell layer than is the case for amphibians. Our morphological identification of ganglion cells in quokka has been supported by HRP labeling and by optic axon counts (Beazley and Dunlop, 1983; Braekevelt *et al.*, 1986); presumed ganglion cell counts are also close to axon estimates in kangaroo (Dunlop *et al.*, 1987a). Similarly, for other marsupials, HRP studies and optic axon estimates have generally supported counts of morphologically identified ganglion cells (opossum: Hokoc and Oswaldo-Cruz, 1978, 1979; Rapaport *et al.*, 1981; Kirby *et al.*, 1982; opossum: Freeman and Tancred, 1978; Freeman and Watson, 1978). Despite controversy as to ganglion cell estimates in cat (Stone, 1965, 1978; Hughes, 1975), currently quoted values are intermediate between those originally cited from cresyl stained material (discussed by Williams *et al.*, 1986).

In developing retinas, cells of the ganglion cell layer appear immature in cresyl-stained material. Cells lack clumped Nissl substance and are fairly regular in outline; in addition, all cell somas in this layer fall within a narrow size range. We do not consider that it is possible to identify immature marsupial ganglion cells without a labeling technique such as HRP (Dunlop and Beazley, 1985b). A similar difficulty seems to apply for other developing mammals. For example, in E47 cat, ganglion cell estimates of 860,000 (Stone *et al.*, 1982) and 300,000 (Wong and Hughes, 1987) remain tentative until labeling studies are undertaken.

### 9.4. MATURE GANGLION CELL TOPOGRAPHY

### 9.4.1. Amphibians

For both *Heleioporus* and *Limnodynastes* (Fig. 9.3), a temporally situated area centralis is embedded within a visual streak (Dunlop and Beazley, 1981; Coleman *et al.*, 1984). In addition, *Heleioporus* possesses a second high-density

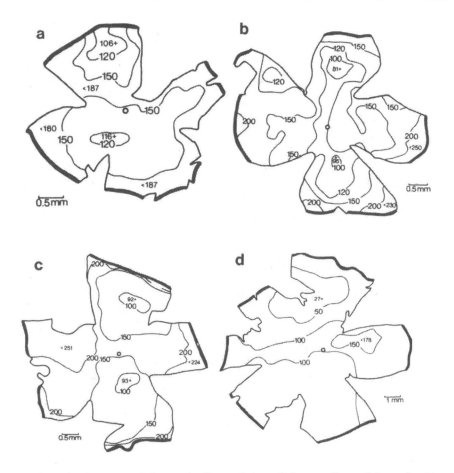

**Figure 9.3.** Isodensity maps of the total cell population of the ganglion cell layer for *Limnodynastes:* (a) tadpole, (b) at metamorphic climax, (c) juvenile at 2 months after metamorphosis and an adult. (a), (b), and (c) are right eyes, (d) is a left eye. Numbers refer to cells/$(100 \ \mu m)^2$. The optic nerve head is denoted by a circle, v = ventral. Reproduced from Fig. 5, Coleman *et al., Journal of Embryology and Experimental Morphology* **83:**119–135 (1984).

patch within the nasal arm of the streak, a specialization that may reflect the burrowing habits of this species.

### 9.4.2. Marsupials

For both quokka and kangaroo, ganglion cells are organized as a temporally situated area centralis and visual streak (Fig. 9.4). However, the streak is much more pronounced in kangaroo, probably reflecting its more open habitat with a clear view of the horizon.

In both species, nonganglion cells do not share the ganglion cell topography, being approximately even in quokka and displaying only a weak visual streak in kangaroo. The majority of these nonganglion cells are probably

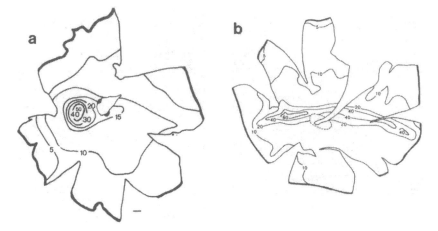

**Figure 9.4.** Isodensity maps of ganglion cells in right retinas of adult quokka (a) and kangaroo (b). Numbers refer to cells/(100 μm)$^2$. Ventral is down. The optic nerve head is split and is hatched in (a) and is shown as a broken line in (b). Scale = 1 mm for (a) and 2 mm for (b). Part (a) is reproduced from Beazley and Dunlop, *Journal of Comparative Neurology* **216:**211–231 (1983); Part (b) is from Dunlop *et al.*, *Vision Research* **27:**151–164 (1987).

displaced amacrine cells. Neuroglia are present in smaller numbers and many are concentrated within a few hundred microns of the retinal periphery.

## 9.5. DEVELOPMENT OF GANGLION CELL TOPOGRAPHY

### 9.5.1. Amphibians

We have studied cell distributions in animals at three developmental stages: tadpoles, metamorphic climax, and juveniles 2 months later. For both *Limnodynastes* and *Heleioporus*, the distribution of cells changes dramatically during development. The topography is fairly uniform in the tadpole, except for a tendency for higher densities to be found around almost the entire retinal periphery. Between metamorphic climax and juvenile stages, high-density patches form at the nasal and temporal poles and densities start to fall dorsally and ventrally thereby leading to formation of a visual streak; the area centralis does not become apparent until more than 2 months after metamorphosis.

The postmetamorphic development of retinal ganglion cell topography is probably a feature common to most frogs. We have documented the transformation for *Xenopus laevis* (Dunlop and Beazley, 1984b) and have observed a similar principle in the Australian frogs *Hyla moorei, Neobatrachus pelobatoides,* and *Heleioporus albopunctatus* (Dunlop and Beazley, 1981). Furthermore, Bousfield and Pessoa (1980) have demonstrated an accentuation of cell density gradients in *Hyla raniceps* as the eyes continue to grow.

## 9.5.2. Marsupials

To avoid misidentification of cell types in developing quokka and kangaroo, we initially chose to describe the distribution of the total cell population of the ganglion cell layer. For both species, cells are initially approximately uniform in their distribution (Figs. 9.5 and 9.6). An area centralis is first apparent at P57-60 in quokka and a weak visual streak has formed by P57 in kangaroo. At later stages, cell densities stabilize first in the area centralis. Peripheral cell densities remain higher than in the adult for several months (Beazley and Dunlop, 1983; Dunlop *et al.,* 1987).

Analyses of the total cell population of the ganglion cell layer leave open the possibility that an area centralis and a visual streak of ganglion cells are masked by the presence of nonganglion cells. We therefore labeled ganglion

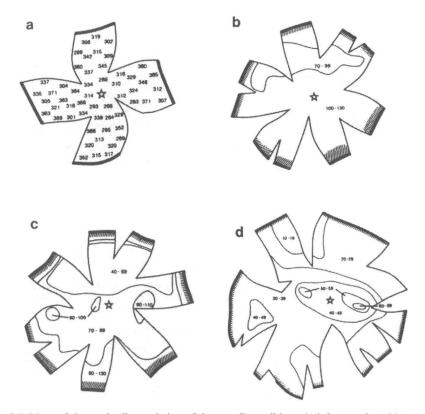

**Figure 9.5.** Maps of the total cell population of the ganglion cell layer in left eyes of quokka: (a) P30, (b) P50, (c) P60, (d) P106. Numbers refer to cells/(100 μm)². Hatching represents densities of 130–200 cells/(100 μm)². Raw data are shown for P30 as densities were too uniform to allow isodensity maps to be drawn. The optic nerve head is denoted by a star; ventral is down. Scale = ×8 for P30 and ×5 for other retinas. Modified from Fig. 2, Dunlop and Beazley, *Journal of Comparative Neurology* **264:**14–23 (1987).

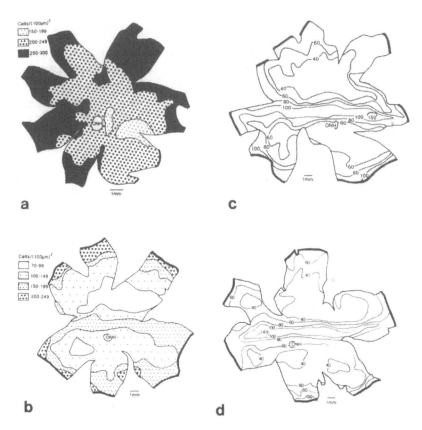

**Figure 9.6.** Isodensity maps of the total cell population of the ganglion cell layer in kangaroo at P39 (a), P57 (b), P84 (c), and P104 (d). P39, P57, and P104 are left eyes; P84 is a right eye. Ventral is down; ONH = optic nerve head. Reproduced from Fig. 3, Dunlop *et al., Vision Research* **27**:151–164 (1987).

cells with HRP injected into the optic tract and visual centers at different stages of development in quokka (Dunlop and Beazley, 1985b). Not all cells contain HRP but, as we argued earlier in this chapter (Section 9.3), axon counts do not indicate whether all ganglion cells become labeled. Indeed, it is likely they did not. From our ultrastructural studies, we found growth cones in the optic nerve immediately behind the eye up to P50 in quokka (Fig. 9.2). If these growth cones arise from axons which enter the optic nerve late, then their cell bodies will remain unlabeled after injection of a tracer into the brain.

Nevertheless, at P20, P30, and P40, stages before the appearance of the area centralis and visual streak in the total cell population, 86%, 79%, and 70%, respectively, of cells in the ganglion cell layer become HRP-labeled (Braekevelt *et al.,* 1986) and their distribution (Fig. 9.7) is approximately uniform (Dunlop and Beazley, 1985b). These findings indicate that, even if we did not identify all ganglion cells by their uptake of HRP, the majority are labeled. Therefore at these early stages there could not be a pronounced

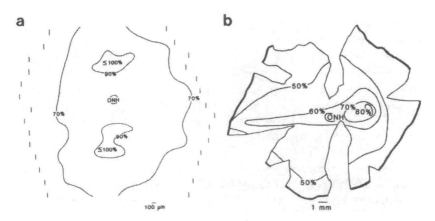

**Figure 9.7.** Maps of percentage of the total cell population of the ganglion cell layer that labeled with HRP in quokkas at P40 and in adulthood. The P40 retina is reconstructed from sections of a right eye whereas the adult is a whole mount of the left eye. Ventral is down; ONH = optic nerve head. Modified from Fig. 5, Dunlop and Beazley, *Developmental Brain Research* **23:**81−90 (1985).

gradient of ganglion cell density which remains undetected by our labeling procedure.

## 9.6. EVENTS UNDERLYING CHANGES IN GANGLION CELL TOPOGRAPHY

### 9.6.1. Cell Addition to the Ganglion Cell Layer

In the studies described here we examined the possibility that differential live cell densities arise by greater addition of cells to regions of the ganglion cell layer in which the area centralis and visual streak will form than elsewhere.

#### 9.6.1.1. Amphibians

Cell number in the ganglion cell layer increases with age. For example, in *Limnodynastes* tadpoles, axon counts indicate that there are approximately 70,000 ganglion cells compared with 200,000 in the juvenile and over 300,000 for the adult (Coleman *et al.*, 1984); corresponding values for *Heleioporus* are approximately 70,000, 180,000, and 440,000 (Dunlop and Beazley, 1981).

In *Limnodynastes,* we examined patterns of cell addition to the ganglion cell layer by counting mitotic figures in normal material and after animals were treated with colchicine to allow a build-up of cells in metaphase. Other animals were injected with $^3$H-thymidine and killed up to 24 hr later. Results indicate that, at each developmental stage, mitosis is confined to the ciliary margin and that more cells are added at the nasal and temporal poles than dorsally or ventrally (Coleman *et al.*, 1984).

However, from these data we cannot predict the proportion of newly generated cells at the ciliary margin which will enter the ganglion cell rather than other retinal layers. We therefore labeled dividing cells by injection of $^3$H-thymidine and allowed animals to develop for several weeks. Densely labeled cells form a ring at a distance from the retinal margin. More $^3$H-thymidine labeled cells are found nasally and temporally than at the other poles. The disparity suggests that differential cell addition to the ganglion cell layer plays a major role in the formation of high cell densities across the nasotemporal axis.

A similar conclusion may be drawn from a $^3$H-thymidine study using *Xenopus laevis* between metamorphic climax and 2 months after metamorphosis. Tay and his colleagues (1982) reported more dividing cells temporally than nasally. The finding is compatible with a wholemount analysis showing that a greater proportion of temporal as compared to nasal retina contained high densities of cells (Dunlop and Beazley, 1984b).

### 9.6.1.2. Marsupials

Total cell numbers in the ganglion cell layer reach a maximum of 500,000 at P45–P50 for quokka (Dunlop and Beazley, 1987) and 2,000,000 for kangaroo at P57 (Dunlop *et al.*, 1987). These values are approximately 30% above adult counts. Similarly, estimates of ganglion cells labeled with HRP injected into visual centers and optic tracts in quokka show a peak at P50 (Braekevelt *et al.*, 1986). Since cell numbers in the ganglion cell layer fall during area centralis and visual streak formation, it is unlikely that cell addition to the ganglion cell layer underlies this change.

This conclusion is supported by a $^3$H-thymidine study (Harman and Beazley, in press) to determine when cells of the ganglion cell layer in quokka complete their final division. We injected $^3$H-thymidine at different ages and examined animals at P100, when ganglion cell topography is well established. Ganglion cells were identified by morphological criteria or by labeling with HRP.

Our results (Fig. 9.8) indicate that centrally located ganglion cells are generated in the first 5 postnatal days. At P7, cells at a more peripheral location are born and by P10 cell addition is panretinal. From P12 to P25, cells in a peripheral ring complete their divisions. When each ganglion cell class (reviewed by Perry, 1982) is analyzed separately, a further level of complexity may be revealed. In cat also, there is an overall central to peripheral pattern of generation for presumed ganglion cells. Furthermore, X cells tend to be generated before Y cells (Walsh *et al.*, 1983; Walsh and Polley, 1985). These authors describe a "rough spiral" pattern amongst $^3$H-thymidine labeled cells. We have seen a similar tendency in quokka but its significance is unknown.

Sengelaub and his co-workers (1986) report that in hamster ganglion cells tend to be born before nonganglion cells in the ganglion cell layer, although

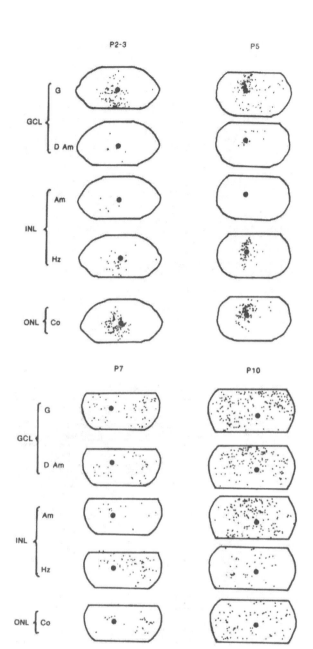

**Figure 9.8.** Maps showing ³H-thymidine labeled cells in the ganglion cell layer (GCL), inner nuclear layer (INL), and outer nuclear layer (ONL) of quokkas sacrificed at P100. Animals were injected with ³H-thymidine at P2-Pd, P5, P7, P10, P12, P15, P25, P40, P50, P55, P60, and P70. Each dot represents a ³H-thymidine labeled cell. Presumed cell types are indicated: g = ganglion cells, D Am = displaced amacrine cells, Am = amacrine cells, H or Hz = horizontals, Co = cones, Glia = neuroglia, B = bipolar cells, M = Müller glia, and Ro = rods. We have reconstructed curved sections as a Mercator's projection, which results in an accentuation of the axis parallel to the plane of section (nasotemporal). The location of the optic nerve head is shown as a solid circle, ventral is down, and temporal is to the left. The absence of label at P40 does not reflect a failure of the technique as many cells were ³H-thymidine labeled in extraocular tissue and in the brain of this animal (Harman and Beazley, 1988).

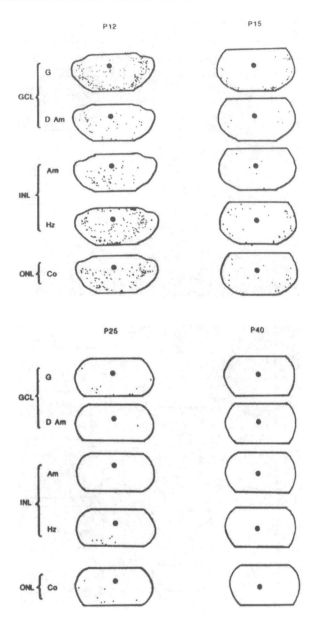

**Figure 9.8.** (*Continued*)

displaced amacrine cells were not distinguished from neuroglia. In quokka, the periods of ganglion cell and displaced amacrine cell generation seem to overlap considerably. Neuroglia, located in a far peripheral ring, are the only cells in the ganglion cell layer generated after P50 (Fig. 9.8).

It is possible that some ganglion cells, despite their early generation, do not enter the ganglion cell layer until after P50. These cells might then be-

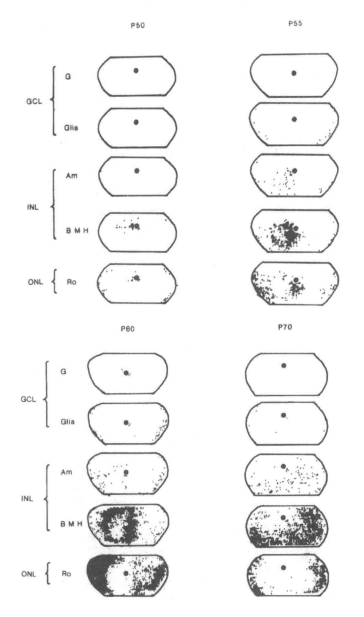

**Figure 9.8.** (*Continued*)

come incorporated into the developing area centralis and visual streak. To test this possibility in quokka, we compared patterns of labeled cells for animals injected with [3]H-thymidine at equivalent stages but examined at either P50 or P100. The patterns suggest that ganglion cells do not enter the ganglion cell layer in regions of the developing area centralis and visual streak between P50 and P100 (compare Figs. 9.8 and 9.9).

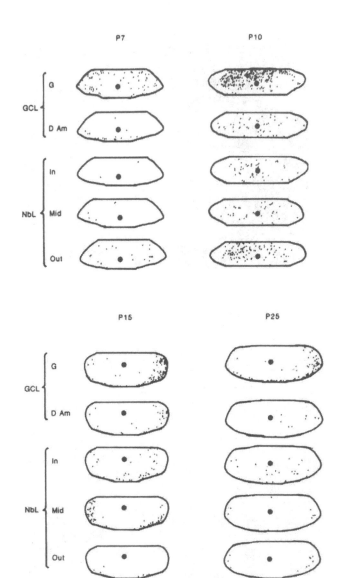

**Figure 9.9.** Maps showing [3]H-thymidine labeled cells in quokkas sacrificed at P50. Animals were injected with [3]H-thymidine at P7, P10, P15, and P25. The neuroblastic layer (NbL) has been divided in three parts of equal depth; however, we cannot determine whether these demarcations correspond to the inner and outer parts of the INL and the ONL. Conventions are as in Fig. 9.8. The number of [3]H-thymidine labeled cells is probably underestimated for the NbL as cells within it are small and it is sometimes difficult to determine whether they are [3]H-thymidine labeled. Comparison of the above maps with Figure 9.8 shows that ganglion cells destined to form the area centralis are generated by P25 and enter the ganglion cell layer before P50 and not between P50 and P100. Since the Mercator's projection distorts retinal shape, particularly in young retinas, it is inappropriate to use such reconstructions of sectioned material to assess the extent of asymmetric retinal growth.

In summary, our ³H-thymidine studies show that, in quokka, ganglion cells are generated and enter the ganglion cell layer before the area centralis and visual streak are formed. Differential cell addition to the ganglion cell layer does not underlie formation of the area centralis and visual streak.

### 9.6.2. Cell Transformation

Hinds and Hinds (1978, 1983) have raised the possibility that during development some ganglion cells lose their axons to become displaced amacrine cells. This issue has yet to be substantiated, although a study by Perry and his co-workers (1983) has been cited as evidence against it. However, for both amphibians and marsupials, we have described the emergence of an area centralis and visual streak within the total cell population of the ganglion cell layer. Therefore, transformation of cell type cannot fully explain the emerging cell topography.

### 9.6.3. Cell Death

We examined patterns of cell death during development of the ganglion cell layer. If cell death plays a part in shaping live cell density and all cells die at similar rates ( Hughes , 1961), more dying cells would be expected in dorsal and ventral retina, destined to be of low live cell density, compared to the presumptive area centralis or visual streak.

#### 9.6.3.1. Amphibia

Although live cell numbers continue to rise throughout life in amphibians, it is possible that this net overall gain is achieved by the rate of cell addition exceeding that for cell death. However, we consider it unlikely that cells die at the stages we have studied as dying cells are not seen in the ganglion cell layer. This apparent absence is probably not due to our inability to recognize dying cells in frog as these are readily apparent after optic nerve crush (Humphrey and Beazley, 1985; Beazley et al., 1986). The suggestion that cell death does not occur at the stages we have examined is supported by studies for Rana in which dying cells were seen only at the earliest stages of eye development (Glucksmann, 1940).

#### 9.6.3.2. Marsupials

Live cell numbers in the ganglion cell layer peak at P50 and P57 for quokka and kangaroo, respectively, and then undergo a 30% reduction to be close to adult values by P87 and P104 (Dunlop et al., 1987; Dunlop and Beazley, 1987). Dying cells are present during both the rise and fall in live cell numbers. Peaks in counts for dying cells, at approximately 1% of the live cell population, are found at the same time as maximal live cell counts. By P87 in quokka and P104 in kangaroo, dying cells represent less than 0.2% of the live cell population.

It is not yet possible to determine whether a pyknotic profile is a degenerating ganglion cell. We consider that from P50 in quokka many dying cells are ganglion cells, since the number of live ganglion cells falls at this time whereas counts of nonganglion cells do not (Dunlop and Beazley, 1985b). Furthermore, our $^3$H-thymidine studies suggest that a considerable proportion of ganglion cells generated between P10 and P25 have died between P50 and P100 (compare Figs. 9.8 and 9.9).

However, to circumvent the problem of the identity of dying cells, their distributions were compared to the topography of the total live cell population (Dunlop et al., 1987; Dunlop and Beazley, 1987). The topography of live cells largely mirrors changes among ganglion cells, since displaced amacrine cells are more uniformly spaced (Dunlop and Beazley, 1985b). Dying cell counts are expressed per 100 live cells to allow comparison between regions of differing live cell density.

From the earliest stages examined as wholemounts, P30 for quokka and P39 for kangaroo, dying cells are seen over the entire retinal surface but their density changes with time. At P30 in quokka, most dying cells per 100 live cells are situated around the optic nerve head but by P50 the wave of death has reached the retinal periphery (Fig. 9.10). A similar distribution develops by P57 in kangaroo. Detailed analyses indicate that in quokka, there are fewer dying cells per 100 live cells in temporal compared to other central retinal regions before and during the emergence of the area centralis. Similarly, in kangaroo, values for dying cells per 100 live cells tend to be lower in the presumptive area centralis and visual streak than elsewhere. Moreover, a high peripheral density of dying cells at the time that live cell topography becomes nonuniform (Fig. 9.10), suggests that cell death plays a role in shaping live cell density gradients around the entire retinal circumference (Dunlop et al., 1987; Dunlop and Beazley, 1987).

Live cell density gradients continue to be accentuated in both quokka and kangaroo after the cessation of cell death. It would seem therefore that cell death probably plays a role in the establishment of live cell density gradients but that dying cells are not present for a sufficiently lengthy period to be the sole explanation for changing live cell topography.

Our conclusion that cell death plays a part in establishing live cell distributions is in accord with findings for hamster (Sengelaub and Finlay, 1982), cat (Wong and Hughes, 1987b), and human (Provis, 1987). Other reports for rat (Dreher et al., 1984), cat (Stone et al., 1984), and rabbit (Stone et al., 1985) have described a more uniform distribution of dying cells. It would be interesting to sample the retina of these species at a higher frequency than in the previous studies, since dying cell numbers always represent a small percentage of the live cell population and any nonuniformity in cell death is easily missed. Furthermore, gradients in dying cell densities may be more readily detected if dying cell distributions had been expressed as a proportion of the live cell population.

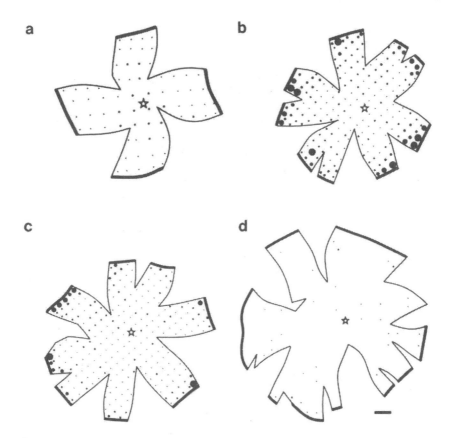

**Figure 9.10** Maps of dying cells in the ganglion cell layer for those quokka retinas shown as live cell maps in Fig. 9.5. P30 (a), P50 (b), P60 (c), P106 (d). Dots, in ascending size, represent numbers of dying cells/100 live cells : 0–0.49, 0.5–0.99, 1.0–1.49, 1.5–1.99, 2.3, 3+. Conventions as in Fig. 9.5. Scale = 1.5 mm for a , 1.0 mm for b–d. Modified from Fig. 2, Dunlop and Beazley, *Journal of Comparative Neurology* **264:**14–23 (1987).

To summarize, cell death has not been demonstrated during normal development in the amphibian ganglion cell layer. However, in marsupials there is a differential distribution of dying cells suggesting that cell death plays a part in sculpting ganglion cell densities.

### 9.6.4. Differential Growth of the Retina

Retinal area increases dramatically with age for both amphibians (Dunlop and Beazley, 1981; Coleman *et al.,* 1984) and marsupials (Beazley and Dunlop, 1983; Dunlop *et al.,* 1987). Changing cell topography would result if areal growth were asymmetric, with least expansion in regions destined to have high cell density. Areal enlargement can be resolved into two orthogonal linear components, namely, radial and tangential growth (Mastronarde *et al.,*

1984). Differences in the extents of radial and/or tangential growth could result in changing cell topography. If the dorsal and ventral axes became longer than the nasal and temporal ones, the initially even distribution of ganglion cells would be transformed into one with high densities across the nasotemporal axis. Differential tangential growth could underlie changing cell topography if the retina underwent greater extension in regions other than those destined to become the area centralis and visual streak.

### 9.6.4.1. Amphibians

To analyze radial growth, we compared the lengths of the dorsal, ventral, nasal, and temporal axes at each stage of development for *Limnodynastes*. Lengths are equivalent up to 2 months postmetamorphosis but by 10 months the nasal axis has undergone a greater extension than the temporal one (Fig. 9.3). Our data suggest that differential axial elongation does not underlie the emergence of visual streak. However, the late formation of the area centralis may result from a transient reduction in the growth of the temporal axis.

We have not undertaken a mathematical analysis of tangential growth patterns in amphibians. The calculations (Mastronarde *et al.*, 1984) assume that the number of cells in the ganglion cell layer remains constant, whereas as we have described in this chapter (section 9.6.1.1), this value increases throughout life. A technique whereby growth patterns of pigment epithelium can be visualized in pigment-chimeric eyes (Hunt *et al.*, 1987) may prove valuable to chart tangential growth in amphibian eyes.

### 9.6.4.2. Marsupials

Analyses of quokka (Dunlop, unpublished observations) and kangaroo retinas (Dunlop *et al.*, 1987) indicate that the increases in lengths of the nasotemporal and dorsoventral axes are approximately comparable throughout development. The results argue that radial growth is not differential and therefore is not a major factor underlying the changes in live cell topography. To examine differential tangential growth for both quokka and kangaroo, we compared numbers of cells in retinal segments at P100 and in adults. We followed a modification of the approach adopted to analyze growth of the kitten eye (Mastronarde *et al.*, 1984). We did not examine retinas younger than P100 as it would not have been possible to allow for the contribution of cell death to changes in live cell density. Our data suggest strongly that there is less tangential growth in the developing area centralis and visual streak than in other retinal regions (Fig. 9.11). Similarly, in postnatal cat, differential tangential growth has been shown to accentuate live cell topography (Mastronarde *et al.*, 1984).

In summary, differential radial growth is probably not an important factor in establishing cell density contours in amphibians or mammals. A possible exception is the formation of the area centralis after metamorphosis in frog. Results for marsupials indicate differential tangential growth is a major component influencing cell topography.

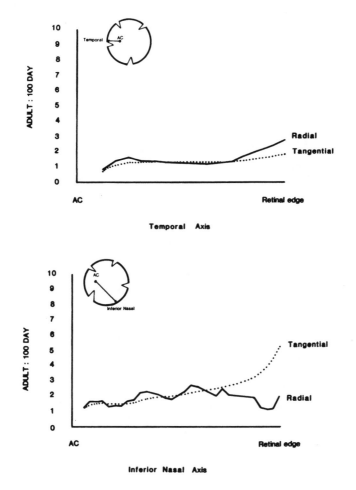

**Figure 9.11.** Curves of tangential and radial expansion, derived from a spherical model of the eye, illustrating patterns of retinal growth between P100 and adult in kangaroo. Sectors from the area centralis to temporal (upper figure) and area centralis to inferonasal (lower figure) peripheries (see insets) show that tangential and radial growth take place to approximately equal extents for temporal retina. However, there is more tangential than radial growth for regions in inferonasal retina outside the area centralis and visual streak.

## 9.7. FACTORS LEADING TO DIFFERENTIAL RETINAL GROWTH IN MARSUPIALS

### 9.7.1. Cell Addition to the Inner and Outer Nuclear Layers

As part of our $^3$H-thymidine studies of developing quokka retina, we examined the number and distribution of dividing cells in the inner and outer nuclear layers, as well as in the ganglion cell layer.

Animals were injected with $^3$H-thymidine at different developmental stages, and retinas were examined at P100. We have tentatively identified the cell types from cresyl-stained sections by the morphology of cells and their location. The exception was displaced ganglion cells, which were identified with HRP and found to be generated at the same time as their orthotopic counterparts. This result, and a comparable finding for mouse (Drager, 1985), argue against the suggestion (Bunt and Minckler, 1977; Linden, 1987) that displaced ganglion cells are generated late and, as a result, are prevented from completing their migration into the ganglion cell layer.

It is clear that cell division takes place in two phases (Fig. 9.8). In the first phase, which proceeds from center to periphery, ganglion cells and displaced amacrine cells, amacrine cells, horizontal cells, and cones are generated by P25. Furthermore, at any particular location, $^3$H-thymidine labeled cells are found in all three retinal layers (Fig. 9.8). This equivalence might lead one to predict that cell types born in the first phase in each layer will be subject to similar changes in their distribution during any subsequent differential growth of the retina. To investigate this possibility, we are currently determining whether the topography of cones matches that of ganglion cells in adult quokka, as is the case as in adult cat (Stone, 1965, 1978; Steinberg *et al.*, 1973) and adult monkey (Perry and Cowey, 1985).

The second phase of cell division is quite different from the first (Fig. 9.8). Different cell types are generated, including bipolar cells, Müller glia, and rods. Furthermore, unlike the first phase of $^3$H-thymidine labeled cells, the second originates in centrotemporal retina, where the area centralis will form, and then extends as a ring centered on the area centralis until it reaches the periphery. From figure 9.8, it is clear that there is a disparity in retinal eccentricity between $^3$H-thymidine labeled cells in the inner and outer nuclear layers. This result suggests that many cells of the inner nuclear layer complete their divisions later than cells lying in the equivalent region of the outer nuclear layer.

In accord with our findings, $^3$H-thymidine studies for cat have shown that cell division in the inner and outer nuclear layers continues into postnatal life (Johns *et al.*, 1979) whereas ganglion cells are generated by midgestation (Walsh and Polley, 1985). Furthermore, as for quokka, cells destined to lie in the inner nuclear layer of cat are amongst the last cells to complete their divisions. However, the studies in cat have not been sufficiently comprehensive to determine whether there are two phases of retinal cell generation such as we have described in quokka.

In another series (Harman and Beazley, 1987) we analyzed patterns of mitosis in quokka retina by counting mitotic figures. As for other mammals (Stone *et al.*, 1985; Beazley *et al.*, 1987), mitoses are largely confined to the ventricular surface of the retina. The density of mitotic figures averaged across the retina is bimodal with peaks at P3–P7 and P43 (Fig. 9.12). This result strongly supports the concept of two phases of cell birth as suggested by our $^3$H-thymidine studies. Moreover, after this second peak, mitosis ceases first in centrotemporal retina at the time the area centralis becomes apparent

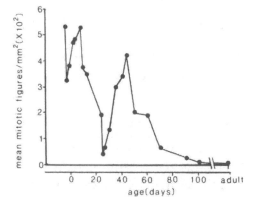

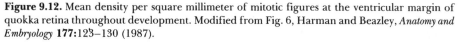

**Figure 9.12.** Mean density per square millimeter of mitotic figures at the ventricular margin of quokka retina throughout development. Modified from Fig. 6, Harman and Beazley, *Anatomy and Embryology* **177**:123–130 (1987).

(Fig. 9.13) and as the outer plexiform layer begins to develop (Harman and Beazley, 1987). As development proceeds, mitosis gradually becomes confined to more peripheral regions. Similarly, an expanding "cold spot" in mitotic activity has been reported for cat (Rapaport and Stone, 1982) and a "cold streak" for rabbit (Stone *et al.*, 1985).

A mitotic figure may represent a cell that will undergo subsequent divisions or one that is about to withdraw from the mitotic cycle. A cell is seen as densely labeled with $^3$H-thymidine only if it does not undergo any subsequent divisions before the animal is sacrificed (Fujita, 1962). Therefore, it is possible to deduce whether mitotic figures will undergo subsequent divisions by comparing the patterns of mitosis and of $^3$H-thymidine labeled cells in a particular retinal region. For example, in quokka at P50 and P60 mitotic figures extend to the retinal periphery. However, if an animal is injected with $^3$H-thymidine

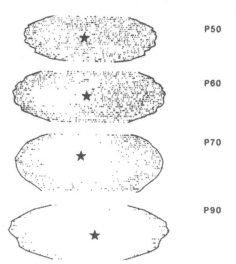

**Figure 9.13.** Maps showing mitotic figures at the ventricular surface of the developing quokka retina at P50, P60, P70, and P90. Each dot represents a mitotic figure. We have reconstructed curved sections as a Mercator's projection which results in an accentuation of the axis parallel to the plane of section (nasotemporal). The location of the optic nerve head is shown as a star, ventral is down, and temporal is to the left. A correction has been applied to allow for shrinkage during processing. Scale = ×4. Modified from Fig. 5, Harman and Beazley, *Anatomy and Embryology* **177**:123–130 (1987).

at P55 and examined at P100, cells become labeled in a ring surrounding the emerging area centralis; there are no $^3$H-thymidine labeled cells near the periphery. The difference between the distributions of mitotic figures and of $^3$H-thymidine labeled cells indicates that cells are continuing to divide more peripherally but are withdrawing from the mitotic cycle in the ring surrounding the cold spot.

In the second phase of cell generation, cells being added to the inner and outer nuclear layers must be inserted between cells born in the first generative phase. If more cells are added to some parts of the retina than others, it seems likely that a differential expansion of the retina will take place. The more extended period for cell division in peripheral regions compared to the centrotemporal region would therefore be expected to stretch peripheral retina to a greater extent than the area centralis. As ganglion cells are born exclusively in the first phase of cell generation, our hypothesis provides at least a partial explanation for their density falling least in the area centralis and most peripherally between P50 and P100.

### 9.7.2. Changes in the Size, Shape, or Packing of Cells

Despite the probable importance of cell addition to differential retinal expansion, accentuation of ganglion cell topography continues after mitosis has ceased. At these later stages, areal expansion may reflect enlargement of cell bodies and/or their processes. Further studies will be necessary to determine which cell types undergo these changes. However, the role of ganglion cells is likely to be minimal since retinal area increases normally in postnatal rat despite total destruction of the ganglion cell population by optic nerve section (Beazley *et al.*, 1987).

ACKNOWLEDGMENTS. L.D.B. and S.A.D. are, respectively, Principal Research Fellow and Senior Research Officer of the NH&MRC(Australia). A.M.H. and L.-A.C. hold Commonwealth of Australia postgraduate research scholarships. Studies were funded by the NH&MRC (Australia), M.D.R.A. (Western Australia), OPSM, and TVW Telethon Foundation. We are grateful to Professor S. D. Bradshaw for assistance from the Zoology Department, University of Western Australia, and in particular to R. McNeice for care of the quokka colony. We also thank W. A. Longley for analyses of some kangaroo retinas, M. Palm for computer programming, S. Hopwood and H. Jurkiewicz for photographic help, and I. Smits and S. Redmond for technical assistance.

### 9.8. REFERENCES

Beach, D. H., and Jacobson, M., 1979, Patterns of cell proliferation in the retina of the clawed frog during development, *J. Comp. Neurol.* **183**:603–614.

Beazley, L. D., 1984, Formation of specific neuronal connections in the visual system of lower vertebrates, in *Current Topics in Research on Synapses,* Vol. 1 (D. G. Jones, ed.), pp. 53–119, Alan R. Liss, New York.

Beazley, L. D., and Dunlop, S. A., 1983, The evolution of an area centralis and visual streak in the marsupial *Setonix brachyurus, J. Comp. Neurol.* **216:**211–231.

Beazley, L. D., Darby, J. E., and Perry, V. H., 1986, Cell death in the retinal ganglion cell layer during optic nerve regeneration for the frog *Rana pipiens, Vision Res.* **26:**543–556.

Beazley, L. D., Perry, V. H., Baker, B., and Darby, J. E., 1987, An investigation into the role of ganglion cells in the regulation of division and death of other retinal cells, *Dev. Brain Res.* **33:**169–184.

Bousfield, J. D., and Pessoa, V. F., 1980, Changes in ganglion cell density during postmetamorphic development in a neotropical tree frog, *Hyla raniceps, Vision Res.* **20:**501–510.

Braekevelt, C. R., Beazley, L. D., Dunlop, S. A., and Darby, J. E., 1986, Numbers of axons in the optic nerve and of retinal ganglion cells during development in the marsupial *Setonix brachyurus, Dev. Brain Res.* **25:**117–125.

Brooke, R. N. L., Downer, J. De C, and Powell, T. P. S., 1965, Centrifugal fibers to the retina in the monkey and cat, *Nature (London)* **207:**1365–1367.

Bunt, A. H., and Minckler, D. S., 1977, Displaced ganglion cells in the retina of the monkey, *Invest. Ophthalmol. Vis. Sci.* **16:**95–98.

Bunt, S. M., and Lund, R. D., 1981, Development of a transient retino-retinal pathway in hooded and albino rats, *Brain Res.* **211:**399–404.

Bunt, S. M., Lund, R. D., and Land, P. W., 1983, Prenatal development of the optic projection in albino and hooded rats, *Dev. Brain Res.* **6:**149–168.

Campbell, G., Ramoa, A. S., and Shatz, C. J., 1987, Transient features of ganglion cell morphology during development of the cat retina, *Invest. Ophthalmol. Vis. Sci. (Suppl.)* **28:**286.

Coleman, L.-A., Dunlop, S. A., and Beazley, L. D., 1984, Patterns of cell division during visual streak formation in the frog *Limnodynastes dorsalis, J. Embryol. Exp. Morphol.* **83:**119–135.

Coleman, L.-A., Harman, A. M., and Beazley, L. D., 1987, Displaced retinal ganglion cells in the wallaby *Setonix brachyurus, Vision Res.* **27:**1269–1277.

Drager, U. C., 1985, Birth dates of retinal ganglion cells giving rise to the crossed and uncrossed optic projections in the mouse, *Proc. R. Soc. Lond. (Biol.)* **224:**57–77.

Dreher, B., Potts, R. A., Ni, S. Y. K., and Bennett, M. R., 1984, The development of heterogeneities in distribution and soma sizes of rat retinal ganglion cells, in *Development of Visual Pathways in Mammals*, Vol. 9 (J. Stone, B. Dreher, and D. H. Rapaport, eds.), pp. 39–57, Alan R. Liss, New York.

Dunlop, S. A., and Beazley, L. D., 1981, Changing retinal ganglion cell distribution in the frog *Heleioporus eyrei, J. Comp. Neurol.* **202:**221–236.

Dunlop, S. A., and Beazley, L. D., 1984a, Development of the area centralis and visual streak in mammals, in *Development of Visual Pathways in Mammals*, Vol. 9 (J. Stone, B. Dreher, and D. H. Rapaport, eds.), pp. 75–88, Alan R. Liss, New York.

Dunlop, S. A., and Beazley, L. D., 1984b, A morphometric study of the retinal ganglion layer and optic nerve in postmetamorphic *Xenopus laevis, Vision Res.* **5:**417–427.

Dunlop, S. A., and Beazley, L. D., 1985a, Cell distributions in the retinal ganglion cell layer of adult Leptodactylid frogs after premetamorphic eye rotation, *J. Embryol. Exp. Morphol.* **89:**159–173.

Dunlop, S. A., and Beazley, L. D., 1985b, Changing distribution of retinal ganglion cells during area centralis and visual streak formation in the marsupial *Setonix brachyurus, Dev. Brain Res.* **23:**81–90.

Dunlop, S. A., and Beazley, L. D., 1986, A retino-retinal projection contributes minimally towards high axon numbers in the marsupial *Setonix brachyurus, Soc. Neurosci. Abstr.* **12:**3413.

Dunlop, S. A., and Beazley, L. D., 1987, Cell death in the developing retinal ganglion cell layer of the wallaby *Setonix brachyurus, J. Comp. Neurol.* **264:**14–23.

Dunlop, S. A., Longley, W. A., and Beazley, L. D., 1987, Development of the area centralis and visual streak in the grey kangaroo *Macropus fuliginosus, Vision Res.* **27:**151–164.

Dunlop, S. A., Coleman, L.-A., Harman, A. M., and Beazley, L. D., 1988, Development of the marsupial primary visual system, in *The Developing Marsupial—A Model for Biomedical Research* (C. H. Tyndale-Biscoe and P. A. Janssens, eds.), pp. 117–131, Springer-Verlag, Heidelberg.

Fite, K. V., 1976, *The Amphibian Visual System. A Multidisciplinary Approach*, Academic Press, New York.

Freeman, B., and Tancred, E., 1978, The number and distribution of ganglion cells in the retina of the brush-tailed possum, *Trichosurus vulpecula, J. Comp. Neurol.* **177:**557–568.

Freeman, B., and Watson, C. R. R., 1978, The optic nerve of the brush-tailed possum, *Trichosurus vulpecula:* Fibre diameter spectrum and conduction latency groups, *J. Comp. Neurol.* **179:**739–752.

Fujita, S., 1962, Kinetics of cellular proliferation, *Exp. Cell Res.* **28:**52–60.

Glucksmann, A., 1940, Development and differentiation of the tadpole eye, *Br. J. Ophthalmol.* **25:**154–178.

Harman, A. M., and Beazley, L. D., 1987, Patterns of cytogenesis in the developing retina of the wallaby *Setonix brachyurus, Anat. Embryol.* **177:**123–130.

Harman, A. M., and Beazley, L. D., 1988, Postnatal cytogenesis in the dorsal lateral geniculate nucleus (LGNd) and superior colliculus (SC) of the marsupial, *Setonix brachyurus* (quokka), *Neurosci. Lett. (Suppl.)* **30:**S73.

Harman, A. M., and Beazley, L. D., 1988, Generation of retinal cells in the wallaby, *Setonix brachyurus, Neurosci.*, in press.

Hendrickson, A., and Kupfer, C., 1976, The histogenesis of the fovea in the macaque monkey, *Invest. Ophthal.* **15:**746.

Hinds, J. W., and Hinds, P. L., 1978, Early development of amacrine cells in the mouse retina: An electron microscopic, serial section analysis, *J. Comp. Neurol.* **179:**277–300.

Hinds, J. W., and Hinds, P. L., 1983, Development of retinal amacrine cells in the mouse embryo: Evidence for two modes of formation, *J. Comp. Neurol.* **213:**1–23.

Hokoc, J. N., and Oswaldo-Cruz, E., 1978, Quantitative analysis of the opossum's optic nerve: An electron microscopic study, *J. Comp. Neurol.* **178:**773–782.

Hokoc, J. N., and Oswaldo-Cruz, E., 1979, A regional specialization in the opossum's retina: Quantitative analysis of the ganglion cell layer, *J. Comp. Neurol.* **183:**385.

Hughes, A., 1961, Cell degeneration in the larval ventral horn of *Xenopus laevis, J. Embryol. Exp. Morphol.* **9:**269.

Hughes, A., 1975, A quantitative analysis of cat retinal ganglion cell topography, *J. Comp. Neurol.* **163:**107–128.

Hughes, A., 1977, The topography of vision in mammals, in *Handbook of Sensory Physiology*, Vol. VII/5 (F. Crescitelli, ed.), pp. 615–756, Springer-Verlag, Berlin.

Hughes, A., 1985, New perspectives in retinal organization, in *Progress in Retinal Research*, Vol. 4 (N. N. Osborne and G. J. Chader, eds.), pp. 243–313, Pergamon Press, Oxford.

Humphrey, M. F., and Beazley, L. D., 1985, Retinal ganglion cell death during optic nerve regeneration in the frog *Hyla moorei, J. Comp. Neurol.* **236:**383–402.

Hunt, R. R., Cohen, J. S., and Mason, B. J., 1987, Cell patterning in pigment-chimeric eyes in *Xenopus:* Germinal cell transplants and their contributions to growth of the pigmented retinal epithelium, *Proc. Natl. Acad. Sci. U.S.A.* **84:**3302–3306.

Itaya, S. K., 1980, Retinal efferents from the pretectal area in the rat, *Brain Res.* **201:**436–441.

Jacobson, M., 1960, The representation of the visual field on the optic tectum of the frog: Evidence for the presence of an area centralis retinae, *J. Physiol.* **154:**31–32.

Jacobson, M., 1962, The representation of the retina on the optic tectum of the frog. Correlation between retinotectal magnification factor and retinal ganglion cell count, *Q. J. Exp. Physiol.* **47:**170–178.

Jacobson, M., 1976, Histogenesis of the retina of the clawed frog with implications for the pattern of development of retino-tectal connections, *Brain Res.* **103:**541–545.

Johns, P. R., 1977, Growth of the adult goldfish eye, III. Source of new retinal cells, *J. Comp. Neurol.* **176:**343–358.

Johns, P. R., and Easter, S., 1977, Growth of the adult goldfish eye. II. Increase in retinal cell number, *J. Comp. Neurol.* **176:**331–342.

Johns, P. R., Rusoff, A. C., and Dubin, M. W., 1979, Postnatal neurogenesis in the kitten retina, *J. Comp. Neurol.* **187:**545–556.

Kirby, M. A., Clift-Forsberg, L., Wilson, P. D., and Rapisardi, S. C., 1982, Quantitative analysis of the optic nerve of the North American opossum (*Didelphis virginiana*): An electron microscope study, *J. Comp. Neurol.* **211:**318–327.

Kuwabara, T., and Wiedman, T. A., 1974, Development of the prenatal rat retina, *Invest. Ophthalmol. Vis. Sci.* **13:**725–739.

Lia, B., Williams, R. W., and Chalupa, L. M., 1987, Formation of retinal ganglion cell topography during prenatal development, *Science* **236:**848–851.

Linden, R., 1987, Displaced ganglion cells in the retina of the rat, *J. Comp. Neurol.* **258:**138–143.

Mann, I., 1964, *The Development of the Eye*, 3rd ed., British Medical Association, London.

Mastronarde, D. N., Thibeault, M. A., and Dubin, W. M., 1984, Non-uniform postnatal growth of the cat retina, *J. Comp. Neurol.* **228:**598–608.

Meyer, R. L., 1978, Evidence from thymidine labelling for continuing growth of retina and tectum in juvenile goldfish, *Exp. Neurol.* **59:**99–111.

Perry, V. H., 1981, Evidence for an amacrine cell system in the ganglion layer of the rat retina, *Neuroscience.* **6:**931–944.

Perry, V. H., 1982, The ganglion cell layer of the mammalian retina, in *Progress in Retinal Research* (N. N. Osborne and G. J. Chader, eds.), pp. 53–80, Pergamon Press, Oxford.

Perry, V. H., and Cowey, A., 1985, The ganglion cell and cone distributions in the monkey's retina: Implications for central magnification factors, *Vision Res.* **12:**1795–1810.

Perry, V. H., Henderson, Z., and Linden, R., 1983, Postnatal changes in the retinal ganglion cell and optic axon populations in the pigmented rat, *J. Comp. Neurol.* **219:**356–368.

Peterson, E. H., and Ulinski, P. S., 1979, Quantitative studies of retinal ganglion cells in a turtle, *Pseudemys scripta elegans*. I. Number and distribution of ganglion cells, *J. Comp. Neurol.* **186:**17–42.

Provis, J. M., 1979, The distribution and size of ganglion cells in the retina of the pigmented rabbit: A quantitative analysis, *J. Comp. Neurol.* **185:**121–139.

Provis, J. M., 1987, Patterns of cell death in the ganglion cell layer of the human fetal retina, *J. Comp. Neurol.* **259:**237–246.

Rapaport, D. H., and Stone, J., 1982, The site of commencement of maturation in mammalian retina: Observations in the cat, *Dev. Brain Res.* **5:**273–279.

Rapaport, D. H., Wilson, P. D., and Rowe, M. H., 1981, The distribution of ganglion cells in the retina of the North American opossum (*Didelphis virginiana*), *J. Comp. Neurol.* **199:**465–480.

Saxen, L., 1954, The development of the visual cells, *Ann. Acad. Sci. Fenn. A* **23:**1–93.

Scalia, F., and Teitelbaum, I., 1978, Absence of efferents to the retina in the frog and toad, *Brain Res.* **153:**340–344.

Schnyder, H., and Kunzle, H., 1984, Is there a retinopetal system in the rat? *Exp. Brain Res.* **56:**502–508.

Sengelaub, D. R., and Finlay, B. L., 1982, Cell death in the mammalian visual system during normal development. I. Retinal ganglion cells, *J. Comp. Neurol.* **204:**311–317.

Sengelaub, D. R., Dolan, R. P., and Finlay, B. L., 1986, Cell generation, death and retinal growth in the hamster retinal ganglion cell layer, *J. Comp. Neurol.* **246:**527–543.

Sidman, R. L., 1961, Histogenesis of mouse retina studied with thymidine-$^3$H, in *The Structure of the Eye* (G. K. Smelser, ed.), pp. 487–506, Academic Press, New York.

Steinberg, R. H., Reid, M., and Lacy, P. L., 1973, The distribution of rods and cones in the retina of the cat (*Felis domesticus*), *J. Comp. Neurol.* **148:**229–248.

Stone, J., 1965, A quantitative analysis of the distribution of ganglion cells in the cat's retina, *J. Comp. Neurol.* **124:**337–352.

Stone, J., 1978, The number and distribution of ganglion cells in the cat's retina, *J. Comp. Neurol.* **180:**753–772.

Stone, J., Egan, M., and Rapaport, D. H., 1985, The site of commencement of retinal maturation in the rabbit, *Vision Res.* **25:**309–317.

Stone, J., Maslim, J., and Rapaport, D. H., 1984, The development of the topographical organisation of the cat's retina, in *Development of Visual Pathways in Mammals* (J. Stone, B. Dreher, and D. H. Rapaport, eds.), pp. 3–21, Alan R. Liss, New York.

Stone, J., Rapaport, D. H., Williams, R. W., and Chalupa, L., 1982, Uniformity of cell distribution in the ganglion cell layer of the prenatal cat retina: Implications for mechanisms of retinal development, *Dev. Brain Res.* **2:**231–242.

Straznicky, K., and Gaze, R. M., 1971, The growth of the retina in *Xenopus laevis:* An auto-radiographic study, *J. Embryol. Exp. Morphol.* **26:**67.

Tay, D., Hiscock, J., and Straznicky, C., 1982, Temporo-nasal asymmetry in the accretion of retinal ganglion cells in late larval and post metamorphic *Xenopus, Anat. Embryol.* **164:**75–83.

Walsh, C., and Polley, E. H., 1985, The topography of ganglion cell production in the cat's retina, *J. Neurosci.* **5:**741–750.

Walsh, C., Polley, E. H., Hickey, T. L., and Guillery, R. W., 1983, Generation of cat retinal ganglion cells in relation to central pathways, *Nature (London)* **302:**611.

Williams, R. W., Bastiani, M. J., Lia, B., and Chalupa, L. M., 1986, Growth cones, dying axons and developmental fluctuations in the fibre population of the cat's optic nerve, *J. Comp. Neurol.* **246:**32–69.

Wong, R. O. L., and Dunlop, S. A. D., 1988, Dendritic development of retinal ganglion cells in the marsupial *Setonix brachyurus,* quokka, *Neurosci. Lett. (Suppl.)* **30:**S141.

Wong, R. O. L., and Hughes, A., 1987a, Developing neuronal populations of the cat retinal ganglion cell layer, *J. Comp. Neurol.* **262:**433–495.

Wong, R. O. L., and Hughes, A., 1987b, The role of cell death in the topogenesis of neuronal populations in the cat retinal ganglion cell layer, *J. Comp. Neurol.* **262:**496–511.

Wong, R. O. L., Wye-Dvorak, J., and Henry, G. H., 1986, Morphology and distribution of neurons in the retinal ganglion cell layer of the adult Tammar wallaby—*Macropus eugenii, J. Comp. Neurol.* **253:**1–12.

# Developmental Heterochrony and the Evolution of Species Differences in Retinal Specializations

# 10

## Kenneth C. Wikler and Barbara L. Finlay

### 10.1. INTRODUCTION

The vertebrate retina is relatively stable across phylogeny in the classes and types of cell that compose its radial organization. Mechanistic studies of development have described how aspects of retinal organization common to all retinas emerge, such as the control of neurogenesis of particular cell types, competitive control of cell survival and dendritic organization in the development of retinal lamination, and the mechanics of directed axon outgrowth. However, vertebrate eyes also differ markedly between species in overall size, shape, and resolving power as well as in the number and arrangement of cells in the retina. Thus, studies of retinal neurogenesis must account for the development of species differences in eye conformation and retinal organization and address the evolutionary regulation of these developmental programs.

In this chapter, we consider some of the factors that produce species differences in retinal organization, with emphasis on the role of alterations in the timing and duration of developmental events. The feature of retinal organization of interest for this discussion is the cellular topography of the ganglion cell layer, where variations in cell density produce an area centralis or a horizontally extended visual streak. The three basic mechanisms that interact in development to produce topographic specializations, differential cell generation, cell death, and retinal growth are well known. Comparison of

**Kenneth C. Wikler** • Section of Neuroanatomy, Yale University, School of Medicine, New Haven, Connecticut 06510. **Barbara L. Finlay** • Department of Psychology, Cornell University, Ithaca, New York 14853.

the developmental programs of animals from species with different adult retinal specializations will allow us to determine which of these basic mechanisms may be varied to produce these specializations, and how they may be varied.

Alteration in the total duration and rate of development and the relative timing of growth in different organ systems are thought to be two of the principal ways by which species evolve (Gould, 1977). We examine various ways in which alterations of developmental timing may be implicated in the evolutionary production of topographic specializations in the retina, by examining the development of retinal topography in animals of closely related species that present interesting contrasts in duration of development, eye size, and morphology.

## 10.2. DIFFERENCES BETWEEN SPECIES IN RETINAL ORGANIZATION

### 10.2.1. The Nature of Variation between Vertebrates in Retinal Topography

Vertebrate species, although possessing the same basic organization of cell and fiber layers in the retina, differ significantly in both the number of cells in a given nuclear layer and also in the distribution of those cells across the retinal surface (Wall, 1942; Hughes, 1977). Mammalian species differences in the distribution of cells across the retinal surface appear to be the morphological features of the retina that are best correlated with species differences in visual acuity (Stone, 1983). For example, although the gerbil and the domestic cat have a comparable number of cells in the retinal ganglion cell layer, the cat's center–periphery gradient in cell density is much more pronounced than that of the gerbil and, correspondingly, the cat's visual acuity is approximately four times that of the gerbil (Stone, 1965, 1978; Emerson, 1980; Hughes, 1977; Wikler *et al.*, 1988).

Researchers differ in their categorization of mammalian retinas. For example, Stone (1983) proposes that the mammalian retina has two structurally and functionally distinct features. These features are the area centralis, an area of increased retinal ganglion cell density localized at the point of binocular fixation, and the visual streak, an area of increased cell density located along a horizontal axis of the retina. He proposes that this fundamental dual area centralis and visual streak organization is found in all mammalian species and that different species emphasize these retinal features to different degrees. For example, although human and nonhuman primates as well as carnivore species have a pronounced area centralis, examination of the distribution of all retinal ganglion cells (or in some cases subtypes of retinal ganglion cells) suggests that cell density is also higher along the horizontal axis in these species (Stone, 1983). Similarly, even species which possess a very pronounced visual streak, like the rabbit, have a detectable elevation in cell density within the temporal region of the streak.

Hughes (1977) proposes another categorization of retinal organization that takes into account both the absolute density and relative distribution of retinal ganglion cells. Species such as guinea pigs, mice, and weasels have retinas with low cell density and a low center–peripheral density gradient across the retinal surface. Tree shrews and squirrels possess a high density of retinal ganglion cells with a small center–peripheral gradient. Primates have very steep center–peripheral gradients. Cats and most carnivores are described as having a central or temporal area of increased cell density superimposed on a visual streak. Hughes includes herbivores such as deer and cattle which also possess this dual retinal organization. Finally, animals that have a very pronounced visual streak with very minimal evidence of an area centralis include the rabbit and ground squirrel.

Hughes (1977) and later Stone (1983) also describe an "anakatabatic area" or "vertical streak" organization where cell density is greater along the vertical axis extending dorsal or ventral to the horizontal meridian. This retinal type has been described in grazing species such as horses and deer. Hughes suggests that this area may be a specialization of functional importance for the viewing of the visual field between the animal's feet and its point of fixation (Hughes, 1977).

Although categorizations of species differences in retinal organization differ, the area centralis and visual streak are found in different degrees of elaboration across all species. Most important for the present analysis is that within any particular mammalian radiation, different types of retinal organization can be seen with more or less pronounced central specializations. For example, within the carnivore radiation the ferret has a 10–12 : 1 center–periphery gradient in cell density whereas the cat has a 100 : 1 ratio (Stone, 1965, 1978; Henderson, 1985). Within the rodent radiation, the Syrian hamster has a 1.5 : 1 difference between the visual streak and the retinal periphery, whereas the very closely related Mongolian gerbil possesses a 6 : 1 center–periphery difference (Sengelaub et al., 1986; Wikler et al., 1988).

Why this diversity in retinal topography has evolved is a central question in this research, and explanations have been advanced at both functional levels—that is, the optimal design of a visual information processing machine—and at ecological levels—that is, how a retinal design best fits a visual/behavioral niche. Also, developmental studies have offered mechanistic accounts for the production of species differences in retinal organization. We discuss later how retinal design might also reflect developmental processes and constraints, but we first review the primary functional and ecological accounts of retinal organization.

### 10.2.2. Functional Explanations for Variations in Retinal Topography

Although the upper limit on visual acuity is set by the photoreceptor matrix, visual acuity is further constrained by the local density of retinal ganglion cells: there are not enough "private channels" available for the outer nuclear layers. There is a reasonably close correspondence between the distribution of retinal ganglion cells, photoreceptors, and other cells, although exceptions to this relationship indicate that this is not an invariant rule of retinal organization (Hughes, 1977). Together these observations indicate the importance of the number and distribution of retinal ganglion cells for species differences in visual acuity.

Hughes (1977) suggests that the limited amount of "neural apparatus" available in the central nervous system is the constraint responsible for the uneven distribution of retinal ganglion cells across the retina. For example, if the peak retinal ganglion cell density in the cat retina were found across the entire retinal surface, the area of the cat's striate cortex would have to be five times that of the human visual cortex in order to accommodate proportionately the increased number of output cells in its retina. Hughes concludes that brain space can be conserved by designating only a local retinal area to subserve high visual resolution of restricted parts of the visual field.

### 10.2.3. Ecological Explanations for Variations in Retinal Topography

Ecological approaches to the understanding of species differences in retinal organization attempt to explain why different structures have evolved in response to certain environmental demands and do provide important information as to the possible functional importance of differences in retinal morphology across species. Although comparisons of species' ecological habitats and retinal type show general trends within mammalian radiations, few invariant relationships emerge. For example, the visual streak was believed to be a specialization that evolved for grazing animals with laterally placed eyes which provide a panoramic visual field. The visual streak may be a retinal specialization that enables individuals to detect the presence of predators on the horizon without the need for eye movements (Stone, 1983). However, Hughes (1977) points out that, contrary to this hypothesis, predator species with laterally placed eyes, like the crocodile, also possess a pronounced visual streak.

The fact that any mapping of retinal topography onto a particular ecological habitat eventually encounters a species with anomalous retinal organization is not surprising in that there is no reason to expect homologous retinal structures to subserve similar functional requirements across all species. Furthermore, differences in retinal topography across species are not necessarily linked to differences in current adaptive significance. Such differences may be the result of each species' particular phylogenetic and ontogenetic history (Stone, 1983). That is, the original ecological pressures that acted to select a particular retinal morphological feature may be unrelated to its current adaptive utility. Also, as discussed below, developmental programs for the production of retinal organization may act to constrain the possible system variations evident across species.

### 10.2.4. Mechanisms of Development as an Explanation for Variation in Retinal Topography

Chapter 9 in this volume describes the basic developmental mechanisms that have been shown to contribute to the production of inhomogeneities in cell density across the retinal surface. These mechanisms, differential neurogenesis, differential neuronal death, and differential retinal stretch, are each discussed briefly below.

### 10.2.4.1. Differential Neurogenesis

Retinal center–periphery gradients in cell density may result from a "hot spot" of neurogenesis during development such that a greater production of cells regionally could correspond to areas of elevated cell densities in maturity. A variety of differences in early cell density can be seen, with cells evenly distributed early in development (rat: Perry *et al.*, 1983; cat: Stone *et al.*, 1982), cell density highest in the peripheral margins of the retina (hamster: Sengelaub *et al.*, 1986), or with cells in greater concentration in the center of the retina (rabbit: Stone *et al.*, 1985; Robinson *et al.*, 1986; human: Provis, 1987; Provis *et al.*, 1985a,b). However, when compared to adult distributions in these species, all these early distributions of cells in the retinal ganglion cell layer early in development are relatively flat. The variability in the early pattern of cell generation in the retina across species may thus provide a potential basis for some of the differences in adult topographic organization, but the small magnitude of these differences indicates that they must be amplified by acting in concert with other developmental mechanisms.

### 10.2.4.2. Differential Cell Death

Evidence for the sculpting of central retinal specializations through the preferential loss of cells in the peripheral margins of the retina has been found for several species. As discussed by Beazley and co-workers (Chapter 9 in this volume), the preferential loss of cells from the periphery has been shown for the hamster (Sengelaub *et al.*, 1986), kangaroo and quokka (Chapter 9 in this volume), cat (Wong and Hughes, 1987), and human (Provis, 1987). However, there is no evidence to support a preferential cell loss in rabbit (Robinson *et al.*, 1986), rat (Horsburgh and Sefton, 1987), or the ferret (Henderson *et al.*, 1988).

### 10.2.4.3. Differential Retinal Growth

In every species investigated to this point, the greater expansion of the retinal periphery contributes to the development of adult retinal topography (Stone *et al.*, 1982; Dreher *et al.*, 1984; Henderson *et al.*, 1988; Mastronarde *et al.*, 1984; Sengelaub *et al.*, 1986; Robinson, 1987).

Two models, not necessarily mutually exclusive, have been proposed to explain the mechanism(s) for the preferential expansion of the retinal periphery during development. First, the retina changes its conformation over development (Mastronarde *et al.*, 1984). The retina initially occupies a large percentage of the sphere of the eye and retracts during development to occupy only a hemisphere. The resulting dilation of the retinal margin produces a thinning of peripheral cell densities (termed retinal "flattening"). The second hypothesis offered to explain differential retinal expansion is the "balloon" model (Sengelaub *et al.*, 1986; Robinson, 1987). This hypothesis suggests that differences in retinal elasticity combined with intraocular pressure result in the differential stretching of the retina as it increases in area. Differential retinal elasticity might be achieved via spatial gradients in neurogenesis, cell migration, or the elaboration of retinal plexiform layers (Rapaport and Stone, 1982; Stone *et al.*, 1985). Such differences in elasticity due to maturational gradients early in development are supported by the findings of Kelling *et al.* (1989), which indicate that retinal segments from the periphery of the developing kitten retina show a greater degree of deformation than retinal segments from the putative area centralis when subjected to equivalent changes in pressure.

The development of retinal topography is thus a multifactorial process. The development of retinal specializations are dependent on a combination of factors whose weighting is varied in animals with different retinal organizations. This explanation for species differences in visual system organization, however, does not address evolutionary issues concerning the means by which developmental programs are altered over phylogenetic time to produce these system differences.

## 10.3. TOWARD A FULL EXPLANATION OF THE DEVELOPMENT AND EVOLUTION OF RETINAL SPECIALIZATIONS

These three preceding accounts, functional, ecological, and developmental, are not mutually exclusive, and a full description of the nature of retinal topographic organization will include elements of all these types of explanation. However, the contrast of explanations being drawn here is not only one of "levels:" ecology or function is the ultimate reason for a particular retinal conformation and developmental processes are the mechanism that produce the evolutionarily desired conformation. More than that, available processes of development may entirely constrain what type of retinal conformation is possible. Also, pleiotropic effects of selection for maturational rate of some other character may produce secondarily a retina of a particular conformation. In these cases development is an "ultimate" cause of retinal conformation equal to that of function and ecology.

In order to address the interaction of developmental mechanisms with phylogenetic processes, we would like to describe empirically how the developmental program for the production of retinal specializations is regulated over evolutionary time. To describe evolutionary scenarios it is essential that we compare relatively closely related species that shared a common stem ancestor in the recent phylogenetic past. Also, as presented below, our comparison of retinal development between related species specifically focuses on processes that may act to regulate or orchestrate mechanisms such as neurogenesis, cell death, and retinal growth in both development and evolution. Can regulatory factors, such as alterations in overall body size or changes in the regulation of developmental timing, act to regulate the importance of these developmental mechanisms and produce adult species differences in retinal organization?

## 10.4. HETEROCHRONY AND THE EVOLUTION OF RETINAL SPECIALIZATIONS

### 10.4.1. Alterations in Developmental Timing

Regulation of timing in development may be an orchestrating process that is important in visual system evolution. Before a discussion of how alterations in developmental duration and timing might be employed to produce differences in retinal conformation, the range over which developmental durations vary should be considered. Regularities by which ocular development conforms to developmental schedules of varying length have been investigated (Dreher and Robinson, 1988) and form the basis for a discussion of developmental heterochrony.

If the absolute time of gestation plus time to eye opening is compared across mammalian vertebrates, the variation is immense, from a minimum of approximately 20 days found in several species of myomorphic rodents, to a maximum of 650 days in elephants (Eisenberg, 1981). Within various orders to be considered here the range is less but still significant. For myomorph and sciuriomorph rodents, for example, the gestation plus eye-opening length ranges from 30 days to approximately 75 days. Primates range from a low of 55 days to a maximum of 255 days. In the following sections, we discuss how retinal neurogenesis must accommodate the period of developmental duration. First, we discuss prior analyses of heterochronic changes in other developing systems as a model for this approach.

### 10.4.2. Heterochrony: The Role of Timing in Development and Evolution

The role of heterochrony in morphological evolution has received considerable attention by both developmental and evolutionary biologists in recent years (Gould, 1977; Bonner, 1980; Goodwin *et al.*, 1983; Raff and Kaufman, 1983; Black, 1986; Finlay *et al.*, 1987). Gould (1977) has formally defined heterochrony as "the retardation or acceleration of developmental events in a descendent's ontogeny relative to that of its ancestor." Gould contends that the alteration in the regulation of timing events during development is an important mechanism for evolutionary change and describes heterochronic scenarios in insect, amphibian, rodent, and human populations (Gould, 1977). Gould has linked heterochronic changes to the influence of ecology on reproductive strategies. That is, fluctuating and unstable ecological conditions are associated with species that invest little parental care and have large litters. Species of this type have accelerated developmental programs for rapid production of young (r selection). Stable environments enable species to use the strategy of producing relatively few young that are slow to mature (k selection).

Heterochronic processes are considered to be powerful and simple mechanisms for producing relatively rapid evolutionary change (Alberch and Alberch, 1981). In contrast to the gradual selection of individual adult features over phylogenetic time, alterations in the regulation of developmental timing may be a "genetically inexpensive" means by which to produce systemwide changes in structure and function.

The control parameters relating to heterochronic mechanisms and processes have recently been systematically addressed (Alberch et al., 1979). The morphological expressions of growth changes in phylogeny are called either peramorphosis (when developmental events are accelerated) or paedomorphosis (when development is retarded). The developmental control parameters for heterochronic change include the onset age of growth or the offset signal for growth, a limiting size or shape, or age for growth, the rate of growth, and the initial size when the growth period begins (Alberch et al., 1979). For example, a paedomorphic species could be produced by decreasing its growth period (progenesis) or by slowing the rate of growth (neoteny) relative to that of its ancestor's. A peramorphic species can arise by extending its growth period or by accelerating its rate of growth (acceleration) relative to its ancestor's. For our discussion, alterations in onset, rate, and duration of growth in retinogenesis are all considered as potential examples of heterochronic change.

Allometric studies of comparative brain size and body size, and rates of somatic growth versus brain growth have long indicated that the rate of brain growth reliably differentiates major vertebrate radiations (Jerison, 1973; Eisenberg, 1981; Sacher, 1982; Calder, 1984; Mann et al., 1988). These allometric studies provide strong support for the idea that the rate of maturation of large, multicomponent structures is an important variable in understanding speciation and is genetically controlled (but see Mann et al., 1988). If change in developmental rates of the whole organism or of gross structures is often the means by which selection of particular traits occurs, understanding the pleiotropic effects of changes in maturational rates will be critical to any level of discussion of evolution of nervous systems, from gross morphology to organizational characteristics. Understanding the constraints and necessary covariations produced by heterochronic variation of any particular stem form in vertebrate radiations will be critical to any explanation of the morphology of any particular feature in extant organisms. However, at a cellular or organizational level, research concerning the role of heterochrony in the evolution of the nervous system has been confined to the invertebrate nervous system of the nematode (Ambros and Horwitz, 1984). An evaluation of heterochrony in the evolution of the mammalian visual system is begun here and an empirical approach to this question is outlined that contrasts with and complements the ecological and functional approaches discussed above.

## 10.5. ACCOMMODATION OF RETINAL NEUROGENESIS INTO DEVELOPMENTAL SCHEDULES OF VARIABLE DURATION

Relatively little is known of how the basic events of retinogenesis, such as neurogenesis, axon outgrowth, cell death, and retinal growth, are fit into varying developmental durations. A study that surveyed six commonly studied mammalian vertebrates (rat, cat, rabbit, quokka, macaque, and human) found that the time of peak axon number and the end of rapid axon loss in the optic nerve occurred at a surprisingly constant percentage of the caecal period (gestational length plus time to eye opening) (Dreher and Robinson, 1988). The end of rapid axon loss occurred at about 74% of the closed eye period. Because the absolute duration of these events varies significantly, the authors suggest that this may indicate a necessary proportional duration and onset of events in retinogenesis, including at least neurogenesis, axon outgrowth, and cell growth. However, the particular collection of species used in this study was not optimal for comparison of closely related species that vary on dimensions of interest, since systematic data of this kind are not available for a large number of mammalian vertebrates. Thus, these conclusions are interesting but must remain speculative.

Similarly, proportional extension of development of this same type has also been observed in allometric studies of the development of gross morphology in rodent species (Creighton and Strauss, 1986). Taken together, these studies suggest a simple relationship between developmental duration and the development of adult morphological features. For example, doubling an animal's gestation length might be expected to result in a lengthening of developmental processes that results in a doubling of retinal cell number and eye size. However, two sources of information suggest that the relationship between developmental duration, developmental mechanisms, and adult morphology is neither direct nor simple.

First, regardless of species differences in the length of the developmental period, many fundamental features of development of neuronal function (e.g., the length of the cell cycle, the rate of process outgrowth, and the duration of an action potential) are unlikely to change proportionately with gestation length. This finding suggests that certain aspects of neuronal development and function are less susceptible to alterations in timing than are others.

Second, the relationship between eye size and developmental duration is not invariant. Whole eye size is significantly correlated with body size (Hughes, 1977), which in turn is related to gestational length (Eisenberg, 1981). Eyes range in axial length from about 1.5 mm in shrews to 60 mm in baleen whales. However, at any particular body size, variation in eye size is from three- to fivefold, dependent particularly on the species' ecological niche (the diurnal/nocturnal dimension is particularly important for eye size). These studies indicate that within a given body size and gestational length there is a significant degree of variability in the development of eye size across species.

These observations together suggest that developmental processes are not uniformly acted on by changes in developmental duration. In the following sections we offer at a cellular level an empirical analysis of how retinal cell number and distribution *do* change as a function of alterations in three regulatory processes: alterations in developmental duration, in size scaling, and finally in rates of development. We can infer potential evolutionary scenarios for the production of species differences in retinal organization through a detailed comparison of retinogenesis of closely related species that vary in these three regulatory parameters.

## 10.6. HETEROCHRONIC CHANGES IN OCULAR DEVELOPMENT AND VARIATION IN RETINAL CELL NUMBER AND TOPOGRAPHY

We have recently compared the development of two very closely related cricetine rodents, the hamster (*Mesocricetus auratus*) (Sengelaub *et al.*, 1986) and the gerbil (*Meriones unguiculatus*) (Wikler *et al.*, 1988), and also two less closely associated carnivores, the ferret (*Mustela putorius furo*) and the domestic cat (*Felis domesticus*) (Henderson *et al.*, 1988). These two pairs each present a strong contrast in the relative development of the visual streak and area centralis and allow several comparisons of the relationship of gestational length and eye size to retinal organization. In addition, enough is known about both eye conformation and neurogenesis in the rat (*Rattus rattus*) and mouse (*Mus musculus*) to permit a preliminary comparison of these two species as well (see Fig. 10.1).

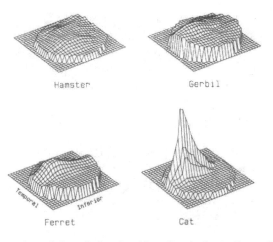

**Figure 10.1.** Representation of the relative densities of retinal ganglion cells in the hamster, gerbil, ferret, and cat retina. To produce these plots, for each species' retina separately, the lowest local cell density is given the value 1, and cell densities elsewhere are given as multiples of the lowest density and are expressed as a height. The information on retinal topography is taken from published reports (hamster, Sengelaub *et al.*, 1986; gerbil, Wikler *et al.*, 1988; ferret, Henderson *et al.*, 1988; cat, Stone, 1983), smoothed, and represented as a surface.

### 10.6.1.  Comparison of the Hamster and Gerbil: Variation in Duration of Development

The hamster has one of the shortest conception to eye-opening durations found in *Rodentia* (30 days), and the gerbil one of the longest (45 days). They differ in the adult number of cells in the ganglion cell layer (hamster, about 145,000; gerbil, 300,000), somewhat in adult eye size (Fig. 10.2), and most importantly in the relative development of the area centralis/visual streak (the hamster has a 1.5/1 difference from center to periphery while the gerbil has a 6/1 difference). Their visual acuity and visuomotor behavioral complexity contrast similarly: the gerbil's acuity for sine wave gratings is about three times that of the hamster, and the gerbil shows predictive tracking and ballistic capture of prey not seen in hamsters (Ingle, 1981). Gerbils are active both diurnally and nocturnally, while hamsters are active predominantly in twilight hours.

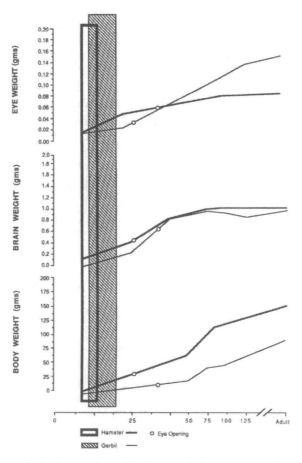

**Figure 10.2.** Eye growth, brain growth, and body growth in hamsters and gerbils expressed relative to duration of retinal ganglion cell genesis and the gross maturational markers of eye opening and sexual maturity.

We then asked what differences in development might account for these morphological and functional differences in adult nervous system organization. We investigated gross allometric measures such as growth of brain, body, and eye size (Wikler *et al.*, 1988). We also determined for both animals the duration of retinal neurogenesis and the topography of cells generated for the ganglion cell layer on each day of cell generation (Sengelaub *et al.*, 1986; Wikler *et al.*, 1988).

A summary of this comparison is shown in Fig. 10.2. Hamsters reach two milestones of general maturation considerably earlier than gerbils: eye opening occurs at about 30 days postconception in hamsters as compared to 45 days in gerbils. Sexual maturity occurs correspondingly earlier in hamsters than in gerbils. The rate of change in body weight of gerbils is always less than that of hamsters. Interestingly, the mature brain weights of hamsters and gerbils are nearly identical. The similarity of the growth of the brain between these species is striking, particularly when contrasted with development of adult body weight or the onset of sexual maturity. The change in eye weight (as a measure of eye volume), retinal area, and the principal axes of the eye show similar maturational patterns (Wikler *et al.*, 1988). The gerbil eye does not become larger than the hamster eye until approximately the fiftieth postconceptional day and also shows an extended late growth. Therefore, the developmental timetables for attainment of adult brain weight, eye size, and body weight appear to be dissociable. These data suggest again that changes in the duration of development are not acting uniformly across all systems.

When does the area centralis/ visual streak emerge in the gerbil? The center–periphery disparity has reached approximately half of its adult magnitude by postconceptional day 37 (Fig. 10.2). At that point, the hamster and gerbil eyes have the same radial dimensions and volume, so the absolute growth of the eye or the absolute amount of differential stretch cannot account for the difference between the hamster's relatively flat distribution and the gerbil's visual streak. What developmental events prior to postconceptional day 37 set up this center–periphery disparity?

First, retinal neurogenesis is prolonged in the gerbil: represented in Fig. 10.3 is the duration over which retinal ganglion cells are produced. Retinal ganglion cells are produced over 6–8 prenatal days in the gerbil, and over 3 prenatal days in the hamster (Sengelaub *et al.*, 1986; Wikler *et al.*, 1988). It appears that the rate of cell production does not differ between the two species. That is, the number of ganglion cells produced in the gerbil is approximately twice that produced by the hamster, which is in direct proportion to the length of retinal ganglion cell neurogenesis.

The spatial pattern of cell production differs significantly between these species. In Fig. 10.3, the spatial distributions of the cells undergoing their final division on the first, middle, and last day of retinal ganglion cell neurogenesis are shown for both hamster and gerbil. In the gerbil, ganglion cells leave the mitotic cycle beginning from the area of the prospective area centralis. (At no point in early development are there greater total numbers of cells centrally in either species; in fact, the opposite is true.) The gerbil retina

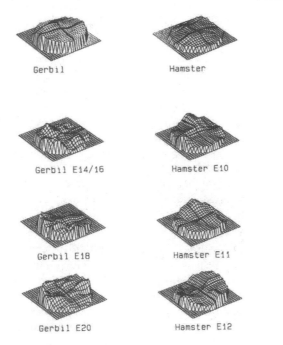

**Figure 10.3.** Spatial gradients in the generation of retinal ganglion cells for the first, middle, and last days of retinal ganglion cell genesis for the hamster (left) and gerbil (right). Data are taken from Sengelaub *et al.* (1986) and Wikler *et al.* (1988), smoothed, and reexpressed as in Fig. 10.1.

shows a center-to-periphery pattern of maturation, while the hamster shows a less pronounced superior temporal-to-inferior nasal progression.

This pattern of early maturation of the center of the retina is associated with the stabilization of the high cell density area of the central retina. As discussed previously, differential retinal elasticity is believed to be the substrate for differential retinal expansion. This substrate difference is associated with spatial gradients in retinal maturation. Central areas that mature earlier than the peripheral margins of the retina are thicker and less elastic early in development (Kelling *et al.*, 1989). After this early maturational period, the hamster and gerbil retina are then subjected to approximately equivalent amounts of absolute retinal growth over the next 10 days of development. Cell density of the central retina in the gerbil remains high whereas cell density is reduced in the retinal periphery. The same process occurs in the hamster, but to a much lesser extent. This difference in the development of retinal topography may be related to species differences in the spatial pattern of early retinal maturation.

Interpretation of the significance of this center–periphery gradient depends on what pattern of neurogenesis (center–periphery in the gerbil or flat in the hamster) is viewed as the "stem" form and what is derived. Many evolutionary scenarios implicitly or explicitly cast the simple or undifferentiated form as the stem form, but clearly this need not be the case. The hamster could be derived from a more gerbil-like stem form in the following plausible, but entirely speculative, scenario. The stem, gerbil-like form possessed a center-to-periphery pattern in retinogenesis that is characteristic of every vertebrate lineage that has been examined (fish—see Chapter 11 in this volume; amphibians—see Chapter 9 in this volume; mammals—see Chapter 1 in this volume). (In support of this speculation, the hamster is unusual in that it possesses the flattest gradient of ganglion cell density reported in rodents, and possibly in all mammals.) The hamster was phylogenetically selected for extreme rapidity of development and as a result of its unusually compressed development period failed to express spatial gradients in retinogenesis. Secondarily, the hamster lost regional differentiation of retinal cell density (through differential retinal growth) dependent on expression of these spatiotemporal gradients. This type of scenario is particularly interesting in that the particular retinal morphology exhibited by the hamster is thus a consequence of a gross alteration in the animal's developmental rate, and not a result of direct selection for a particular retinal morphology. Conversely, animals selected for extended neurogenesis might develop highly differentiated central specializations as a direct consequence of their extended development and not because of direct selection for a well-differentiated eye.

Not every difference in the morphology of the hamster and gerbil retina can be attributed to differences in the relative rate or duration of development. The initial shape of the eye is different in the two species. The gerbil's eye is initially ovoid and extended along its nasotemporal axis, while the hamster eye is spherical (Wikler *et al.*, 1988). In adulthood, the eyes of both species are spherical. This may directly contribute to the development of the pronounced asymmetric visual streak in gerbils, extended on its nasotemporal axis. The superior–inferior axis of the gerbil eye is preferentially extended over development and correlates with a decrease in cell density in the inferior and superior margins of the retina. Although heterochronic differences as investigated here in development can potentially account for two major differences in the hamster and gerbil retina (cell number and the presence of a central specialization), it does not account for a third difference, the presence of asymmetry in the central specialization. The production of a pronounced visual streak in the gerbil retina may be directly related to shape, rather than timing differences, in early retinal development.

### 10.6.2. Comparison of the Rat and the Mouse: Size Scaling

The laboratory rat (*Rattus rattus*) and the house mouse (*Mus musculus*) are representatives of two closely related genera. The retinal organization and the retinogenesis of these species have been studied in great detail by a variety of investigators and permit a comparison similar to that made above for hamster versus gerbil though details of the topography of cells generated at various conceptional days are not known. In general, the topography of ganglion cells and the entire organization of the eye are strikingly similar in the mouse and the rat. The mouse eye is approximately half the linear size of the rat's (Hallet, 1987), and with appropriate linear scaling, the gross features of the eyes can be superimposed (Remtulla and Hallet, 1985). Mean spacing of photoreceptors per square millimeter is very similar, as is the mean spacing of ganglion cells (Hallet, 1987). The distribution of ganglion cells is very similar, with approximately a 5/1 maximum to minimum disparity from center to periphery reported for the rat (Stone *et al.*, 1983; Dreher *et al.*, 1985) and a 4–5/1 maximum to minimum reported for the mouse (Drager and Olsen, 1981). The arrangement of these cells is intermediate between the hamster and gerbil (Fig. 10.1): a temporal area of highest density, extended along the horizontal axis, with lowest density superiorly. The rat with a larger eye and similar cell spacing has approximately 119,000 ganglion cells, whereas the mouse possesses 50,000 cells. Correspondingly, the measured maximal acuity of the mouse is approximately half that of the rat.

What are the differences in development that produce two retinas of strikingly similar conformation but that differ in areal size by a factor of 2? Interestingly, duration of development is quite similar between these species. The time from conception to eye opening is approximately 35 days in both rats and mice. The duration of retinal ganglion cell neurogenesis is also similar, about 6–7 days (Sidman, 1961; Webster, 1985). The topography of the cell distribution on generation has not been described quantitatively for either animal, though both are described as having a gross center-to-periphery progression by birthdate when they are examined as adults (Sidman, 1961; Morest, 1970; Bruckner *et al.*, 1976; Drager and Olsen, 1981; Drager, 1985).* A comparison of the size of the retina at various ages (as in Fig. 10.2) has not been done, and it would be very interesting to see at what point in development the twofold size difference in retinal area appears. We would hypothesize, given the comparable developmental duration between these species, that the size difference in the retina appears early on, producing a germinal zone of twice the area in the rat. This larger initial germinal zone would then undergo an identical number of cell divisions and an identical scaled pattern of differential areal growth thereafter.

Taken together with the hamster/gerbil comparison, this comparison underlines the fact that eye size alone, or the absolute amount of retinal growth, is not associated with a particular retinal conformation. Rather, the effects of eye growth, particularly differential areal expansion of the retina during development, must be understood in the context of initial retinal conditions. This qualification is important to the further comparison of the process of development of the cat and the ferret.

---

*In the hamster and gerbil, the initial topography of retinal ganglion cells generated on the same day does not resemble the topography seen at adulthood: presumably the same would be true for the mouse and rat.

### 10.6.3. Comparison of the Ferret and the Cat: The Effects of Rate of Eye Growth

The ferret and the cat are both carnivores. However, they are of different families (ferret: Canoidae, Mustelidae; cat: Feloidae; Felidae) and are therefore considerably more distantly related than the two rodent pairs discussed above and thus might be expected to differ on many more developmental dimensions. Their ganglion cell number does not differ as much as eye size or visual capacity might indicate: the adult ferret retina has about 90,000 retinal ganglion cells compared to 150,000 cells in the cat. However, the close similarity of their time from conception to eye opening (ferret, 72 days; cat, 75 days) and period of retinal ganglion cell neurogenesis (both, postconception days 20–40; see Chapter 1, Table 1.1, in this volume; Greiner and Weidman, 1980; Greiner, 1981), combined with the major difference in the topography of their ganglion cell distributions, makes them ideal candidates for comparison.

Differential retinal expansion is an important mechanism for the production of central retinal specializations in both the ferret (Henderson *et al.*, 1988) and the cat (Lia *et al.*, 1987; Wong and Hughes, 1987). However, the most striking feature differentiating ferret and cat eyes is simple size: mature retinal area in the ferret is about 60 mm² while the cat's retinal area is nearly 10 times larger at 550 mm². The rate of retinal growth in the ferret and the cat (taken from Henderson *et al.*, 1988, and Mastronarde *et al.*, 1984) is presented in Fig. 10.4. Growth of the cat eye is both accelerated in rate and increased in duration relative to that of the ferret.

The duration of retinogenesis and expression of spatial gradients in neurogenesis (see Chapter 1 in this volume) are similar in the ferret and cat. We therefore speculate that extended or amplified eye growth (given the appropriate early gradients in neurogenesis) will lead inevitably to an exaggerated area centralis or visual streak. Both conditions, differential retinal expansion coupled with growth of a larger eye, must be true. Note from the prior species comparisons that large eye size alone does not necessarily lead to a highly

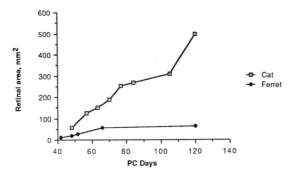

**Figure 10.4.** Relative growth of retinal area in the ferret and the cat. From Henderson *et al.* (1988), Mastronarde *et al.* (1984), and Stone *et al.* (1982).

specialized area centralis (rat/mouse). Also note that even with its small eye size the ferret has a very respectable area centralis. The ferret's center–periphery disparity is larger than that of the gerbil, which may be related to species differences in the duration of retinogenesis (12 days RGC production in the gerbil, 20 days in the ferret).

### 10.6.4. Summary of Cross-Species Comparisons

We offer the following summary of these results as a hypothesis to guide future analyses and choice of species:

1. The developmental mechanism responsible for substantial species differences in retinal topography is differential retinal expansion. The absolute duration or rate of retinal growth, given proper initial conditions, may account for some species differences in retinal topography.
2. The initial conditions producing differential retinal growth, however, are various. The initial shape of the eye (Wikler et al., 1988), conformational change of the eye (Mastronarde et al., 1984), early spatial gradients in retinal maturation (Robinson, 1987; Wikler et al., 1988), and gradients in cell death (Sengelaub et al., 1986; Provis, 1987; Wong and Hughes, 1987) are likely candidates to set up conditions predisposing the retina to differential retinal expansion.
3. Alterations in developmental timing as well as differences in overall eye size are regulatory means by which the mechanisms described above can be differentially weighted. The developmental program for the production of a particular type of retinal organization might be more or less susceptible to either of these regulatory processes.
4. Species-specific differences in retinal organization are related to the functional utility of the retina, the particular life history of a species, as well as developmental parameters that constrain changes in organization. Also, selective pressures for features that may seem unrelated to visual system function—rates of development (r versus k selected species) and differences in body or eye size—may act developmentally to produce alterations in retinal function and structure.

These results suggest several avenues for investigation. A species with a relatively long period of neurogenesis and a relatively flat distribution of ganglion cells could test the proposal that the expression of maturational gradients across the retina is contingent on long durations of neurogenesis and that these gradients are directly associated with differential growth. Species with bifoveate or "anakatabatic" retinal organizations would provide interesting extensions for this model. Finally, an account of how varying cross-retinal distributions of particular cell types (such as X and Y cells) are produced, as discussed by Beazley et al. (Chapter 9 in this volume), must be included.

## 10.7. SUMMARY AND CONCLUSIONS

Heterochrony has traditionally been viewed as an alteration in either the rate of duration of developmental timing in gonadal versus somatic systems which produces morphological diversity; the most classic example is that of the evolution of morphologically juvenile salamanders that are reproductively mature as a result of an acceleration of gonadal development. We argue that the concept of heterochrony may also be applied usefully to neural systems and can account for several aspects of visual system diversity. It is also evident that heterochrony is not the only means by which retinal diversity can evolve. However, we are beginning to see how changes in the rate and timing of development produce particular adult retinal morphologies, and the regularities of even this initial empirical evidence give promise of an eventual developmental grammar of phylogenetic change in eye form.

ACKNOWLEDGMENTS. This chapter summarizes work supported by NSF grants 79-19491, and NIH grants KO4 NS 00783 and RO1 NS 19245 to B. Finlay. We particularly thank Vincent Tamariz for graphics innovations and Gail Perez for her contributions to the analysis of rodent allometry. We thank Dale Sengelaub for his continuing intellectual and editorial contributions and Brad Miller for his useful criticisms of this manuscript.

## 10.8. REFERENCES

Alberch, P., and Alberch, A., 1981, Heterochronic mechanisms of morphological diversification and evolutionary change in the neotropical salamander, *Bolitoglossa occidentalis* (Amphibia: Plethodontidae), *J. Morphol.* **167:**249–264.

Alberch, P., Gould, S. J., Oster, G. F., and Wake, D. B., 1979, Size and shape in ontogeny and phylogeny, *Paleobiology* **5:**296–317.

Ambros, V., and Horvitz, H. R., 1984, Heterochronic mutants of the nematode *Caenorhabditis elegans*, *Science* **226:**409–416.

Black, I. B., 1986, Trophic molecules and the evolution of the nervous system, *Proc. Natl. Acad. Sci. U.S.A.* **83:**8249–8252.

Bonner, J. T. (ed.), 1980, *Evolution and Development*, Springer-Verlag, New York.

Bruckner, G., Mares, V., and Biesold, D., 1976, Neurogenesis in the visual system of the rat: An autoradiographic investigation, *J. Comp. Neurol.* **166:**245–256.

Calder, W. A., 1984, *Size, Function and Life History*, Harvard University Press, Cambridge, MA.

Creighton, G. K., and Strauss, R. E., 1986, Comparative patterns of growth and development in cricetine rodents and the evolution of ontogeny, *Evolution* **4:**94–106.

Drager, U., 1985, Birthdates of cells giving rise to the crossed and uncrossed optic projections in the mouse, *Proc. R. Soc. London Ser. B* **224:**57–77.

Drager, U. C., and Olsen, J., 1981, Ganglion cell distribution in the retina of the mouse, *Invest. Ophthalmol. Vis. Sci.* **20:**285–293.

Dreher, B., and Robinson, S. R., 1988, Development of the retinofugal pathway in birds and mammals: Evidence for a common "time table," *Brain Behav. Evol.* **31:**325–392.

Dreher, B., Potts, R. A., Ni, S. Y. K., and Bennett, M. R., 1984, The development of heterogeneities in distribution and soma sizes of rat retinal ganglion cells, in *Development of Visual Pathways in Mammals* (J. Stone, B. Dreher, and D. Rapaport, eds.), pp. 39–57, Alan R. Liss, New York.

Dreher, B., Sefton, A. J., Ni, S. Y. K., and Nisbett, G., 1985, The morphology, number, distribution and central projections of Class I retinal ganglion cells in albino and hooded rats, *Brain Behav. Evol.* **26:**10–48.

Eisenberg, J. F., 1981, *The Mammalian Radiations: An Analysis of Trends in Evolution, Adaptation and Behavior,* The University of Chicago Press, Chicago.

Emerson, V. F., 1980, Grating acuity of the golden hamster: Effects of stimulus orientation and luminance, *Exp. Brain Res.* **38:**43–52.

Finlay, B. L., Wikler, K. C., and Sengelaub, D. R., 1987, Regressive events in brain development and scenarios for vertebrate brain evolution, *Brain Behav. Evol.* **519:**102–117.

Goodwin, B. C., Holder, N., and Wylie, C. C. (eds.), 1983, *Development and Evolution,* Cambridge University Press, Cambridge.

Gould, S. J., 1977, *Ontogeny and Phylogeny,* Harvard University Press, Cambridge, MA.

Greiner, J. V., 1981, Histogenesis of the ferret retina, *Exp. Eye Res.* **33:**315–332.

Greiner, J. V., and Weidman, T. A., 1980, Histogenesis of the cat retina, *Exp. Eye Res.* **30:**439–453.

Hallet, P. E., 1987, The scale of the visual pathways of mouse and rat, *Biol. Cybern.* **57:**275–286.

Henderson, Z., 1985, Distribution of ganglion cells in the retina of adult pigmented ferret, *Brain Res.* **358:**221–228.

Henderson, Z., Finlay, B. L., and Wikler, K. C., 1988, Development of ganglion cell topography in ferret retina, *J. Neurosci.* **8:**1194–1205.

Horsburgh, G. H., and Sefton, A. J., 1987, Cellular degeneration and synaptogenesis in the developing retina of the rat, *J. Comp. Neurol.* **263:**553–556.

Hughes, A., 1977, The topography of vision in mammals of contrasting lifestyle: Comparative optics and retinal organization, in *Handbook of Sensory Physiology,* Vol. VII, 5th ed. (F. Crescitelli, ed.), pp. 613–756, Springer-Verlag, Berlin.

Ingle, D. J., 1981, New methods for analysis of vision in gerbils, *Behav. Brain Res.* **3:**151–175.

Jerison, H. J., 1973, *Evolution of the Brain and Intelligence,* Academic Press, New York.

Kelling, S. T., Sengelaub, D. R., Wikler, K. C., and Finlay, B. L., 1989, Differential Elasticity of the Immature Retina: a Contribution to the Development of the Area Centralis? *Vis Neurosci* **2,** in Press.

Lia, B., Williams, R. W., and Chalupa, L. M., 1987, Non-uniform growth of the fetal retina can account for the prenatal development of regional specialization in the ganglion cell layer of the cat, *Science* **236:**848–851.

Mann, M. D., Glickman, S. E., and Towe, A. L., 1988, Brain/body relationships among myomorph rodents, *Brain Behav. Evol.* **31:**111–124.

Mastronarde, D. M., Thibeault, M. A., and Dubin, M. W., 1984, Non-uniform postnatal growth of the cat retina, *J. Comp. Neurol.* **288:**598–608.

Morest, D. K., 1970, The pattern of neurogenesis in the retina of the rat, *Z. Anat. Entwicklungsgesch.* **131:**45–67.

Perry, V. H., Henderson, Z., and Linden, R., 1983, Postnatal changes in retinal ganglion cell and optic axon populations in the pigmented rat, *J. Comp. Neurol.* **219:**356–368.

Provis, J. M., 1987, Patterns of cell death in the ganglion cell layer of the human fetal retina, *J. Comp. Neurol.* **259:**237–246.

Provis, J. M., van Driel, D., Billson, F. A., and Russell, P. 1985a, Development of the human retina: Patterns of cell distribution and redistribution in the ganglion cell layer, *J. Comp. Neurol.* **233:**429–452.

Provis, J. M., van Driel, D., Billson, F. A., and Russell, P., 1985b, Human fetal optic nerve: Overproduction and elimination of retinal axons during development, *J. Comp. Neurol.* **238:**92–100.

Raff, R. A., and Kaufman, T. C., 1983, *Embryos, Genes, and Evolution,* Macmillan, New York.

Rapaport, D. H., and Stone, J., 1982, The site of commencement of maturation in mammalian retina: Observations in the cat, *Dev. Brain Res.* **5:**273–279.

Remtulla, S., and Hallet, P. E., 1985, A schematic eye for the mouse and comparisons with the rat, *Vision Res.* **25:**21–31.

Robinson, S. R., 1987, Ontogeny of the area centralis in the cat, *J. Comp. Neurol.* **254:**50–71.

Robinson, S. R., Dreher, B., Horsburgh, G. M., and McCall, M. J., 1986, Development of ganglion cell density in the rabbit, *Soc. Neurosci. Abstr.* **12:**985.

Sacher, G. A., 1982, The role of brain maturation in the evolution of the primates, in *Primate Brain Evolution: Methods and Concepts* (E. A. Armstrong and D. Falk, eds.), Plenum Press, New York.

Sengelaub, D. R., and Finlay, B. L., 1982, Cell death in the mammalian visual system during normal development: I. Retinal ganglion cells, *J. Comp. Neurol.* **204:**311–317.

Sengelaub, D. R., Dolan, R. P., and Finlay, B. L., 1986, Cell generation, death, and retinal growth in the development of the hamster retinal ganglion cell layer, *J. Comp. Neurol.* **246:**527–543.

Sidman, R. L., 1961, Histogenesis of mouse retina studied with thymidine-$^3$H, in *Structure of the Eye* (G. K. Smelser, ed.), pp. 487–506, Academic Press, New York.

Stone, J., 1965, A quantitative analysis of the distribution of ganglion cells in the cat's retina, *J. Comp. Neurol.* **124:**337–352.

Stone, J., 1978, The number and distribution of ganglion cells in the cat's retina, *J. Comp. Neurol.* **180:**753–772.

Stone, J., 1983, *Parallel Processing in the Visual System,* pp. 265–325, Plenum Press, New York.

Stone, J., Rapaport, D. H., Williams, R. W., and Chalupa, L., 1982, Uniformity of cell distribution in the ganglion cell layer of prenatal cat retina: Implications for mechanisms of retinal development, *Dev. Brain Res.* **2:**231–242.

Stone, J., Egan, M., and Rapaport, D. H. 1985, The site of commencement of retinal maturation in the rabbit, *Vision Res.* **25:**309–317.

Wall, G. L., 1942, *The Vertebrate Eye and Its Adaptive Radiation,* Hafner, New York.

Webster, M., 1985, Cytogenesis, histogenesis and morphological differentiation of the retina, Ph.D. dissertation, University of New South Wales, N.S.W. Australia.

Wikler, K. C., 1987, Developmental heterochrony and the development of species differences in retinal specializations, Ph.D. dissertation, Cornell University, New York.

Wikler, K. C., Perez, G., and Finlay, B. L., 1988, Neurogenesis in the gerbil retina: a comparative analysis of the effects of developmental duration or retinal conformation, unpublished manuscript.

Wong, R. O. L., and Hughes, A., 1987, Role of cell death in the topogenesis of neuronal distributions in the developing cat retinal ganglion cell layer, *J. Comp. Neurol.* **262:**496–511.

# Fish Vision

# 11

## Russell D. Fernald

> To suppose that the eye with all its inimitable contrivances for adjusting the focus to different distances, for admitting different amounts of light, and for the correction of spherical and chromatic aberration, could have been formed by natural selection, seems, I freely confess, absurd in the highest degree.
>
> Charles Darwin, *The Origin of the Species*

## 11.1. INTRODUCTION

All vertebrate eyes have evolved from those of common underwater living ancestors and consequently are strikingly similar structures which form an image with a single lens. In contrast, invertebrates have a rich variety of eye types which form images in one of three ways: with single lenses, multiple lenses, or mirrors (Land, 1984). Since eyes must obey fundamental optical laws, physical constraints on eye design and structure provide the most straightforward means of understanding the adaptive value of ocular specializations. Using these physical constraints, inferences about the selective forces that have undoubtedly "shaped" eyes can be made with some confidence, particularly in the study of aquatic eyes. In contrast, in the analysis of ocular development there is no corresponding *a priori* knowledge of fundamental constraints to aid in interpreting these developmental processes. Developmental similarities themselves must serve to guide our understanding of these processes. Phylogenetic comparisons offer significant advantages because the modifications that have occurred during evolutionary time are carried in organisms and are most evident during development.

Fish eyes have long played a prominent role in the analysis of retinal function for a variety of reasons. The retina is accessible and, as numerous studies have shown, it is a suitable model for retinal function in higher vertebrates as well as being interesting in its own right (Fernald, 1984). Although a great deal of work has been done on the domestic goldfish, this narrow experimental focus obscures the fact that bony fish form the largest class of

**Russell D. Fernald** • Institute of Neuroscience, University of Oregon, Eugene, Oregon 97403. Written while R.D.F. was the Hilgard Visiting Professor at Stanford University, Stanford, California 94305.

vertebrates, comprising over 25,000 species. Evidence of teleost success can be found in the adaptive radiation of forms occupying an extraordinary range of habitats. Fish eyes and brains variously reflect success in mastering different life history strategies with the unsurprising general rule that the favored sensory modality occupies disproportionately more brain volume (see Geiger, 1956).

## 11.2. PHYSICAL CONSTRAINTS ON EYE DESIGN

As noted above, in striking contrast to the great variety of eyes that have evolved among invertebrate species, there exists essentially a single type of image-forming vertebrate eye. Because the vertebrate eye first evolved in water, it is sensitive to a narrow band of wavelengths of electromagnetic radiation, which are transmitted through water without significant attenuation (Fig. 11.1). The only other band with comparably low transmitted attenuation ($<10^3$ Hz) has been exploited by weakly electric fish. At low frequencies, however, the very long wavelengths make image formation impossible, since any receptor must be large relative to the wavelengths to be detected. As a consequence, electric fish do not form an image directly but instead have detectors distributed all along their bodies from whose response they "compute" an image.

The exploitation of light for vision has proved to be a powerful selective force in all organisms. Light travels in straight lines, can be reflected, and varies in both wavelength (color) and intensity so that objects in the world can

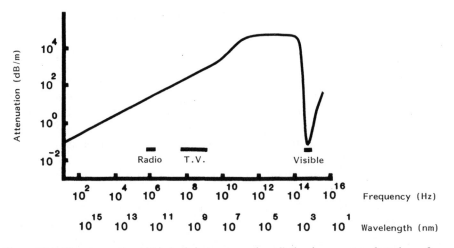

**Figure 11.1.** The attenuation (dB/m) of electromagnetic radiation in seawater plotted as a function of frequency (Hz) and wavelength (nm) of that radiation. Bands of electromagnetic energy used for radio and television transmission are shown, as is the narrow band corresponding to visible light. From Fernald (1987).

be distinguished on the basis of several different properties. Many of the design principles and even apparent errors we recognize in eyes reflect constraints arising directly from the physical properties of light. Natural selection cannot be expected to produce an error-free eye, but rather one just good enough for the animal's particular circumstances (Darwin, 1987). To inform their owners about the visual world, image-forming eyes must both capture light and resolve it into images. These two requirements for vision have shaped the evolution of eyes throughout the animal kingdom. Correspondingly, these requirements also constrain the developmental processes that are responsible for assembling eyes.

## 11.3. WHY STUDY FISH VISION?

There are over 25,000 species of fishes, making them the largest vertebrate class. Nearly 20,000 of these are the ray-finned bony fish or teleosts. This large number of species reflects widespread success in adapting to nearly every kind of habitat imaginable. They thrive in waters ranging from as low as −17 to 38°C, from the water's surface to 9000 m deep, and in currents as different as stagnant swamps and raging torrents. The diversity of habitats has resulted in a spectacular range of adaptations in the physiology, anatomy, and social behavior of fishes.

Studies of teleost cichlid fish species in particular have led to important insights about how selective advantages are reflected in evolutionary change. Baerends and Baerends van der Roon (1950) systematically described the coloration and behavioral patterns in numerous species then available after two major expeditions, one British (Regan, 1920) and one Belgian (Poll, 1956), had made it clear that the African rift lakes rivaled the Galapagos islands as a natural laboratory for evolutionary change. Since then, the thorough examination of the species flocks by Greenwood (1981) has confirmed that explosive speciation occurred in the lakes and left extraordinary diversity in its wake. For example, in Lake Tanganyika there are 126 cichlid species, all endemic, and in lake Malawi, 200 species are present, 99% of which are endemic (Greenwood, 1981). Cichlids have also been studied because of their specialized behavioral adaptations (e.g., Fryer and Iles, 1972) and many unique morphological traits (e.g., Liem and Osse, 1975).

Among the cichlids, *Haplochromis burtoni* is particularly well studied. It is brightly colored and uses a few well-defined chromatic and spatial patterns on its body to communicate with conspecifics (Fig. 11.2 and see Fernald, 1984, for a review). The color patterns can signal immediate intent or behavioral state including the social status of the signaler depending on the social context.

The visual sense is important because social communication among *H. burtoni* depends almost solely on visual signals, although during spawning, chemical cues may be important as well. Audition appears to play no role (Fernald and Hirata, 1975). In their natural habitat, the fish live largely in

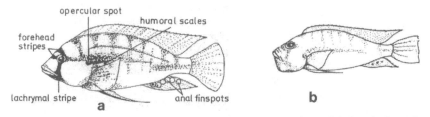

**Figure 11.2.** Coloration patterns of a territorial adult male (a) and an adult female (b) *H. burtoni*. The lachrymal stripe (eye to mouth bar), forehead stripes, and opercular spot of adult males are black. The humoral scales are orange-red and the 5–9 anal fin spots are yellow-orange. The dorsal and caudal fins have rows of small reddish spots between the rays; pectoral fins are transparent and the pelvic fins have a black lower edge. The female is shown with larvae in her buccal cavity. Females and nonterritorial males are nearly colorless, except for a few small anal fin spots. Adult males are 8–10 cm and adult females 5–8 cm. From Fernald and Hirata (1979).

shore pools scattered along the edge of the lake (Fernald and Hirata, 1977a,b). Brightly colored territorial males vigorously defend contiguous territories located in shallow water (<30 cm) above detritus covered sand. In this leklike social system, only males with territories attract mates and females spawn almost exclusively with these males.

Since only about 10% of the males in a shore pool are able to defend and maintain territories and hence are responsible for the majority of the reproduction, male territorial behavior is of central importance for the survival and reproductive success of individual *H. burtoni*. Territorial males are easy to distinguish because of the bright coloration patterns, including blue or yellow body color, yellow-orange egg spots on the anal fin, orange humeral scales, blue lips, and black forehead stripes, opercular spot, eye-to-mouth bar, and pelvic fins. Several of these coloration patterns are under neural control, which allows them to be changed from moment to moment during social encounters (Muske and Fernald, 1987a,b).

The impression that vision is important is amply confirmed by laboratory experiments in which variously colored models are presented to the animals and the responses measured. From a variety of such experiments it is clear that the black eye bar serves as the most important visual signal, indicating territoriality and generally aggressive intent (Leong, 1969; Heiligenberg and Kramer, 1972; Heiligenberg *et al.*, 1972; Fernald, 1977). In fact, the animals can recognize this signal without prior visual experience (Fernald, 1980b), meaning that the recognition processing is present from birth. The importance of this particular signal is reflected in a variety of specialized physiological adaptations which render it more salient (Muske and Fernald, 1987a,b). In addition to the coloration patterns, specific body and eye movements may also provide information to conspecifics (Fernald, 1975, 1980a, 1985b). Observations in the natural habitat revealed the same distinct behavioral patterns found in the laboratory and their frequency of occurrence was similar to what we had previously described (Fernald and Hirata, 1977b).

Taken together, these data illustrate the importance of vision for the behavior of the animals. They move and interact vigorously, obliging rapid decisions and increasing the need for accurate visual information to be available at all times. The adaptations in the visual system which provide this information are correspondingly impressive.

## 11.4. TELEOST RETINAL STRUCTURE

Teleost fish have a retinal structure characteristic of all vertebrates: gelatinous, thin, and transparent. The retina sits at the back of the eye and is comprised of well-defined alternating layers of cell bodies and neuropil, which contains the complex interconnections between cells. The photoreceptors lie in the outermost layer of the neural retina, adjacent to the dark black pigmented epithelium at the back of the eye. There are three different kinds of photoreceptors in the retina of *H. burtoni:* single cones, twin cones, and rods. When seen in cross section, the cone photoreceptors, which are primarily responsible for the daylight acuity, are arranged in a highly ordered array (Fig. 11.3). Each single cone is centered in a square composed of four twin cones. The rod photoreceptors, which account for dim light sensitivity, are interspersed throughout this array. The photoreceptors are thus arrayed somewhat like bowling pins (cones) in a field of grass (rods).

Highly ordered mosaics of cone photoreceptors in fish retinas have been known since the early work of Eigenmann and Shafer (1900) and may be an adaptation to the unusual optical requirements of living underwater (Fernald, 1981b). Since the fish cornea is optically inactive underwater, the focusing power of the fish eye is vested entirely in the spherical lens. As a result, the lens has a very high effective refractive index and a correspondingly short focal length, requiring that the retina lies close to the lens. To achieve reasonable visual acuity in a retina positioned so close to the focusing lens, the photoreceptors must be closely spaced and hence efficiently packed. As might be expected if mosaicism is related to acuity, precise mosaic organization of cone photoreceptors is found in other vertebrate retinas only in regions with high acuity requirements, such as the fovea of humans (Schultze, 1866; Heine, 1901).

Since our field observations suggested that color vision was likely in *H. burtoni,* we measured the wavelength sensitivity of the cone outer receptor photopigments using microspectrophotometry (Fernald and Liebman, 1980). In *H. burtoni,* the short single cone central to each array is maximally sensitive to 454 nm and each member of the twin cones contains a different photopigment with maximal absorbance at 523 and 562 nm. The rod photopigment absorbs maximally at 500 nm, as expected for a rod rhodopsin. Since color resolution depends on the details of the distribution of the color detectors across the retina, we used a histochemical technique, nitro-blue tetrazolium (Marc and Sperling, 1976), to allow their distribution to be visualized. As shown in Fig. 11.4, the twin cones are arranged so that their photopigments have an alternat-

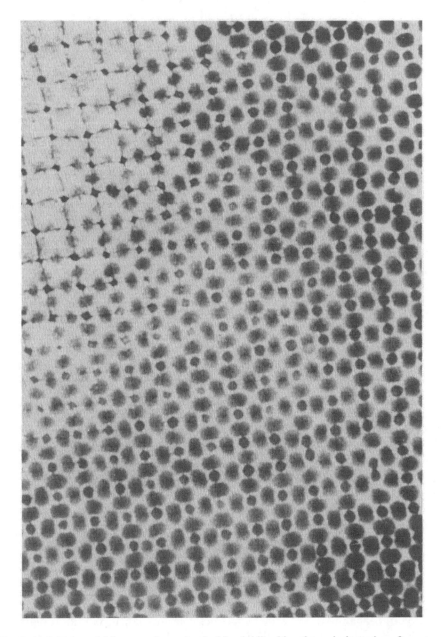

**Figure 11.3.** Tangential 3-μm section stained with toluidine blue through the retina of a mature male *H. burtoni*. Plane of section slightly oblique through the inner segments of the cone photoreceptors.

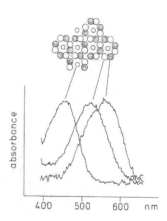

**Figure 11.4.** The schematic illustration at the top shows the *distribution of cone photopigments within the mosaic pattern of the retina of H. burtoni.* The smaller open circles represent the short single cones at the center of each photoreceptor array, and the larger, paired gray circles represent the twin cones, each having a different photopigment. The graph shows the microspectrophotometrically measured absorbance (normalized) of the three different cone pigment types plotted as a function of wavelength. This fourfold rotational symmetry of the cones most closely packs dissimilar pigment types into the retina so that detailed chromatic patterns can be resolved equally well over the entire retinal surface.

ing symmetry about each central single cone (Fernald, 1981b). This distribution optimally packs dissimilar pigment types into each unit area and allows detailed chromatic patterns to be resolved equally well over the entire retina.

## 11.5. GROWTH OF THE FISH EYE

### 11.5.1. Regulation of Eye Growth

The beauty and symmetry of this retinal organization is all the more interesting because it persists in the face of significant enlargement of the eye during growth. Like many animals, fish continue to grow throughout life, although in fish there are unique features about this growth. In particular, the eyes and brain grow by two processes: hyperplasia and cell addition. And, despite enormous growth of these organs, there are no overt consequences detectable in the behavior of the animal. In teleosts, the processes of cell addition are especially interesting because they occur both from well-defined germinal zones (Müller, 1952; Lyall, 1957; Wagner, 1974; Scholes, 1976; Meyer, 1978) and *in situ* (Fernald, 1984). Moreover, the rate of growth is socially regulated making it a particularly useful model system for the general analysis of growth and development of the visual system (see Chapter 2 in this volume).

As *H. burtoni* fry grow, social behavior and, in particular, visual signals from conspecifics *regulate* the growth rate of males (Fernald and Hirata, 1979; Fraley and Fernald, 1982; Davis and Fernald, 1986). For the first 7–8 weeks of life, living in a group facilitates growth of males as compared to brood mates reared in total isolation. After this time, group-reared males that do not form and defend territories grow more slowly than do those with territories. That is, males with territories develop their coloration at a faster rate, weigh more, and have larger and more highly developed gonads than do animals reared under any other conditions. Males reared in total isolation grow and develop at a rate intermediate to the territorial and nonterritorial animals. Thus, the presence of territorial animals sometimes slows the growth rate of

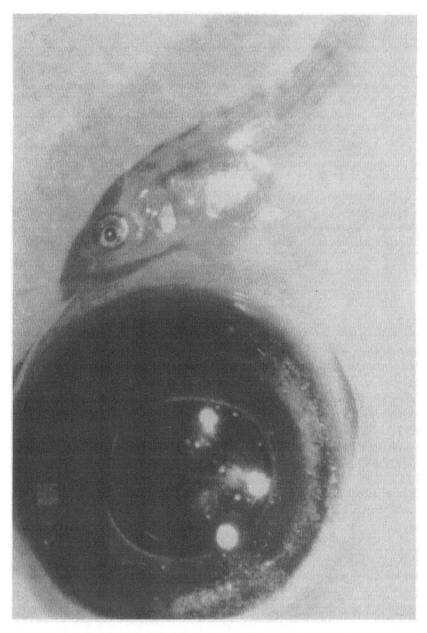

**Figure 11.5.** Under the right environmental and social conditions, *H. burtoni* can grow so quickly that after just over a year its eye is larger than the body of a newly hatched fry. Such rapid growth poses complex problems for the animal's visual system to maintain the good vision necessary for survival. From Fernald (1984).

nonterritorial animals, an effect that may be enhanced when the nonterritorial animals are reared in the presence of larger conspecifics rather than broodmates.

Recently, we demonstrated that this social intimidation slows the growth of soma size in a small group of forebrain neurons implicated in the modulation of gonadal maturation (Davis and Fernald, 1986). We are presently carrying out experiments to determine the stimuli responsible for regulating development.

The consequences of this regulation of growth can be extreme (see Fig. 11.5). Some animals subjected to intimidation by conspecifics will grow more slowly, while remaining appropriately proportioned. This modulation of growth rate is important because it is reflected throughout the organism (see Chapter 2 in this volume).

### 11.5.2. Retinal Change during Eye Growth

In fish, the rate of eye growth relative to growth of the body depends on the importance of vision to the animal. For example, in *H. burtoni*, eye growth is almost directly proportional to body growth, whereas in less visually oriented cyprinids, it is significantly less than directly proportional. Consequently, the absolute amount of growth of the eye can be quite large. For example, the eye of *H. burtoni* may triple in radius during the first 2 months of life (see Fig. 11.5 and Fernald, 1985a). Since the animals continue to behave quite normally, despite this enlargement, we have examined the evolutionary adaptations in the eye which allow these changes to occur.

Specifically, we asked three questions about the effect of change in eye size on vision:

1. Does the quality of the image available to the retina change?
2. Does the photopic visual acuity change?
3. Does the scotopic visual sensitivity change?

### 11.5.3. Image Quality during Eye Growth

#### 11.5.3.1. Lens

The lens is the primary optical element in the fish eye, and in teleost fish, the lens is spherical and symmetrical optically (Fernald and Wright, 1983). Lenses from fish primarily dependent on vision deliver very-high-quality images, despite the fact that spherical aberration is an obvious potential problem (see Fig. 11.6). We have measured the optical quality of lenses of increasing size, assuming that this corresponds to the change that occurs with growth (Fernald and Wright, 1985a). The measured resolving power closely approximates the optimal value predicted from the physically predicted diffraction limits (Airy's disc). Moreover, the lenses produce very good images right up to their edges (see Fernald, 1987, and Fig. 11.7). Since the lens protrudes

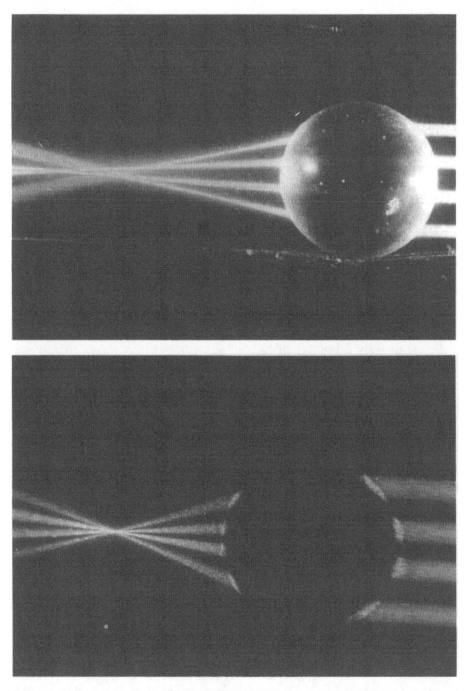

**Figure 11.6.** *Top:* A glass sphere of uniform refractive index ($n \simeq 1.53$) illustrating longitudinal spherical aberration in the focusing of four laser beams. The argon laser beam is at 494 nm. *Bottom:* A freshly excised *H. burtoni* lens suspended by its ligament in oxygenated fish Ringer's solution and illuminated as above. The finely focused cone of light is typical of illuminated fresh fish lenses. From Fernald (1985a).

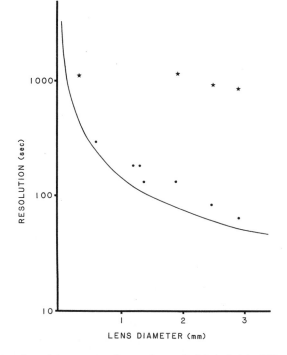

**Figure 11.7.** Measured resolving power of seven lenses (solid circles) in 604 nm light and computed retinal acuity (stars) as a function of lens diameter. Solid line is the graph of the function that describes diffraction-limited resolution of round apertures (Airy's disc). Note that the resolution of fish lens is nearly diffraction limited.

through the pupil, the iris cannot limit images to the central area of the lens, so that producing a good optical image at every point is crucial for the animal's vision.

It has long been presumed that the spherical fish lens is of high quality because there exists a refractive index gradient within it (Maxwell, 1854). For this gradient to correct spherical aberration completely, it would require a very high value in the center of the lens, decreasing continuously and symmetrically with radius in all directions. Maxwell (1854) allegedly first postulated this gradient while contemplating his breakfast herring (cited in Pumphrey, 1961), though the idea appears apparently independently in the writings of Young (1801), Maxwell (1854), and Matthiessen (1882). It clearly also occurred to Brewster because his diary describes his ill-fated attempts at producing a spherical lens with a refractive index gradient to test this hypothesis (Brewster, 1816). The difficulty of making a good spherical lens is seriously complicated by the fact that fishes grow throughout life, so that in addition some mechanism must exist to maintain the correct refractive index gradient as the lens enlarges.

How does the fish preserve the optical quality of its spherical lens despite continual growth? If a spherical lens has no spherical aberration at all and the refractive index has a continuous value from the center to the edge, there is a single unique refractive index gradient that can satisfy this requirement. We have recently estimated the refractive index gradient by the direct measurement of the optical properties of a successively surgically reduced lens and have proposed that there is a central hard core surrounded by a gelatinous outer cortex. In all sizes of lenses, the radius of the core is 0.67% of the radius of the whole lens (Fernald and Wright, 1983). This core appears to have different optical properties than does the entire lens, suggesting that there might be a discontinuous distribution of refractive index across the lens, rather than a continuous one. Even if that were the case, it is unclear how any gradient could be successfully altered during growth.

### 11.5.3.2. Focus

The image available depends not only on the lens quality but also on the state of focus produced by the accommodative apparatus. Since the fish focuses the image by moving the lens, different parts of the eye are focused at different distances from it simultaneously. In most fish with fully lateral eyes, the lens usually moves parallel to the plane of the pupil along an axis of pupillary eccentricity. Thus, in the relaxed eye, the lens lies nearer to the nasal pole so that the temporal part of the retina (nasal visual field) is focused for near vision and the nasal pole retina (temporal visual field) for far vision (Fernald and Wright, 1985b).

Accommodative state is described by the point of focus of incident parallel rays. Emmetropia refers to focus of parallel rays in the plane of the photoreceptors, myopia refers to focus between the lens and retina, and hyperopia is the focus of parallel rays behind the retina. For seeing images clearly, both accommodative state and location of the object of interest are important.

Investigators have used electrical stimulation and drugs to induce accommodative change in fish to allow measurement of the accommodative focus. Beer (1894) first measured the state of refraction of fish eyes, concluding that the fish eye is slightly myopic. However, discussions using optometric terms such as myopia and emmetropia, which are based on the human visual system, may be inappropriate for at least two reasons. The first is that investigators have claimed emmetropia as the desired state in the teleost vision since it is so in humans by definition. However, visibility underwater is limited by backscatter and turbidity so that myopic vision may be quite appropriate for fish. The second difficulty has been that many investigators have not taken into account the fact that the state of accommodation is a function of the angle of view because of the lens movement within the globe. So, measures of accommodation may not be comparable among species and occasionally are not even comparable within a single study (for details see Fernald and Wright, 1985b). Since the magnitude of lens movement varies with fish size, the size of the specimen used will also affect detectability and measurement of accommodative state. Retinoscopic measurements of refractive states have been difficult in fish eyes because of their small size. Furthermore, because the measurements are made underwater, the vergence of the retinoscopic reflection must be corrected for the air–water interface, or the degree of ametropia in either myopic or hyperopic eyes will be overestimated (Hueter and Gruber, 1980).

The most vexed issue in accommodative estimates regards the source of reflection in the eye. In most fish studied there are no retinal blood vessels to be used as landmarks, since the choroid rete provides oxygen to the eye. Estimating or assuming a reflective surface can lead to sizable errors, particularly in small eyes, because the retinal thickness is nearly constant over a wide range of eye sizes. Another error can arise because of chromatic aberration of the optic system (Nuboer and van Genderen-Takken, 1978).

We measured lens movement during accommodation and the corresponding refractive state in a variety of sizes of eyes in *H. burtoni* using two different methods (Fernald and Wright, 1985b). The magnitude of lens movement responsible for accommodation increases with eye size, but with increasing lens diameter, the power of the lens decreases. So the net effect is nearly neutralized. That is, as the animal gets larger, its accommodative mechanism is appropriately scaled to body size. The slight loss of accommodative amplitude with growth (Fernald and Wright, 1985b) means that the distance to the nearest focal point increases from about 2 to 4.5 cm as the animal grows. Since the fish's total length is increasing, this visual near point maintains its same position relative to the body and hence the eyes.

We have calculated accommodative amplitude as a function of retinal location in two different sizes of fish. By making this comparison, two important points are clear. First, larger animals have a smaller total dynamic range of accommodative power corresponding to the higher power of their lenses. Second, active accommodation can only be effective for images falling onto either the nasal or temporal poles of the retina in this fish. This is because the axis of accommodation is in the nasal–temporal plane and lens movement is very nearly parallel to the plane of the pupil. So for most fish it will be true that the state of focus on the dorsal and ventral retinas will be a compromise achieved during active accommodation by lens movement in the nasal–temporal plane. Our various methods for measuring refractive state lead us to propose that the fish eyes are typically emmetropic to slightly myopic, which agrees with the results of recordings from the optic tectum and is appropriate for the fish.

### 11.5.3.3. Visual Field

Since the eye of the animal enlarges so significantly, it is possible that the field of view is also changing, which would have consequences for the changes in the retina during growth. If, for example, the field of view grew smaller as the eye enlarged, then one might imagine a less severe challenge for the retina to preserve its functional integrity.

Fish have periscopic vision because the incompressible spherical lens pro-

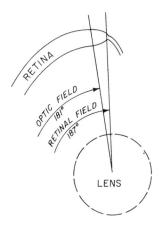

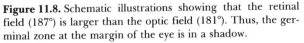

**Figure 11.8.** Schematic illustrations showing that the retinal field (187°) is larger than the optic field (181°). Thus, the germinal zone at the margin of the eye is in a shadow.

trudes through the pupil. In many fish species the pupillary opening is slightly eccentric with a small aphakic space nasal to the spherical lens. The major axis of this eccentricity is collinear with the axis of lens movement and was once hypothesized to serve to enlarge the field of binocular vision (Kahmann, 1936). We have measured the optic field size in *H. burtoni* and found it to be radially symmetric and 181° ± 1.1°, independent of both fish size and state of accommodation (Fernald, 1981a; Fernald *et al.*, 1988). This is shown schematically in Fig. 11.8. From this it appears that the pupillary eccentricity probably exists to allow lens movement and not to increase optic field size. This is made more plausible since the aphakic space is actively reduced during lens movement by iris muscle movement. In the goldfish, which has a round pupil and apparently little or no accommodative movement, the optic field size is approximately the same as that measured here and is also independent of fish size (Easter *et al.*, 1977).

Binocular field size in *H. burtoni* is about 28° and, like the monocular field, independent of fish size. It is interesting that a binocular visual field exists even though there is no brain site that receives the binocular retinal projection (Fernald, 1982).

The constancy of optic field size despite significant increase in retinal size has important implications for visual function as the eye grows.

### 11.5.4. Visual Acuity during Eye Growth

Photopic visual acuity depends on the solid angle viewed by the eye, on the cone photoreceptor spacing across the retina, and on the convergence of the cones onto higher-order processing cells (ganglion), since in bright light usually only the cones participate in phototransduction. Since the solid angle viewed by the eye is independent of eye size, the same volume of space projects onto a larger retinal surface as the eye grows. So to maintain the

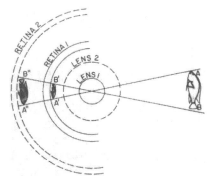

**Figure 11.9.** Sketch showing that a large eye (retina 2) will have a larger image of the same object than does a small eye (retina 1). Since the lens remains very well corrected as it grows, the larger image is equally as sharp as the smaller image.

visual acuity, the density of cone photoreceptors per degree of visual angle must remain the same, as must the ratio of cones to higher-order cells (Fig. 11.9).

In fact, because new cones are added throughout life, the cone density per visual angle increases slightly with eye size, and with that increase the best theoretically achievable acuity increases as well. For example, as *H. burtoni* grows from 8 weeks to 4 months and its eye increases in size from a radius of 0.85 to 2.4 mm, the cone density increases from 6.9 to 14.8 cones/degree (Fernald, 1985a). Because the cell addition accounts for only 40% of the increase in retinal area (Fernald, 1983), the cone density per unit area decreases, while the cone density per visual angle increases (see Fernald, 1985a, for details). Behavioral studies of the sunfish have confirmed that visual acuity improves with fish size (Baerends *et al.,* 1960; Hairston *et al.,* 1982).

### 11.5.5. Visual Sensitivity during Eye Growth

Scotopic visual sensitivity depends on capture of photons by rod photoreceptors, since cones are moved out of the plane of focus in low light due to retinomotor movements. The behaviorally measured scotopic visual threshold is constant over a wide range of eye sizes (Allen and Fernald, 1985; Powers and Bassi, 1981), meaning that the fish can detect very low levels of light energy despite its enlarging eye. As Müller (1952) first showed, the number of rods per unit area remains constant as the fish grows. Since the number of ganglion cells per visual angle, like the cones, increases slightly, the solid angle of visual space per ganglion cell is slightly smaller in larger animals. Clearly, these ganglion cells receive their input from an increasingly large number of rod photoreceptors, and how they preserve their response properties during growth is unknown.

It is now clear, however, that the rod density is maintained, presumably for the purpose of preserving visual sensitivity by the insertion of new rod photoreceptors into the adult retina. The discovery of just how rods are added is described elsewhere in this volume (see Chapter 2). It seems quite reasonable that the selective advantage of maintaining a constant scotopic visual sensitivity is sufficient reason for this evolutionary adaptation.

## 11.6. UNDERSTANDING DEVELOPMENT OF THE EYE

The myriad of novel adaptations which allow the fish eye to grow significantly larger and yet remain fully functional are interesting not only in their own right but because the continued growth of the fish eye offers a novel opportunity to analyze retinal development in an adult animal. Each point along the margin of the eye corresponds in its development to some point in embryonic time: the earliest time corresponding to the outer edge and later times corresponding to points toward the center. Thus, in a single section, the entire ontogeny of the retina can be visualized. To understand the regulation of normal development, identifying tags are needed to allow detailed examination of the time course of development of these cells.

To begin analysis of these processes, we have been developing monoclonal antibodies (Fig. 11.10) and using *in situ* hybridization with antisense mRNA probes (Fernald *et al.*, 1987) to label events at the edge of the retina. These tools should become more than molecular crayons which facilitate description because they can be related to functional events within retinal cells. We anticipate that with these new markers the sequence of developmental events that underlie retinal development will be elucidated.

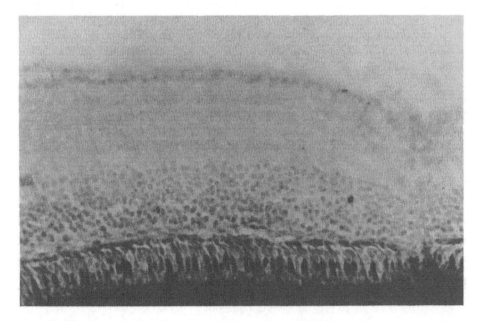

**Figure 11.10.** Monoclonal antibody staining of the *H. burtoni* retina. To visualize the antibody we used HRP with nickel intensification in the DAB reaction product (Vectastain ABC kit, Vector Labs).

## 11.7. SUMMARY

The analysis of vision in *H. burtoni* has led to a number of unexpected insights into adaptations which underlie its visual specialization. At every level of analysis, there are structural, functional, or developmental modifications that have resulted from selective forces during evolutionary time. In many cases, the discoveries in fish have led to similar discoveries in other species, confirming the common ancestry of vertebrate eyes in fishes. We anticipate that further work will offer insights into molecular control mechanisms which may be common throughout vertebrate phylogeny.

ACKNOWLEDGMENTS. This research was supported by NIH grant EY 05051 and the Medical Research Foundation of Oregon. I would like to thank Ms. L. Shelton and Ms. T. Walden-Hansen for technical help and Ms. J. Kinnan for clerical help.

## 11.8. REFERENCES

Allen, E. E., and Fernald, R. D., 1985, Scotopic visual threshold in the African cichlid fish *Haplochromis burtoni, J. Comp. Physiol.* **157:**247–253.

Baerends, G. P., and Baerends-Van Roon, J. M., 1950, An introduction to the ethology of cichlid fishes, *Behaviour (Suppl.)* **1:**1–243.

Baerends, G. P., Bennema, B. E., and Vogelzang, A. A., 1960, Ueber die anderung der sehschar-fe mit dem wachstum bei *Aequidens portalegrensis* (Hensel) (Pisces, cichlidae), *Zool. Jahrb. Allg. Syst. Oko.* **88:**67–78.

Beer, T., 1894, Die accommodation des fischauges, *Pflugers Arch. Gesamte Physiol. Menschen Tiere* **58:**523–650.

Brewster, D., 1816, On the structure of the crystalline lens in fishes and quadrupeds, as ascertained by its action on polarized light, *Philos. Trans. R. Soc. London* 311–317.

Darwin, C., 1987, *Origin of Species*, abridged and introduced by Richard Leakey, Rainbird Publishing Group, London.

Davis, M. R., and Fernald, R. D., 1986, Social environment modulates the development of a forebrain peptidergic nucleus in the cichlid fish, *Haplochromis burtoni, Soc. Neurosci. Abstr.* **11:**1283.

Easter, S. S., Johns, P. R., and Baumann, L. R., 1977, Growth of the adult goldfish eye. I. Optics, *Vision Res.* **16:**469–476.

Eigenmann, C. H., and Shafer, G. D., 1900, The mosaic of single and twin cones in the retina of fishes, *Am. Natural.* **34:**109–118.

Fernald, R. D., 1975, Fast body turns in a cichlid fish, *Nature (London)* **258:**228–229.

Fernald, R. D., 1977, Quantitative behavioral observations of *Haplochromis burtoni* under semi-natural conditions, *Anim. Behav.* **25:**643–653.

Fernald, R. D., 1980a, Optic nerve distention in a cichlid fish, *Vision Res.* **20:**1015–1019.

Fernald, R. D., 1980b, Responses of male *Haplochromis burtoni* reared in isolation to models of conspecifics, *Z. Tierpsychol.* **54:**85–93.

Fernald, R. D., 1981a, Visual field and retinal projections in the African cichlid fish, *Neurosci. Abstr.* **7:**844.

Fernald, R. D., 1981b, Chromatic organization of the cichlid fish retina, *Vision Res.* **20:**1749–1753.

Fernald, R. D., 1982, Retinal projections in the African cichlid fish, *Haplochromis burtoni, J. Comp. Neurol.* **206:**379–389.

Fernald, R. D., 1983, Neural basis of visual pattern recognition in fish, in *Advances in Vertebrate Neuroethology* (J. P. Ewert, R. R. Caparnica, and D. J. Ingle, eds.), pp. 569–580, Plenum Press, New York.

Fernald, R. D., 1984, Vision and behavior in an African cichlid fish, *Am. Sci.* **72:**58–65.

Fernald, R. D., 1985a, Growth of the teleost eye: Novel solutions to complex constraints, *Environ. Biol. Fish.* **13:**113–123.

Fernald, R. D., 1985b, Eye movements in the African cichlid fish, *Haplochromis burtoni, J. Comp. Physiol.* **156:**199–208.

Fernald, R. D., 1987, Aquatic adaptations in fish eyes, in *Sensory Biology of Aquatic Animals* (J. Atema, R. R. Fay, A. N. Popper, and W. N. Tavolga, eds.), Chap. 19, Springer-Verlag, New York.

Fernald, R. D., and Hirata, N., 1975, Non-intentional sound production in a cichlid fish (*Haplochromis burtoni*, Gunther), *Experientia* **31:**299–300.

Fernald, R. D., and Hirata, N., 1977a, Field study of *Haplochromis burtoni:* Quantitative behavioral observations, *Anim. Behav.* **25:**964–975.

Fernald, R. D., and Hirata, N., 1977b, Field study of *Haplochromis burtoni:* Habitats and co-habitats, *Environ. Biol.* **2:**299–308.

Fernald, R. D., and Hirata, N., 1979, The ontogeny of social behavior and body coloration in the African cichlid fish, *Haplochromis burtoni, Z. Tierpsychol.* **50:**180–187.

Fernald, R. D., and Liebman, P. A., 1980, Visual receptor pigments in the African cichlid fish, *Haplochromis burtoni, Vision Res.* **20:**857–864.

Fernald, R. D., and Wright, S. E., 1983, Maintenance of optical quality during crystalline lens growth, *Nature (London)* **301:**618–620.

Fernald, R. D., and Wright, S. E., 1985a, Growth of the visual system of the African cichlid fish, *H. burtoni:* Optics, *Vision Res.* **25:**155–161.

Fernald, R. D., and Wright, S. E., 1985b, Growth of the visual system of the African cichlid fish, *H. burtoni:* Accommodation, *Vision Res.* **25:**163–170.

Fernald, R. D., McDonald, R., and Korenbrot, J., 1987, Light–dark cycle of opsin mRNA production in toad and fish, *Invest. Ophthalmol. Visual Sci. (Suppl.)* **28**(3):184.

Fernald, R. D., Wright, S. E., and Shelton, L., 1988, Growth of the visual system of the African cichlid fish, *H. burtoni:* Optic field and retinal field, in press.

Fraley, N. B., and Fernald, R. D., 1982, Social control of development rate in the African cichlid fish, *Haplochromis burtoni, Z. Tierpsychol.* **60:**66–82.

Fryer, G., and Iles, T. D., 1972, in *Cichlid Fishes of the Great Lakes of Africa,* Oliver and Boyd, Edinburgh.

Geiger, W., 1956, Quantitative untersuchungen uber das gehirn der knochenfische, mit besondere Berucksichtigung seines relativen wachstums, *Acta Anat.* **26:**121–163; **27:**324–350.

Greenwood, P. H., 1981, Species flocks and explosive evolution, in *Chance, Change and Challenge—The Evolving Biosphere* (P. H. Greenwood and P. L. Foley, eds.), pp. 61–74, Cambridge University Press and the British Museum, London.

Hairston, N. G., Li, K. T., and Easter, S. S., 1982, Fish vision and the detection of planktonic prey, *Science* **218:**1240–1242.

Heiligenberg, W., and Kramer, U., 1972, Aggressiveness as a function of external stimulation, *J. Comp. Physiol.* **77:**332–340.

Heiligenberg, W., Kramer, U., and Schultz, V., 1972, The angular orientation of the black eye-bar in *Haplochromis burtoni* (cichlidae, pisces) and its relevance to aggressivity, *Z. Vergl. Physiol.* **76:**168–176.

Heine, C., 1901, Demonstration der zapfenmosaiks der menschlichen fovea, *Dtsch. Ophthalmol. Ges. Ber. Vers.* **19:**265–266.

Hueter, R. E., and Gruber, S. H., 1980, Retinoscopy of aquatic eye, *Vision Res.* **20:**197–200.

Kahmann, H., 1936, Uber das foveale sehen der wirbeltiere. I. Uber die fovea centralis und die fovea lateralis bei einigen wirbeltieren, *Albrecht von Graefe's Arch. Ophthalmol.* **135:** 265–276.

Land, M., 1984, Crustacea, in *Photoreception and Vision in Invertebrates* (M. A. Ali, ed.), pp. 401–438, Plenum Press, New York.

Leong, C. Y., 1969, Quantitative effect of releasers in the attack readiness of the fish *Haplochromis burtoni*, Z. Vergl. Physiol. **65**:29–50.

Liem, K. F., and Osse, J. W. M., 1975, Biological versatility, evolution and food resources, exploitation in African cichlid fishes, Am. Zool. **15**:427–454.

Lyall, A. H., 1957, The growth of the trout retina, Q. J. Microsc. Sci. **98**:101–110.

Marc, R. E., and Sperling, H. G., 1976, Color receptor identities of goldfish cones, *Science* **191**:487–489.

Matthiessen, L., 1882, Uber die beziehungen, welche zwishen dem brechungsindex des kernzentrums der krystalllinse und den dimensionen des auges bestehen, *Pflugers Arch. Ges. Physiol.* **27**:510–523.

Maxwell, J. C., 1854, Some solutions of problems, *Camb. Dubl. Math. J.* **8**:188–195.

Meyer, R. L., 1978, Evidence from thymidine labelling for continuing growth of retina and tectum in juvenile goldfish, *Exp. Neurol.* **59**:99–111.

Müller, H., 1952, Bau und wachstum der Netzhaut des guppy, *Lebistes reticulatus, Abt. Allg. Zool. Physiol. Tiere* **63**:275–324.

Muske, L. E., and Fernald, R. D., 1987a, Control of teleost social signal: Neural basis for differential expression of a color pattern, *J. Comp. Physiol.* **160**:89–97.

Muske, L. E., and Fernald, R. D., 1987b, Control of teleost social signal: Anatomical and physiological specializations of chromatophores, *J. Comp. Physiol.* **160**:99–107.

Nuboer, J. F. W., and van Genderen-Takken, H., 1978, The artifact of retinoscopy, *Vision Res.* **18**:1091–1096.

Poll, M., 1956, Poissons cichlidae, *Result Sci. Explor. Hydrobiol. Lake Tanganika (1946–47)* **3**(5b):1–619.

Powers, M. K., and Bassi, C. J., 1981, Absolute visual threshold is determined by the proportion of stimulated rods in the growing goldfish retina, *Neurosci. Abstr.* **7**:541.

Pumphrey, R. J., 1961, Concerning vision, in *The Cell and Organism* (J. A. Ramsey, ed.), pp. 193–208, Cambridge University Press, Cambridge.

Regan, C. T., 1920, The classification of the fishes of the family cichlidae. I. The Tanganyika genera, *Annu. Mag. Natl. Hist.* **5**:33–53.

Scholes, J. H., 1976, Neuronal connections and cellular arrangement in the fish retina, in *Neural Principles in Vision* (F. Zettler and R. Weiler, eds.), pp. 63–93, Springer-Verlag, Berlin

Schultze, M., 1866, Zur anatomie und physiologie der retina, *Arch. Microsk. Anat. Entwicklungsmech.* **2**:165–174.

Wagner, H. J., 1974, Development of the retina of *Nannacara anomala*, with reference to regional variations in differentiation, *Z. Morphol. Tiere* **79**:113–131.

Young, T., 1801, On the mechanism of the eye, *Philos. Trans.* **92**:23–88.

# Development of Accommodation and Refractive State in the Eyes of Humans and Chickens

# 12

## Howard C. Howland and Frank Schaeffel

### 12.1. INTRODUCTION

The optical quality of the eyes of some higher vertebrates (mainly primates and birds) is so good that aberrations induced by the image-forming structures are not the limiting factor of optical resolution. Instead, diffraction of light at the pupil aperture prohibits a further improvement of the optical resolution at the particular eye size. To attain such optical quality, considerable precision in the arrangement of the ocular structures is required. It can be calculated that for an emmetropic human eye with a depth of focus of about 0.5 D (Campbell, 1957), the retina may not move out of its proper position by more than about 100 μm without detectable loss of resolution. (Emmetropia is the technical name for that condition where a distant image is in good focus on the retina with resting accommodation.)

It has long been questioned whether genetic control of eye growth is sufficient to result in such precision. It is obvious that the tuning of focal length to image surface is often disturbed in humans, resulting in myopia (nearsightedness) or hyperopia (farsightedness). Recently, it was found that proper regulation of eye growth can be interrupted experimentally by abnor-

---

**Howard C. Howland and Frank Schaeffel** • Section of Neurobiology and Behavior, Cornell University, Ithaca, New York 14853.

mal visual input. Deprivation of *form vision* in early postnatal development, by lid suturing or by fitting the animal with translucent plastic occluders, results in myopia in monkeys, tree shrews, cats, and chickens.

Due primarily to the work of Wallman and co-workers (Wallman and Adams, 1987; Wallman *et al.*, 1978, 1981a,b, 1987) and of Hodos and co-workers (Hodos and Keunzel, 1984; Hodos *et al.*, 1985), the chick eye has received increasing attention as an animal model for *form deprivation* myopia. Its high optical quality requires a precise regulation of eye growth. The axial length of the chick eyes increases from about 8.6 mm at the age of 1 day to about 12 mm at the age of 40 days. Related to its rapid growth, experimental manipulations produce rapid changes in refractive state. Experimental myopia due to deprivation of form vision develops at a rate of 1–2 diopters/day.

When investigating this kind of myopia, the most surprising finding of Wallman *et al.* (1987) was that it can be induced selectively and solely in those parts of the visual field that have been deprived. In addition, it has been shown by Troilo and Wallman (1987) that myopia can also be induced after optic nerve section. Both observations lead to the conclusion that the increased growth in the posterior part of the globe is mediated by local mechanisms in the retina rather than by central factors, such as changes in the accommodative tonus. However, it has not yet been shown whether, during the deprivation of form vision, mechanisms are triggered in the eye that are also responsible for regulation of growth during normal visual experience.

We would expect that an "emmetropizing" mechanism is sensitive to a refractive error present in the eye and that it is able to trigger a subsequent growth response to correct for it. It was found that the chicken eye shows the phenomenon of emmetropization even in the presence of external optical disturbances (Schaeffel *et al.*, 1988). The growth of the eye and its parts is such that the eye tends to maintain a focused image of the world on its retina, despite the imposition of negative or positive lenses. The result clearly shows that emmetropizing mechanisms are acting in the chicken eye.

By analyzing the morphological changes in the eyes subsequent to the treatment with the lenses, we found that cornea and lens were not consistently altered. Instead, the "target" of compensatory growth was a shift in the relative position of the retina with respect to lens and cornea (Schaeffel and Howland, 1988). Thus, the ametropia induced is "axial," similar to that which occurs (for reasons not yet known) most frequently in the human eye.

In this chapter, we regard the chick eye as a model for vertebrate ocular development, and we examine how development of accommodation and refractive state in chicks may be similar to or different from collateral visual development in humans.

We first consider the differences and similarities of human and chicken eyes. After an exposition of some of the photographic methods employed in studying the optics of eyes, we describe some human developmental data and compare it to similar data for chickens. Lastly, we consider how data from chicks may be used in the study of myopia and amblyopia in humans.

## 12.2. DIFFERENCES AND SIMILARITIES IN HUMAN AND CHICK EYES

Human and chick eyes are compared in Table 12.1.

### 12.2.1. Differences

First, we should note that, in avian terms, chicks are precocial while humans are altricial. The fovea of newborn human infants is not completely developed, as Abramov and others have shown (Abramov *et al.*, 1982), whereas the area centralis is formed by day 17 (approximately 4 days before hatching) in the afoveate chick (Morris, 1982).

In humans there is a very large binocular overlap, while chicks display only a small degree of overlap. Normal humans always make coordinated eye movements in order to foveate visual targets. In contrast, chicks can fixate and focus their eyes either in a coordinated fashion or independently (Schaeffel *et al.*, 1986).

As will be discussed later, the focusing of human infant eyes takes considerable time after birth to develop, whereas chicks have their full accommodative range at hatching (Schaeffel *et al.*, 1986). Moreover, the basic mechanisms for accommodation in chicks and humans differ, humans depending

**Table 12.1. Similarities and Differences in Human and Chick Eyes**

| Characteristic | Humans | Chicks |
|---|---|---|
| Degree of development at birth | Incomplete | Functional |
| Binocular overlap | Large | Small |
| Divergence of optic axes | 10° | 144° |
| Fovea | Present | Absent |
| Eye movements | Large and binocular | Small and independent |
| Symmetrical binocular focusing | Present at birth, increases with age | Independent focusing |
| Accommodation | Slow development over first 6 months | Complete at birth |
| Corneal accommodation | Absent | Present |
| Scleral ossicles | Absent | Present |
| Anterior (Crampton's) muscle of accommodation | Absent | Present |
| Color vision | Present | Present |
| Visual habit | Diurnal | Diurnal |
| Optical quality | Good | Good |
| Ganglion cells/mm retina | 200,000 (fovea) | >100,000 (center; 20,000 periphery) |

solely on changes in lenticular shape for accommodation (Helmholtz, 1962), while approximately half of the change of refractive power in the accommodated chick eye is due to changes in corneal curvature (Schaeffel and Howland, 1987; Troilo and Wallman, 1987).

These differences in accommodative mechanism are accompanied by fundamental morphological differences. Humans have no bones in the eye, whereas the chick eye possesses scleral ossicles which give great rigidity to the globe immediately behind the cornea. Chicks possess an extra muscle of accommodation (Crampton's muscle; Walls, 1967). Both the scleral ossicles and Crampton's muscle may play functional roles in corneal accommodation.

### 12.2.2. Similarities

Both humans and chicks have color vision, their visual habits are diurnal, and their optics are of good quality as judged by the crispness of retinoscopic reflexes. Commensurate with the high optical quality, both human and chick retinas have high densities of retinal ganglion cells in the fovea or area centralis (Walls, 1967).

Perhaps most importantly, both kinds of eye must solve the fundamental problem of emmetropization. Namely, they must keep real-world images within their accommodative ranges as the eyes grow. And both human eyes (Howland, 1982a) and chick eyes (Schaeffel and Howland, 1987) do grow significantly over the lives of the animals.

### 12.3. METHODS FOR STUDYING THE OPTICS OF EYES

Before we describe our results on the refractive development of human and chicken eyes, it is worth considering the various methods for studying the optics of eyes and their relative advantages and disadvantages. Many of these methods have either been invented or further developed in our laboratory.

### 12.3.1. Retinoscopy

For the past century or so, retinoscopy has been the "objective" method of choice for determining where an eye is focused. The method is not as objective as some would have it, because the observations depend to a certain extent on the judgment of the retinoscopist. In this method, a light is shone into an eye from a distance of approximately 0.5 m. The light source is caused to move up and down while the retinoscopist observes the image of this light on the subject's retina as seen through the subject's pupil from a position almost collinear with the light source itself. If the subject's eye is focused behind the retinoscopist (relative hyperopia), the retinoscopist will see an upright virtual image of the source behind the subject's retina. If, however, the subject is focused in front of the retinoscopist (relative myopia), the latter will see a real image of the source located between the subject and himself.

Generally, the retinoscopist moves the light up and down in one meridian and notes the corresponding "with" or "against" motion of the reflex, depending on whether the subject is relatively hyperopic or myopic and the retinoscopist, accordingly, is viewing a virtual or real image. Usually, the retinoscopist "neutralizes" the reflex by inserting ophthalmic lenses in front of the eye of the subject until the motion of the reflex is neither "with" nor "against," at which point the refracted meridian of the eye of the subject is focused at the distance of the retinoscopist.

### 12.3.1.1. Problems of Retinoscopy

As may be seen from the above description, retinoscopy has certain drawbacks which are inherent in the method.

1. Only one meridian of one eye may be refracted at a time. It takes a second or two to observe the motion of the reflex, and the accommodation of the eye may change between refractions of different reflexes. This may make it difficult to determine the astigmatism of an eye or the difference in refraction between two eyes.
2. The placement of a lens in front of the eye may cause the subject to attempt to focus on the lens itself.
3. With uncooperative subjects, it may be difficult to determine where the eye is fixated and hence to determine which meridian is being refracted.
4. The layer from which the light is reflected in the fundus may not be the layer of sharp focus on the retina (Glickstein and Millodot, 1970; Murphy *et al.*, 1983). The discrepancy between the actual refractive state of the eye and the apparent one due to this difference in layers has been termed *the small eye artifact*. It is generally thought that human infants show a small eye artifact of a fraction of a diopter (Howland, 1982a). It should be noted that this artifact is also inherent in all the methods discussed below.

### 12.3.2. Photographic Refractive Methods

There are three different photographic methods currently employed to refract eyes. The advantage of all these methods is that they can be used to refract both eyes of a subject simultaneously. Also, the refractive state can be monitored continuously if video equipment is used.

### 12.3.2.1. Point Spread Methods of Photorefraction

Two of the earliest methods of photographic refraction, which have come to be called *orthogonal* and *isotropic* photorefraction, respectively, are basically point spread methods for measuring defocus. The double pass point spread (or, more often, line spread) has been measured in experimental setups to assess the optical quality of an eye in optimal focus (Campbell and Gubisch, 1966; Howland and Roehler, 1977; Nalwalk and Howland, 1986). However, ignoring problems with optical quality, the same method may be used to assess degree of defocus of an eye, provided only the diameter of the pupil is known. In these photorefractive methods, a point of light emanating from a fiber optic light guide located in the center of a camera lens is imaged on the subject's retina. Light from this image returns in the direction of the source and hence to the camera. If the point is in perfect focus on the subject's retina, virtually all the light from the retinal image will be reimaged at the source, and very little light returned from the subject's eye to the camera will be captured by the surrounding camera lens. However, if the subject's eye is not accurately focused at the plane of the source, the returned light will spread around the fiber optic light guide and hence be captured by the camera lens. This light returned to the plane of the source is known technically as the *double pass point spread* or simply, *point spread.*

In orthogonal photorefraction (Howland and Howland, 1974), a set of ancillary cylinder lenses forms the point spread into a cross-shaped pattern at the film plane, wherein the length of the cross arms record the width of the point spread at the camera lens, and together with the pupil diameter, the degree of defocus of the eye in corresponding meridians.

In isotropic photorefraction (Howland *et al.,* 1983), the camera itself is slightly defocused relative to the subject, and the diameters of the blur circles of the subject's pupils at the film plane indicate the corresponding diameters of the point spread at the camera lens and hence, by computation, the degree of defocus of the subject's eyes.

**Problems with Point Spread Methods.** Point spread methods have the advantage over retinoscopy of providing an objective record of the refraction and of refracting both eyes in multiple meridians simultaneously, but they are often incapable of recording the sign of defocus (hyperopic or myopic) in a single photograph. Occasionally, due to the chromatic aberration of the eye, if photographs are recorded on color film, it may be possible to detect the sign of defocus from the color fringes in a single photograph (Howland, 1982b). Otherwise, to determine the sign of defocus, a comparison must be made between photographs taken with the camera focused behind and in front of the subject.

### 12.3.2.2. Photoretinoscopic Methods

A third photographic method of refracting the eye is called *photoretin-oscopy, paraxial* or *eccentric* photorefraction (Kaakinen, 1979, 1981a,b; Howland, 1980, 1985; Bobier and Braddick 1985; Kaakinen and Ranta-Kemppainen, 1986; Norcia *et al.,* 1986). In this method the retinoscopic reflex is photographically recorded when the light source is in a known position in relation to the camera. When the camera is accurately focused on the subject's pupils, the photographs record simultaneously the subject's face, the corneal reflexes of the light source (which are useful in establishing the refracted axis), and the retinoscopic reflexes, as well as the pupil diameters. From a knowledge of the pupil diameters and the eccentricity of the light source, the degree of defocus may be computed (Howland, 1984, 1985; Bobier and Braddick, 1985), and from the position of the reflexes in the pupils, the sign or signs of defocus of the eyes are clearly established.

We have developed an infrared (IR) photoretinoscopic technique using IR-emitting diodes of different eccentricities. We find it works very well on avian subjects (Schaeffel *et al.,* 1987). Usually, this has been used with an IR-sensitive video camera together with a frame grabber and a video image digitizer board. The use of IR light adds two advantages to normal photoretinoscopy. Because the eye is not sensitive to the wavelength used (890 nm), (1) there is no light-induced pupillary constriction, which improves the precision of measurement, and (2) the subject is not aware it is being measured. Several new measurements were possible as a result of these two features of IR photoretinoscopy (Schaeffel *et al.,* 1986; Schaeffel and Howland, 1987).

**Problems with Photoretinoscopic Methods.** A drawback of photoretinoscopic methods, as usually practiced, is that only one axis of each eye is refracted in a single photograph, although some researchers have used dual light sources simultaneously (Kaakinen 1981b; Kaakinen and Ranta-Kemppainen, 1986). Also, the requirements of accurate focus are much more stringent than those of point spread methods.

### 12.3.3. Photokeratoscopy

Another photographic method that is useful in the study of eyes is photokeratoscopy. This is not strictly a refractive method in that it is used to measure corneal curvature from which the refractive power of the cornea may be inferred. The technique as we have employed it (Howland and Sayles, 1985; Schaeffel and Howland, 1987) involves illuminating the eye with a circular series of point sources of light which are then reflected from the subject's cornea. These corneal reflections are then photographed and their diameters measured. The more curved the cornea, the closer together will be the reflected points. In order to achieve accuracy with the method, the distance between the subject and the point sources must be carefully controlled, as must be the camera-to-subject distance.

### 12.4. HUMAN REFRACTIVE DEVELOPMENT

Over the past 10 years or so, our laboratory has attempted to characterize the refractive development of human infants and young children, with a view toward being able to detect unusual refractions. We are now attempting to determine if an "unusual" refractive development is related to the subsequent development of amblyopia.

Arguing both from animal studies of cortical development (for a review, see Boothe *et al.*, 1985) and from the experience of clinicians (Ingram and Walker, 1979; Ingram *et al.*, 1986), our major hypothesis has been that the first few years of life contain the "critical period" for the development of amblyopia, and that the length of this critical period may be 2 or 3 years.

From our photorefractions, we have assembled a picture of accommodative and refractive development that has the following features.

### 12.4.1. Newborn Infants

Newborn infants present considerable problems for the vision researcher. They are intermittently awake and asleep, their heads are very unstable, and it is very difficult to judge in which direction they are looking since their motor coordination is poor.

We found that newborns are often astigmatic. Four of the 15 infants we examined between 0 and 9 days of age had greater than 1 D of astigmatism (Howland *et al.*, 1978). Approximately half of these infants were able consistently to focus a camera at 0.75 m distance, while only three were able to focus a camera at 1.5 m, and then only occasionally (Braddick *et al.*, 1979). It appears that newborns tonically accommodate to near distances and only occasionally focus at distances over 1.0 m.

### 12.4.2. Infants between 1 and 3 Months of Age

During this period, the ability of infants to focus distant targets increases as does the astigmatism measured by photorefraction. The latter increases to a level where almost 70% of the infants show astigmatisms of greater than 1 D (Howland *et al.*, 1978; Howland and Sayles, 1984, 1987).

It may well be that this astigmatism, whose principal axes are usually horizontal and vertical, is responsible for the oblique effect in adults.

### 12.4.3. Infants between 3 and 6 Months of Age

From about 3 to 6 months we see a rapid increase in ability to focus the photographer's face. From 3 months on it is easy to attract an infant's gaze to almost any point in the refraction room. We can show this by taking pictures of the first corneal reflexes and studying their location in the pupil when the infant's gaze has been attracted to points at increasing angles from the camera (Howland and Sayles, 1987).

### 12.4.4. Infants and Children between 6 Months and 4 Years of Age

Between 6 months and 4 years of age, some of the trends which have been established during the first 6 months of life continue while others reverse. In particular, the amount of astigmatism decreases steadily over the first 5 years of life (Howland and Sayles, 1984, 1987). The symmetry of the refraction increases, and the difference in both spherical and cylindrical magnitudes decreases over this period.

At about 10 or 11 months, infants begin to show a fear of strangers that can become so strong it may prevent retinoscopy, though it does not usually affect photorefraction. In some children, this fear appears to turn into anger with increasing age. In the "terrible twos," these children become very difficult subjects. Here bribery works better than threats, as tears are the enemy of the refractionist.

### 12.4.5. The Nature of Infant Astigmatism

We have looked at the nature of astigmatism by using a portable photokeratoscope with a short depth of field and taking many pictures in hopes of getting some in focus. As a result of these studies (Howland and Sayles, 1985), we found that early astigmatism is primarily corneal in nature, but that as the infant grows older and the astigmatism reduces, it becomes partly corneal, partly lenticular.

Meanwhile, many studies show that the contrast acuity increases steadily with age. Davida Teller's rule of thumb for human infants is that the contrast acuity in cycles per degree is about equal to the age in months (Boothe *et al.,* 1985).

Beginning at about 2.5 years of age, we can collect visual acuity data from children using conventional subjective techniques such as broken wheel acuity cards or symbol charts. We have found that no single acuity test or presentation routine seems optimal for all children. Furthermore, normal 6 year old children may still test out as 20/30 or worse without a refractive correction, particularly when we use multiple letters in a line or letters bunched together.

### 12.4.6. Summary of Human Refractive Development

Thus, we see that many of the infant mechanisms are immature at birth. For the first few months or so, fixation and focusing are inconsistent and somewhat chaotic. From about 3 months on, fixation is consistent, and by 6 months, so is focusing. Thereafter, all the asymmetries and inaccuracies in the visual system tend to diminish till the fourth or fifth year of life. The work of Dobson and Teller (1978) shows that contrast sensitivity is steadily increasing over this period. Conventional visual acuity may lag somewhat behind contrast sensitivity and not reach adult levels until 7 years (Blum *et al.,* 1959).

### 12.5. CHICK REFRACTIVE DEVELOPMENT

Chicks are precocial, a fact that is also true of their vision. Using infrared photoretinoscopy, we have shown that they accommodate through their full range of 17 D on the first day after hatching (Schaeffel *et al.,* 1986).

Incidentally, it should be noted that the eyes most frequently accommodate and fixate independently. They do, however, have a frontal field of binocular overlap and they are capable of fixating and accommodating to the same point simultaneously (Schaeffel *et al.*, 1986).

Over the first 50 days of life, the axial length of the eye and the posterior nodal distance increase from 8 to 11 and from 5 to 7 mm, respectively. At the same time, the mean pupil diameter increases far more proportionally from about 2.5 to 5.5 mm. This is to say that the $f/\#$ decreases with increasing age.

Other developmental differences appear in the near pupillary response (the constriction of the pupil that accompanies accommodation) and the response of the pupil to illumination. Young chicks have a poorly developed near pupillary response and show less constriction than 50 day old chicks to lights of the same intensity.

### 12.5.1. Mode of Accommodation

Surprisingly, we found that accommodation in the bird eye is due in part (sometimes in very large part) to changes in corneal radius (Schaeffel and Howland, 1987). This, of course, is not the case in humans. We accommodate or focus our eyes by changing the radii of curvature of our lenses—those of us that still can—but chickens rely about half on lenticular accommodation and half on corneal accommodation. Pigeons appear to rely totally on corneal accommodation.

Corneal accommodation can be measured by recording the diameter of a pattern of lights in object space reflected from the cornea. (This is the method we used to measure infant astigmatism; Howland and Sayles, 1985.) To be successful, however, one must very carefully control the distance between the eye and the array of lights. We accomplished this with chickens by simultaneously recording (1) the photorefractive reflex, (2) the keratoscopic reflex (from which the radius of curvature of the cornea was measured), and (3) the cornea of the chick aligned against a fiducial mark (Schaeffel and Howland, 1987).

### 12.5.2. Growth of the Eye with Ophthalmic Lenses

Because of the many papers showing that the chick eye became myopic when it was grown with occluders (Wallman *et al.*, 1978, 1981a,b, 1987), we thought there was some chance that we could control the growth of the eye with lenses. We hoped to show that we could make the eye not only myopic with negative lenses but also hyperopic with positive lenses.

One of us (F.S.) designed a set of lens hoods for the chickens, using five different sizes for each chick to allow for their growth. The hoods were held in place with Velcro fasteners. Velcro fasteners were also used to attach the lenses to the hoods. This allowed removal for cleaning, which was done every 1–2 hr during the day.

We found that the resting refractive state of the chickens was rapidly altered in a direction, but not necessarily magnitude, that would compensate for the effect of the lens (Schaeffel *et al.*, 1988). When the chickens were sacrificed, we measured their posterior nodal distances and found that in seven of eight cases, the distances increased on the visual axis and in the temporal retina. The correlation between the difference in posterior nodal distances and difference in lens powers is significant for both the optic axis and the 20° in the temporal retina.

These results suggest that there is a closed feedback loop regulating the growth of the eye and that this loop uses some information associated with the location of the focal plane of the eye relative to the retina. One such source of information is the accommodative effort needed to focus images in the real world onto the retina. We are currently attempting to test the hypothesis that the growth of the eye is related to accommodation in the chick.

We have initiated experiments (together with D. Troilo and J. Wallman) using chicks whose Edinger Westphal (EW) nucleus has been ablated and who have been raised with positive and negative lenses. Our preliminary results show that the eyes of these chicks, in the absence of any accommodation, can also emmetropize (though with greater variance), compensating for the negative or positive lenses. The path of emmetropization, however, is quite different from that of normal chicks wearing lenses. These EW ablated chicks first become quite hyperopic and then approach their steady state, hyperopic or myopic, depending on the power of the applied lens. These results may be interpreted most easily by assuming that there are two feedback loops acting in parallel to emmetropize the eye, one that employs accommodation, another that does not.

## 12.6. HOW TO USE CHICKS TO STUDY HUMAN VISION

Emmetropization is selectively advantageous in both human and animal eyes because both are relatively large-aperture, high-resolution eyes with fine retinal mosaics and low cone/ganglion cell ratios that grow significantly over their lifetimes.

Can we expect, however, that the same mechanisms operate to cause emmetropization in these eyes, given the many differences in the eyes which we have noted in Table 12.1 and given that their common ancestor lived some 200 million years ago? (As contrasted with the common ancestor of apes and humans, who lived some 20 million years ago.) Moreover, even if the mechanism or mechanisms are the same, what have they to do with relevant human diseases such as myopia and amblyopia?

In the case of myopia, it may be argued that emmetropization must have been a feature of some of the earliest vertebrate eyes, and that some mechanisms must have been well developed long before the lineage of birds and primates parted company. So it may be that some of the mechanisms of emmetropization in bird and human eyes are identical. Indeed, any mechanism of emmetropization is of interest in itself, simply because it shows what is possible in human eyes.

### 12.6.1. The Relationship of Emmetropization and Myopia

Occlusion experiments with chicks involve an asymmetrical open feedback loop. Namely, when the eye is occluded, no change in growth of the eye will improve the image on the retina. Nevertheless, the blur of the retinal image seems to be a sufficient signal for the eye to grow in length and to continue to grow. It is as if, for lack of other information, it is assumed that the blur of the image must be due to a hyperopic defocus, and the correction is in the direction of increased myopia.

Our experiments on chicks, however, show that blur need not always result in this runaway growth. Presumably, under our closed loop conditions of raising normal chicks with lenses, the sign of the perturbation may be inferred by the eye, and growth of the length of the eye may be augmented or retarded accordingly.

Therefore, the mechanism(s) that causes occlusion myopia in chicks operating in open loop conditions may be identical to that, or those, which causes emmetropization of the eye with lenses.

### 12.6.2. Schemes for Emmetropization

Let us look at some possible schemes for emmetropization.

#### 12.6.2.1. Scheme I: Accommodative

In the first scheme we assume the following:

1. The growth of the eye is programmed so that it will always be somewhat hyperopic. Without other feedback, this hyperopia would be of varying degrees.
2. Defocus in the eye is interpreted as a signal that the eye should accommodate.
3. The act of accommodation induces growth of the length of the eye.
4. The ratio of growth to accommodation is such that the length of the eye is always maintained on the side of hyperopia, perhaps by a nonlinear mechanism.

#### 12.6.2.2. Scheme II: Nonaccommodative

In the second scheme we assume the following:

1. The growth of the eye is programmed so that it will initially be hyperopic but will increase its axial length relative to its corneal curvature until a focused image appears on the retina.
2. Focused images on the retina (perhaps via action of neurotransmitters

or neuromodulators from retinal cells) lead to a cessation or inhibition of growth.

3. The relation between a focused image and inhibition is such that the growth of the eye is halted on the hyperopic side of emmetropia.

Note that the two schemes are not incompatible and that there is evidence to suggest that both may be working simultaneously in the normal chick, while Scheme II may work alone in the EW ablated chicks discussed above.

It may also be argued that Scheme II, by itself, would be inadequate in chick with normal accommodation, as the act of accommodation would always keep an image in focus until the end of the accommodative range were reached, thus prematurely halting the growth of the eye at high hyperopic values.

### 12.6.3. Relationship of Emmetropization Schemes to Myopia

How might these schemes of emmetropization apply to humans? "School" myopia, for which there is abundant evidence (Curtin, 1985), may be due to either or both of these emmetropization schemes controlling the growth of the eye to adjust it to the targets most frequently focused, namely, textbooks. It could be that the relative and absolute gains, or amplification factors, of these mechanisms may vary from person to person. Furthermore, the academic myopia so obvious in the classroom before the advent of contact lenses is simply due to the normal process of emmetropization, which kept our ancestor's eyes focused on the horizon.

But what of early onset, pathological myopia? This could result from a failure of either or both schemes. In Scheme I there may be a failure in the first point; namely, the preprogrammed hyperopia fails, and the eye is initially myopic rather than hyperopic. The assumption that accommodation will result in a focused image is false and leads to a reversal of sign in the feedback loop, which then explodes in pathological myopia.

In Scheme II a failure of the preprogrammed hyperopia could lead to the same result. In addition, it could be that focused images no longer cause a cessation of growth, and the preprogrammed axial growth is never stopped.

Perhaps, given our present knowledge of emmetropizing mechanisms, the most significant insight we can gain from these studies of chicks is that there may be more than one feedback loop operating in the emmetropization of human eyes. This possibility opens a wider field of potential pathologies arising from failures of these feedback mechanisms than was previously envisioned.

### 12.6.4. Amblyopia

Finally, let us turn our attention back to human amblyopia and ask what chick studies can teach us about that. At first glance the answer might appear to be—nothing.

Amblyopia is most often monocular. It is frequently viewed as a disturbance in the normal balance of access to cortical cells in the visual cortex by pathways originating from the two eyes. Two possible causes for this imbalance are anisometropia and strabismus. Thus, the afoveate chick with little intraocular overlap would appear to be a poor choice for an animal model of human amblyopia.

However, the chick does possess frontal vision, and we have seen it accommodate in a coordinated binocular fashion. Moreover, as Pettigrew and Konishi (1976) have shown in the barn owl, the visual wulst bears remarkable similarities to the visual cortex of the cat. In fact, ocular dominance histograms of owls whose eyes have been deprived of form vision resemble those of kittens who have been similarly treated.

So perhaps it is not totally absurd to look again at the chick as a possible animal model in the area of amblyopia. Can we make a chick amblyopic, and, if we did, could we detect the amblyopia? These are questions which we think are worth considering.

ACKNOWLEDGMENTS. This work was supported by grant EY-02994 from the National Eye Institute, National Institutes of Health, Bethesda, Maryland, to H.C.H. and by a U.S. Dept. Agr. Res. Hatch grant to H.C.H.

## 12.7. REFERENCES

Abramov, I., Gordon, J., Hendrickson, A., Dobson, V., and LaBossiere, E., 1982, The retina of the newborn human infant, *Science* **217:**265–267.

Blum, H. L., Peters, H. B., and Bettman, J. W., 1959, *Vision Screening for Elementary Schools: The Orinda Study,* University of California Press, Berkeley.

Bobier, W. R., and Braddick, O. J., 1985, Eccentric photorefraction: Optical analysis and empirical measures, *Am. J. Optom. Physiol. Opt.* **62:**614–620.

Boothe, R., Dobson, V., and Teller, D., 1985, Postnatal development of vision in human and nonhuman primates, *Annu. Rev. Neurosci.* **8:**495–545.

Braddick, O., Atkinson, J., French, J., and Howland, H. C., 1979, A photorefractive study of infant accommodation, *Vision Res.* **19:**1319–1330.

Campbell, F. W., 1957, The depth of field of the human eye, *Opt. Acta* **4:**157–164.

Campbell, F. W., and Gubisch, R. W., 1966, Optical quality of the human eye, *J. Physiol. (London)* **186:**558–578.

Curtin, B. J., 1985, *The Myopias: Basic Science and Clinical Management,* Harper & Row, New York.

Dobson, V., and Teller, D. Y., 1978, Visual acuity in human infants: A review and comparison of behavioral and electrophysiological studies, *Vision Res.* **18:**1469–1483.

Glickstein, M., and Millodot, M., 1970, Retinoscopy and eye size, *Science* **168:**605–606.

Helmholtz, H. von, 1962, *Helmholtz's Treatise on Physiological Optics* (translation of the 3rd ed. of *Handbuch der Physiologischen Optik,* 1909), Dover, New York.

Hodos, W., and Kuenzel, W. J., 1984, Retinal-image degradation produces ocular enlargement in chicks, *Invest. Ophthalmol. Vis. Sci.* **25**(6):652–659.

Hodos, W., Fitzke, F. W., Hayes, B. P., and Holden, A. L., 1985, Experimental myopia in chicks: Ocular refraction by electroretinography, *Invest. Ophthalmol. Vis. Sci.* **26:**1423–1430.

Howland, H. C., 1980, The optics of static photographic skiascopy, Comments on a paper by K. Kaakinen: A simple method for screening of children with strabismus, anisometropia or ametropia by simultaneous photography of the corneal and fundus reflexes, *Acta Ophthalmol.* **58:**221–228.

Howland, H. C., 1982a, Infant eyes: Optics and accommodation, *Current Eye Res.* **2:**217–224.

Howland, H. C., 1982b, Optical techniques for detecting and improving deficient vision, in *Optics in Biomedical Sciences* (G. von Bally and P. Greguss, eds.), pp. 188–196, Springer-Verlag, Berlin.

Howland, H. C., 1984, The optics of retinoscopy and photoretinoscopy: Results from ray tracing, *J. Opt. Soc. Am.* **1**(12):1268.

Howland, H. C., 1985, Optics of photoretinoscopy: Results from ray tracing, *Am. J. Optom. Physiol. Opt.* **62**:621–625.

Howland, H. C., and Howland, B., 1974, Photorefraction, a technique for the study of refractive state at a distance, *J. Opt. Soc. Am.* **64**(2):240–249.

Howland, H. C., and Rohler, R., 1977, Photographic measurement of linespread of human eye, *J. Opt. Soc. Am.* **67**:1407.

Howland, H. C., and Sayles, N., 1984, Photorefractive measurements of astigmatism in infants and young children, *Invest. Ophthalmol. Vis. Sci.* **25**:93–102.

Howland, H. C., and Sayles, N., 1985, Photokeratometric and photorefractive measurements of infant astigmatism, *Vision Res.* **25**(1):73–81.

Howland, H. C., and Sayles, N., 1987, A photorefractive characterization of focussing ability of infants and young children, *Invest. Ophthalmol. Vis. Sci.* **28**:1005–1015.

Howland, H. C., Atkinson, J., Braddick, O., and French, J., 1978, Infant astigmatism measured by photorefraction, *Science* **202**:331–333.

Howland, H. C., Braddick, O., Atkinson, J., and Howland, B., 1983, Optics of photorefraction: Orthogonal and isotropic methods, *J. Opt. Soc. Am.* **73**(12):1701–1708.

Ingram, R. M., and Walker, C., 1979, Refraction as a means of predicting squint or amblyopia in preschool siblings of children known to have these defects, *Br. J. Ophthalmol.* **63**(4):238–242.

Ingram, R. M., Walker, C., Wilson, J. M., Arnold, P. E., and Dally, S., 1986, Prediction of amblyopia and squint by means of refraction at age 1 year, *Br. J. Ophthalmol.* **70**:12–15.

Kaakinen, K., 1979, A simple method for screening of children with strabismus, anisometropia or ametropia by simultaneous photography of the corneal and the fundus reflexes, *Acta Ophthalmol.* **57**:161–171.

Kaakinen, K., 1981a, Photographic screening for strabismus and high refractive errors of children aged 1–4 years, *Acta Ophthalmol.* **59**:38–44.

Kaakinen, K., 1981b, Simultaneous two-flash static photoskiascopy, *Acta Ophthalmol.* **59**:378–386.

Kaakinen, K., and Ranta-Kemppainen, L., 1986, Screening of infants for strabismus and refractive errors with two-flash photorefraction with and without cyclopegia, *Acta Ophthalmol.* **64**:578–582.

Murphy, C. J., Howland, H. C., Kwiecinski, G. G., Kern, T., and Kallen, F., 1983, Visual accommodation in the flying fox (*Pteropus giganteus*), *Vision Res.* **23**:617–620.

Morris, V. B., 1982, An afoveate area centralis in the chick retina, *J. Comp. Neurol.* **210**:198–203.

Nalwalk, J., and Howland, H. C., 1986, Aberroscopic measurements and linespread correlates of monochromatic aberrations of the eye, *Invest. Ophthalmol. Vis. Sci. (Suppl.)* **26**(3):77 (March).

Norcia, A., Zadnick, K., and Day, S., 1986, Photorefraction with a catadioptic lens: Improvement on the method of Kaakinen, *Acta Ophthalmol.* **64**:379–385.

Pettigrew, J. D., and Konishi, M., 1976, Effect of monocular deprivation on binocular neurones in the owl's visual wulst, *Nature (London)* **264**:753–754.

Schaeffel, F., and Howland, H. C., 1987, Corneal accommodation in chick and pigeon, *J. Comp. Physiol.* **A160**:375–384.

Schaeffel, F., and Howland H. C., 1988, Visual optics in normal and ametropic chickens, *Clin. Vis. Sci.* **3** (2): 83–98.

Schaeffel, F., Howland, H. C., and Farkas, L., 1986, Natural accommodation in the growing chicken, *Vision Res.* **26**:1977–1993.

Schaeffel, F., Farkas, L., and Howland, H. C., 1987, Infrared photoretinoscope, *Appl. Opt.* **26**:1505–1509.

Schaeffel, F., Glasser, A., and Howland, H. C., 1988, Accommodation, refractive error and eye growth in chickens, *Vision Res.* **28**:639–657.

Troilo, D., and Wallman, J., 1987, Changes in corneal curvature during accommodation in chicks, *Vision Res.* **27**:241–247.

Wallman, J., and Adams, J. I., 1987, Developmental aspects of experimental myopia in chicks: Susceptibility, recovery and relation to emmetropization, *Vision Res.* **27:**1139–1163.

Wallman, J., Turkel, J., and Tractman, J., 1978, Extreme myopia produced by modest change in early visual experience, *Science* **201:**1249–1252.

Wallman, J., Rosenthal, D., Adams, J. J., Trachtman, J. N., and Romagnano, L., 1981a, Role of accommodation and developmental aspects of experimental myopia in chicks, *Doc. Ophthalmol. Proc. Ser.* **28:**197–206.

Wallman, J., Adams, J. I., and Trachtman, J. N., 1981b, The eyes of young chickens grow towards emmetropia, *Invest. Ophthalmol. Vis. Sci.* **20:**557–690.

Wallman, J., Gottlieb, M. D., Rajaram, V., and Wentzek, L. A., 1987, Local retinal regions control local eye growth and myopia, *Science* **237:**73–77.

Walls, G. L., 1967, *The Vertebrate Eye and Its Adaptive Radiation,* Hafner, New York.

# Index

**283**